D1172227

Bergman Spaces

Mathematical
Surveys
and
Monographs

Volume 100

Bergman Spaces

Peter Duren
Alexander Schuster

American Mathematical Society

2000 *Mathematics Subject Classification.* Primary 30H05, 46E15, 30D55.

For additional information and updates on this book, visit
www.ams.org/bookpages/surv-100

Library of Congress Cataloging-in-Publication Data

Duren, Peter L., 1935–
Bergman spaces / Peter Duren, Alexander Schuster.
p. cm. — (Mathematical surveys and monographs, ISSN 0076-5376 ; v. 100)
Includes bibliographical references and index.
ISBN 0-8218-0810-9 (alk. paper)
1. Bergman spaces. I. Schuster, Alexander, 1968– II. Title. III. Mathematical surveys and monographs ; no. 100.

QA331.7.D87 2004
515′.9—dc22 2003063825

∞ The paper used in this book is acid-free and falls within the guidelines
established to ensure permanence and durability.
Visit the AMS home page at http://www.ams.org/

10 9 8 7 6 5 4 3 2 1 09 08 07 06 05 04

Contents

Preface

Over the last ten years, the theory of Bergman spaces has undergone a remarkable metamorphosis. In a series of major advances, central problems once considered intractable were solved, and a rich theory has emerged. Although progress continues, the time seems ripe for a full and unified account of the subject, weaving old and new results together.

The modern subject of Bergman spaces is a blend of complex function theory with functional analysis and operator theory. It comes in contact with harmonic analysis, approximation theory, hyperbolic geometry, potential theory, and partial differential equations. Our aim has been to develop background material and make the subject accessible to a broad segment of the mathematical community. We hope the book will prove useful not only as a reference for research workers, but as a text for graduate students.

In fact, the book evolved from a rough set of notes prepared for graduate students in a two-week course that one of us gave in 1996 at the Norwegian University of Science and Technology in Trondheim, in conjunction with a conference on Bergman spaces supported by the Research Council of Norway. Since that time we have used successive versions of the manuscript in graduate courses we taught at the University of Michigan (1998), Washington University in St. Louis (1999), and San Francisco State University (2001). The last course was supported by the NSF CIRE (Collaborative to Integrate Research and Education) program. The students in all of these courses were enthusiastic, and their perceptive remarks on the manuscript often led to substantial improvements.

In striving for clear presentations of material, we have had the benefit of expert advice from many friends and colleagues. We are most grateful to Kristian Seip for guiding us to a self-contained account of his deep results on interpolation and sampling. Harold Shapiro showed us an elegant way to develop the biharmonic Green function and helped with other constructions. Dmitry Khavinson fielded a steady barrage of technical questions and offered many useful suggestions on the manuscript. Sheldon Axler made a careful reading of several chapters and gave valuable criticism. Mathematical help of various sorts came also from Marcin Bownik, Brent Carswell, Eric Hayashi, Håkan Hedenmalm, Anton Kim, John McCarthy, Maria Nowak, Stefan Richter, Richard Rochberg, Joel Shapiro, Michael Stessin, Carl Sundberg, James Tung, Dror Varolin, Dragan Vukotić, Rachel Weir, and Kehe Zhu. Special thanks go to Joel Shapiro for permission to base our treatment of cyclic singular inner functions (Section 8.2) on his unpublished notes. Christopher Hammond read large portions of the manuscript with an eagle eye and spotted a number of misprints, minor errors, and obscurities. Anders Björn also made helpful remarks. We want to express our sincere appreciation to all of these people, and others whose names we may have overlooked, for helping to improve this book. Any defects that remain, however, are the authors' responsibility.

We also had the benefit of the earlier book *Theory of Bergman Spaces* by Håkan Hedenmalm, Boris Korenblum, and Kehe Zhu (Springer–Verlag, 2000), which served as a useful reference in our approach to several topics. As may be expected, the two books have considerable overlap, but ours develops more of the prerequisite material. It also treats topics not discussed in the earlier book, and treats some of the same topics in different ways. A few results appear here for the first time. On the other hand, the book of Hedenmalm, Korenblum, and Zhu contains extensive discussions of several topics barely touched upon in our book, such as invertible noncyclic functions and logarithmically subharmonic weights.

Our book is essentially self-contained. It should be accessible to advanced graduate students who have studied basic complex function theory, measure theory, and functional analysis. Prior knowledge of Hardy spaces is helpful, since that theory often serves as a model for Bergman spaces, but the main facts about Hardy spaces are reviewed in two "crash courses" early in the book and later as motivation for corresponding topics in Bergman spaces. A few Hardy space results are actually needed for the theoretical development of Bergman spaces, and proofs are given.

Most of the writing was carried out during summers together in Ann Arbor, where the University of Michigan provided excellent facilities for our work. Thanks also go to the Ann Arbor Diamondbacks, who were an extra incentive for the second-named author to return to Michigan every summer.

Over the last decade, the American Mathematical Society held several Special Sessions on Bergman spaces at national and regional meetings, and sponsored a week-long research conference at Mt. Holyoke College in the summer of 1994. That summer conference, in particular, did much to stimulate further research in the field. We were therefore especially pleased when the AMS agreed to publish our book. We are grateful to Sergei Gelfand of the AMS for his initial vision that encouraged us not to settle for a revised set of lecture notes, but to do the extra work needed to produce a full expository account of the subject. He showed remarkable patience with the slow pace of the resulting project, but pushed us to finish when the end was in sight and helped with the technical aspects of production. We hope our book may be judged a worthy successor to the classic book by Stefan Bergman, which appeared in the same AMS series many years ago.

Peter Duren and Alexander Schuster
September 2003

CHAPTER 0

Overview

The theory of Bergman spaces has evolved from several sources. A primary model is the related theory of Hardy spaces. For $0 < p < \infty$, a function f analytic in the unit disk $\mathbb{D}$ is said to belong to the Hardy space H^p if the integrals $\int_0^{2\pi} |f(re^{i\theta})|^p \, d\theta$ remain bounded as $r \to 1$. It belongs to the Bergman space A^p if the area integral $\int_{\mathbb{D}} |f(z)|^p \, d\sigma$ is finite. It is clear that $H^p \subset A^p$. The space H^∞ consists of the bounded analytic functions.

The structural properties of individual functions in H^p were studied actively in the period 1915–1930, beginning with a classical paper by G. H. Hardy. It was found that each such function has radial limits at almost every point of the unit circle, and that the boundary function cannot vanish on any set of positive measure unless the function vanishes identically. The zero-sets $\{z_k\}$ of functions in H^p were neatly characterized by the Blaschke condition $\sum(1 - |z_k|) < \infty$, a property independent of p. For $1 < p < \infty$ the space H^p was shown to be invariant under harmonic conjugation. Detailed information was obtained about Taylor coefficients. One early impetus for the study of Hardy spaces was their application to the theory of trigonometric series.

With the emergence of functional analysis in the 1930s, H^p spaces began to be viewed as examples of Banach spaces, for $1 \le p \le \infty$. This point of view suggested a variety of new problems and provided effective methods for the solution of old problems. In 1949, a seminal paper of Beurling [2] gave a complete description of the invariant subspaces of the shift operator in the sequence space ℓ^2 by relating it to the operator of multiplication by z in the Hardy space H^2. Beurling coined the terms "inner function" and "outer function" and made essential use of the canonical factorization of an arbitrary H^2 function into a product of inner and outer functions, a result due to Smirnov. At about the same time, S. Ya. Havinson and H. S. Shapiro independently introduced a theory of dual extremal problems in H^p spaces, unifying an existing collection of examples into a coherent framework that could be fully understood only in the setting of functional analysis. Around 1960, Lennart Carleson [2,3] and later Shapiro and Shields [1] solved universal interpolation problems in H^p spaces with the help of functional analysis. A prominent feature of these developments was the interplay of "hard" and "soft" analysis that has given the subject a special appeal.

Meanwhile, Stefan Bergman [1] had developed an elegant theory of Hilbert spaces of analytic functions in planar domains and in higher-dimensional complex space, relying heavily on a reproducing kernel that became known as the Bergman kernel function. Bergman's work focused on spaces of analytic functions that are square-integrable over the given domain with respect to Lebesgue area or volume measure. When attention was later directed to the spaces A^p over the unit disk, it was natural to call them Bergman spaces. As counterparts of Hardy spaces, they presented analogous problems. However, it soon became apparent that Bergman spaces are in many respects much more complicated than their Hardy-space cousins. Functions in Bergman spaces may have wild boundary behavior, their zero-sets are difficult to fathom, there is no obvious analogue of Blaschke products or "inner–outer" factorization, and the invariant subspaces need not be singly generated as they are (according to Beurling's theory) for the Hardy spaces. An attempt to develop a corresponding theory of dual extremal problems for Bergman spaces runs into the serious difficulty of describing the annihilator of A^p as a subspace of L^p. The techniques of functional analysis that had dealt so successfully with interpolation problems in H^p seemed doomed to failure when applied to Bergman spaces. In short, although many problems in Hardy spaces were well understood by the 1970s, their counterparts for Bergman spaces were generally viewed as intractable. There had been some isolated progress, such as the work of Charles Horowitz [1,2,3] and Boris Korenblum [1,2] on zero-sets and zero-divisors of A^p functions, but expectations were low and the prevailing mood was one of pessimism.

All of this changed in the 1990s. In a flurry of important advances, problems previously considered intractable began to be solved. Major breakthroughs attracted other workers to the field and inspired a period of intense research on Bergman spaces and related topics. First came Håkan Hedenmalm's construction [1] of canonical divisors, an analogue of Blaschke products that had clear advantages over the products introduced earlier by Horowitz. According to a classical theorem of F. Riesz, Blaschke products act as isometric zero-divisors for the H^p spaces, and this fact allows many problems for H^p to be reduced to the easier setting of H^2. It was important to find a similar result for Bergman spaces. Hedenmalm's leading idea was to view Blaschke products as solutions to certain extremal problems in H^2 and to pose similar extremal problems in A^2. The resulting extremal functions have unit norm and were found to be contractive zero-divisors of functions in the Bergman space A^2. His work also showed that A^2 spaces have no isometric zero-divisors. Duren, Khavinson, Shapiro, and Sundberg [1,2] then generalized Hedenmalm's results to A^p spaces for $0 < p < \infty$, discovered the surprising relevance of the biharmonic Green function, and developed an integral formula that allowed a partial extension of the theory to more general invariant subspaces.

Around the same time, Kristian Seip [2,3], motivated in part by results in signal analysis and wavelets, characterized the interpolation and sampling sets for the Bergman space A^2 in terms of certain densities. His results were later generalized to A^p for $0 < p < \infty$. Hedenmalm [3] used Seip's results on sampling sets to construct a specific invariant subspace of A^2 that has codimension (or *index*) 2 and is therefore not singly generated (or *cyclic*). It had been known from prior work of Apostol, Bercovici, Foiaş, and Pearcy [1] that the space A^2 has invariant subspaces of arbitrary index, a result that underscored their extremely complex structure and made a complete description seem unlikely. Hedenmalm's explicit construction was then generalized to arbitrary index by Hedenmalm, Richter, and Seip [1]. As they pointed out, one particular reason for studying the invariant subspaces of A^2 is that the general invariant subspace conjecture in Hilbert space reduces to a special question about invariant subspaces of A^2. (The invariant subspace conjecture, still unresolved, is that every bounded linear operator on a separable Hilbert space has a nontrivial invariant subspace.)

In a more positive direction, Aleman, Richter, and Sundberg [1] showed that the invariant subspaces of A^2 have a common structural property that can be viewed as an analogue of Beurling's theorem for H^2. They showed that every invariant subspace M of A^2 is generated by the orthocomplement of zM within M. They also showed that for $0 < p < \infty$ every cyclic subspace of A^p is generated by its canonical extremal function. They used these results to show that the cyclic elements of A^p are precisely the A^p *outer functions* defined earlier by Korenblum [6], who had proved one half of the theorem. The work of Aleman, Richter, and Sundberg also led to a credible analogue of inner–outer factorization for functions in Bergman spaces. The proof of their main structural theorem on invariant subspaces of A^2 was long and technical, but Shimorin [7] found a simpler proof that actually embedded the result in a more general theorem about linear operators in Hilbert space.

Some basic questions remain open. For instance, there is still no complete description of the zero-sets of functions in an A^p space. The invariant subspaces have not been characterized. It is not even known how or whether the structural theorem for A^2 might extend to A^p for other values of p. Much remains to be done on extremal problems in A^p. However, the remarkable progress over the last decade has provided a rich body of material for an expository account of the theory of Bergman spaces. All of the developments outlined above are treated in this book.

The book begins with a chapter on Bergman's general theory of orthogonal systems of functions and kernel functions, plus a general account of biharmonic Green functions. Both topics recur repeatedly throughout the book. Chapter 2 develops the basic properties of A^p as a linear space, and gives further background material that is needed later in the book. Uniformly convex spaces are presented for applications to canonical extremal functions in Chapter 5. The boundedness of the Bergman projection plays a role in the representation of the dual space of A^p for $1 < p < \infty$. The dual space structure of A^1 is more delicate and involves the Bloch spaces, which are introduced here. The chapter also includes an extensive discussion of the pseudohyperbolic metric and the related concept of uniformly discrete sequences, both of which are fundamental to the entire theory. Carleson measures for the Bergman spaces are described in terms of the pseudohyperbolic metric.

In Chapter 3 the focus shifts to analytic properties of individual functions in a Bergman space A^p. Most of the results relate to various measures of growth and the size of Taylor coefficients. The zero-sets of functions in A^p are discussed in Chapter 4. Most of the material here is due to Charles Horowitz. The Horowitz products, which are essentially Blaschke products with appropriate convergence factors, are shown to be bounded zero-divisors for the Bergman spaces. The ordinary Blaschke products with this property are also described. Chapter 5 contains a full account of the modern theory of canonical divisors, which can be viewed as improved versions of Horowitz products, although their analytic structure is less apparent. Their basic properties are developed, and a representation is given in terms of reproducing kernels of weighted A^2 spaces. The integral formula is proved and is used to extend some properties of canonical zero-divisors to canonical extremal functions of arbitrary invariant subspaces of A^p. The concept of an A^p inner function evolves in a natural way from these developments.

Chapters 6 and 7 treat sampling and interpolation sets for A^p spaces. Chapter 6 contains introductory material, explicit examples, dual formulations, and some special theorems. The two main theorems of Kristian Seip, characterizing sampling and interpolation sets in terms of certain densities, are stated and used to derive further consequences. The proofs of Seip's theorems are deferred to Chapter 7, where they are given in full detail.

Chapters 8 and 9 return to the theme of invariant subspaces. Chapter 8 begins with a brief account of Beurling's theorems for H^p spaces, which serve as a kind of model, although the situation is much more complex for Bergman spaces. In fact, most of the chapter points out various ways in which things go wrong for Bergman spaces. A classical singular inner function may or may not generate the whole space A^p ; a criterion in terms of the associated singular measure is presented and proved. This leads to a wider discussion of cyclic elements of A^p. Next come the constructions of invariant subspaces of A^2 with higher index, using some facts about sampling and interpolation sets. The concept of index is then generalized to A^p spaces, and corresponding constructions are given. Chapter 9 reverts to a more positive mode and presents some unexpectedly precise theorems about invariant subspaces of Bergman spaces. The results of Aleman, Richter, and Sundberg are stated and applied to characterize the cyclic elements of A^p and to give the analogue of inner–outer factorization for Bergman spaces. Then comes Shimorin's proof of the theorem of Aleman–Richter–Sundberg that specifies a set of generators for each invariant subspace of A^2. Finally, a surprisingly difficult proof is given for the theorem that every cyclic subspace of A^p is generated by its canonical extremal function. The proof makes heavy use of the integral formula and kernel functions of weighted A^2 spaces, as discussed in Chapter 5, and an explicit description of the subspace generated by an A^p inner function. Here the biharmonic Green function plays a prominent role.

Other topics could well have been included. For instance, the book contains no unified treatment of extremal problems. Some general results have come from attempts to develop a theory of dual extremal problems for Bergman spaces. Partial accounts of this material and related results can be found in papers of Vukotić [3] and Khavinson and Stessin [1].

Bergman spaces with standard weights arise only sporadically in this book, but results on zero-sets and on sampling and interpolation sets extend readily to that more general setting, mainly because the standard weights cooperate well with the pseudohyperbolic metric. On the other hand, the factorization results that come from canonical divisors do not generalize automatically to weighted spaces. Phenomena such as the absence of extraneous zeros depend critically on the choice of parameter in the weight function. These matters are treated in works of Hedenmalm and Zhu [1], Hedenmalm [2], Shimorin [1,2,3,4], and Weir [1,2,3,4]. Functions in standard weighted Bergman spaces arise in an unexpected way when a function in H^p of the polydisk is restricted to the diagonal; that is, when all variables are equated to produce a function of one complex variable. This phenomenon was discovered by Rudin [1] and was further explored in papers of Horowitz and Oberlin [1] and Duren and Shields [3].

Although the cyclic elements of Bergman spaces are characterized as A^p outer functions, it is not easy to identify the A^p outer functions or to describe their analytic properties. Further results on cyclic elements and invertibility, primarily due to Borichev and Hedenmalm [1,2], are discussed in the book of Hedenmalm, Korenblum, and Zhu [2].

There is an extensive literature on linear operators that act on Bergman spaces. In particular, there are many interesting results on composition operators, and on Toeplitz and Hankel operators. These topics are beyond the scope of this book. Some of the material can be found in Axler [2,3] and in the book of Zhu [1]. Treatments of composition operators are given in the books of J. H. Shapiro [6] and Cowen and MacCluer [1].

CHAPTER 1

The Bergman Kernel Function

This book will treat primarily the Bergman spaces A^p over the unit disk

$$\mathbb{D} = \{z \in \mathbb{C} : |z| < 1\}.$$

For $0 < p < \infty$, the space A^p consists of the functions f analytic in $\mathbb{D}$ that belong to the space $L^p(\mathbb{D})$. In other words, the integral $\int_{\mathbb{D}} |f(z)|^p \, dA$ with respect to area measure dA is required to be finite. Actually, the pioneering work of Stefan Bergman (1895–1977) deals mainly with the case $p = 2$ and is concerned with more general domains in the plane or in higher-dimensional complex space. In this Hilbert space setting, orthogonal systems of functions come to the fore and the reproducing kernel plays a prominent role. To set the stage for later developments, we shall begin by discussing some of the basic ingredients of Bergman's theory. The book of Bergman [1] gives further information.

§1.1. Point-evaluation functionals.

Let $\Omega \subset \mathbb{C}$ be an arbitrary domain (an open connected set) in the complex plane. For $0 < p < \infty$, the *Bergman space* $A^p(\Omega)$ consists of all functions f analytic in Ω for which

$$\|f\|_p = \left\{ \int_\Omega |f(z)|^p \, dA \right\}^{1/p} < \infty .$$

The quantity $\|f\|_p$ is called the *norm* of the function f; it is a true norm if $p \geq 1$. In any case the norm has the properties $\|\alpha f\|_p = |\alpha| \|f\|_p$ and $\|f\|_p = 0$ if and only if $f = 0$, but for $p \geq 1$ the *triangle inequality* $\|f+g\|_p \leq \|f\|_p + \|g\|_p$ is satisfied. If $0 < p < 1$, the triangle inequality is replaced by the inequality $\|f + g\|_p^p \leq \|f\|_p^p + \|g\|_p^p$. In particular, $A^p(\Omega)$ is always a linear metric space.

Our first theorem asserts that functions in a Bergman space cannot grow too rapidly near the boundary.

THEOREM 1. *Point-evaluation is a bounded linear functional in each Bergman space $A^p(\Omega)$. More specifically, each function $f \in A^p(\Omega)$ has the property*

$$|f(z)| \leq \pi^{-\frac{1}{p}} \delta(z)^{-\frac{2}{p}} \|f\|_p , \qquad z \in \Omega ,$$

where $\delta(z) = dist(z, \partial\Omega)$ is the distance from z to the boundary.

PROOF. Fix a point $z \in \Omega$ and let $\delta = \delta(z)$. Then the disk

$$D = \{\zeta \in \mathbb{C} \, : \, |\zeta - z| < \delta\}$$

lies in Ω. The integrals

$$\int_0^{2\pi} |f(z + re^{i\theta})|^p d\theta$$

are nondecreasing functions of r (see Duren [5], p. 9), so it follows that

$$|f(z)|^p \le \frac{1}{2\pi} \int_0^{2\pi} |f(z + re^{i\theta})|^p \, d\theta \,, \qquad 0 \le r < \delta \,.$$

Another integration gives

$$\begin{aligned} \pi\delta^2 |f(z)|^p &\le \int_0^{\delta} \int_0^{2\pi} |f(z + re^{i\theta})|^p \, d\theta \, r \, dr \\ &= \int_D |f(\zeta)|^p \, dA \le \int_\Omega |f(\zeta)|^p \, dA = \|f\|_p^p \,, \end{aligned}$$

which is the stated result. □

One immediate corollary is that norm convergence implies locally uniform convergence. In other words, if f_n and f are in $A^p(\Omega)$ and $\|f_n - f\|_p \to 0$ as $n \to \infty$, then $f_n(z) \to f(z)$ uniformly on each compact subset of Ω. Another corollary is that the space $A^p(\Omega)$ is *complete*; every Cauchy sequence converges in norm to some function in $A^p(\Omega)$. Since $L^p(\Omega)$ is complete, we need only show that $A^p(\Omega)$ is a *closed* subspace. But if $f_n \in A^p(\Omega)$ and $\{f_n\}$ converges in norm to a function $f \in L^p(\Omega)$, then some subsequence $\{f_{n_k}(z)\}$ converges to $f(z)$ almost everywhere in Ω. On the other hand, $\{f_n\}$ is a Cauchy sequence in norm, hence a locally uniform Cauchy sequence, so it converges locally uniformly to a function g that is *analytic* in Ω. Thus $g(z) = f(z)$ almost everywhere in Ω, and the limit function f can be identified with a function in $A^p(\Omega)$.

As a consequence, $A^p(\Omega)$ is a Banach space for $p \ge 1$, and $A^2(\Omega)$ is a Hilbert space with inner product

$$\langle f, g \rangle = \int_\Omega f(z) \, \overline{g(z)} \, dA \,.$$

Another important corollary of Theorem 1 should be noted. If a sequence of functions $f_n \in A^p(\Omega)$ is bounded in norm, then it is locally bounded, and so by Montel's theorem it constitutes a normal family; some subsequence converges locally uniformly in Ω to a function in $A^p(\Omega)$. (See for instance Duren [6], Chapter 1.)

The proof of Theorem 1 shows that $A^p(\mathbb{C})$ contains only the zero-function. Thus we shall suppose henceforth that our domain Ω has at least one nondegenerate boundary component, consisting of more than a single point. Then it can be shown that for each point $z \in \Omega$ there is a function $f \in A^p(\Omega)$ for which $f(z) \neq 0$.

Now let $p = 2$. Since each point-evaluation functional $\phi(f) = f(z)$ is bounded, the Riesz representation theorem for Hilbert space guarantees the existence of a unique function k_z in $A^2(\Omega)$ such that $f(z) = \langle f, k_z \rangle$ for every $f \in A^2(\Omega)$. The function $K(z,\zeta) = \overline{k_z(\zeta)}$ is known as the *reproducing kernel*, or the *Bergman kernel function* of the domain Ω. It has the reproducing property

$$f(z) = \int_\Omega f(\zeta)\, K(z,\zeta)\, dA(\zeta)\,, \qquad z \in \Omega\,,$$

for each function $f \in A^2(\Omega)$. Taking $f(\zeta) = k_w(\zeta) = \overline{K(w,\zeta)}$ for some $w \in \Omega$, we conclude that

$$\overline{K(w,z)} = \int_\Omega \overline{K(w,\zeta)}\, K(z,\zeta)\, dA(\zeta) = K(z,w)\,.$$

Thus the kernel function has the symmetry property $K(z,\zeta) = \overline{K(\zeta,z)}$. This shows that $K(z,\zeta)$ is analytic in z and anti-analytic in ζ. Another consequence is the formula

$$K(z,z) = \int_\Omega |K(z,\zeta)|^2\, dA(\zeta) = \|K(z,\cdot)\|_2^2 > 0\,.$$

In view of the last identity, the Schwarz inequality applied to the reproducing formula shows that $|f(z)| \leq \sqrt{K(z,z)}\, \|f\|_2$. For each point $\zeta \in \Omega$, this shows that $\|f\|_2 \geq K(\zeta,\zeta)^{-\frac{1}{2}}$ for all $f \in A^2(\Omega)$ with $f(\zeta) = 1$. In fact, the lower bound is sharp and is uniquely attained by the function $f(z) = K(z,\zeta)/K(\zeta,\zeta)$.

It should be emphasized that among functions in $A^2(\Omega)$, the kernel function $K(z,\zeta)$ is uniquely determined by its reproducing property. This is part of the Riesz representation theorem — if a function $\ell_z(\zeta) = \overline{L(z,\zeta)}$ belongs to $A^2(\Omega)$ and also has the reproducing property, then $f(z) = \langle f, \ell_z \rangle$ and so $\langle f, k_z - \ell_z \rangle = 0$ for every $f \in A^2(\Omega)$. But this implies that $k_z - \ell_z = 0$, so that $K(z,\zeta) = L(z,\zeta)$ for all $z, \zeta \in \Omega$.

§1.2. Orthonormal bases.

Suppose now that $\{\varphi_n\}$ is an orthonormal basis of $A^2(\Omega)$. Then

$$\langle \varphi_n, \varphi_m \rangle = \delta_{nm} = \begin{cases} 0\,, & n \neq m \\ 1\,, & n = m\,, \end{cases}$$

the Kronecker delta. Furthermore, each function $f \in A^2(\Omega)$ has a unique expansion $f = \sum_{n=1}^{\infty} c_n \varphi_n$, convergent in norm and therefore uniformly convergent on compact subsets:

$$f(z) = \sum_{n=1}^{\infty} c_n \, \varphi_n(z) \,, \qquad z \in \Omega \,,$$

where $c_n = \langle f, \varphi_n \rangle$. By Parseval's identity, $\sum_{n=1}^{\infty} |c_n|^2 = \|f\|_2^2$. Choosing the kernel function $f(z) = K(z, \zeta)$, we find that $c_n = \langle K(\cdot, \zeta), \varphi_n \rangle$, or

$$\overline{c_n} = \langle \varphi_n, K(\cdot, \zeta) \rangle = \int_\Omega \varphi_n(w) \, K(\zeta, w) \, dA(w) = \varphi_n(\zeta) \,.$$

We have therefore arrived at the following remarkable theorem.

THEOREM 2. *The kernel function of* Ω *has the representation*

$$K(z, \zeta) = \sum_{n=1}^{\infty} \varphi_n(z) \, \overline{\varphi_n(\zeta)} \,,$$

where $\{\varphi_n\}$ *is an arbitrary orthonormal basis of* $A^2(\Omega)$.

A heuristic verification of this formula proceeds as follows. Let $\widetilde{K}(z, \zeta)$ denote the sum. For each function

$$f(z) = \sum_{m=1}^{\infty} c_m \, \varphi_m(z)$$

in $A^2(\Omega)$, the orthonormality relations $\langle \varphi_n, \varphi_m \rangle = \delta_{nm}$ give

$$\int_\Omega f(\zeta) \, \widetilde{K}(z, \zeta) \, dA(\zeta) = \sum_{m=1}^{\infty} c_m \sum_{n=1}^{\infty} \varphi_n(z) \, \langle \varphi_m, \varphi_n \rangle = \sum_{m=1}^{\infty} c_m \, \varphi_m(z) = f(z) \,.$$

Thus $\widetilde{K}(z, \zeta)$ has the reproducing property and is therefore equal to the kernel function $K(z, \zeta)$.

Theorem 2 is readily extended to weighted Bergman spaces $A_w^2(\Omega)$, with weight function $w(z) \geq 0$ and inner product

$$\langle f, g \rangle_w = \int_\Omega f(z) \, \overline{g(z)} \, w(z) \, dA \,,$$

provided the point-evaluation functionals are uniformly bounded on compact subsets of Ω. Under this hypothesis, the space $A_w^2(\Omega)$ is complete and is therefore a Hilbert space. More generally, the proof for the unweighted case (see Section 1.1) then applies to show that the space $A_w^p(\Omega)$ is complete for $0 < p < \infty$. We shall see in Chapter 5 (Section 5.4) that any weight function of the form $w = |\psi|^q$, where $0 < q < \infty$ and $\psi \in A^q(\Omega)$, gives rise to a weighted Bergman space $A^p_w(\Omega)$ where point-evaluation functionals are uniformly bounded on compact subsets.

Let us now calculate the kernel function of the unit disk $\mathbb{D}$. Observe first that the monomials $1, z, z^2, \dots$ form an orthogonal set in $A^2 = A^2(\mathbb{D})$. An easy computation gives

$$\int_{\mathbb{D}} z^n \, \overline{z^m} \, dA = \int_0^{2\pi} \int_0^1 r^n e^{in\theta} \, r^m e^{-im\theta} \, r \, dr \, d\theta = \frac{2\pi}{n+m+2} \, \delta_{nm} \, .$$

This shows that the functions

$$\varphi_n(z) = \sqrt{\frac{n+1}{\pi}} \, z^n \, , \qquad n = 0, 1, 2, \dots \, ,$$

are orthonormal in A^2. To show that they form a basis, we have to verify that they span the space. Equivalently, we must show that Parseval's relation

$$\sum_{n=0}^{\infty} |\langle f, \varphi_n \rangle|^2 = \|f\|_2^2$$

holds for every $f \in A^2$. But this is equivalent to the identity

$$\|f\|_2^2 = \pi \sum_{n=0}^{\infty} \frac{|a_n|^2}{n+1} \, , \qquad f(z) = \sum_{n=0}^{\infty} a_n \, z^n \, ,$$

which is easily established by similar use of the orthogonality relations. More specifically, let $\mathbb{D}_\rho$ denote the disk $|z| < \rho < 1$, let

$$s_n(z) = \sum_{k=0}^{n} a_k z^k, \qquad n = 1, 2, \dots,$$

denote the partial sums of the Taylor series expansion of f, and compute

$$\int_{\mathbb{D}_\rho} |s_n(z)|^2 \, dA = \int_0^{2\pi} \int_0^{\rho} s_n(re^{i\theta}) \overline{s_n(re^{i\theta})} \, r \, dr \, d\theta = \pi \sum_{k=0}^{n} \frac{|a_k|^2}{k+1} \, \rho^{2(k+1)} \, .$$

But $s_n(z) \to f(z)$ uniformly on $\mathbb{D}_\rho$, so it follows that

$$\int_{\mathbb{D}_\rho} |f(z)|^2 \, dA = \pi \sum_{k=0}^{\infty} \frac{|a_k|^2}{k+1} \, \rho^{2(k+1)} \, .$$

Now let $\rho \to 1$ to arrive at the given formula for $\|f\|_2^2$. Thus the polynomials are dense in A^2. However, they are not dense in $A^2(\Omega)$ for *every* simply connected domain Ω. For instance, they are not dense if Ω is the disk minus a radial slit.

Since $\{\varphi_n\}$ is an orthonormal basis of A^2, Theorem 2 says that the kernel function is

$$K(z,\zeta)=\sum_{n=0}^{\infty}\varphi_n(z)\,\overline{\varphi_n(\zeta)}=\frac{1}{\pi}\sum_{n=0}^{\infty}(n+1)(z\overline{\zeta})^n\,,$$

or

$$K(z,\zeta)=\frac{1}{\pi}\frac{1}{(1-z\overline{\zeta})^2}\,.$$

For an alternate derivation, write

$$K(z,\zeta)=\sum_{n=0}^{\infty}a_n(\zeta)z^n$$

and use the reproducing property of the kernel function to calculate

$$\zeta^m=\int_{\mathbb{D}}z^m K(\zeta,z)\,dA(z)=\sum_{n=0}^{\infty}\overline{a_n(\zeta)}\int_{\mathbb{D}}z^m\overline{z^n}\,dA=\frac{\pi}{m+1}\overline{a_m(\zeta)}$$

for $m=0,1,2,\dots$.

The explicit formula for the kernel function shows that

$$f(z)=\frac{1}{\pi}\int_{\mathbb{D}}\frac{1}{(1-\overline{\zeta}z)^2}f(\zeta)\,dA(\zeta)\,,\qquad f\in A^2\,.$$

Choosing in particular $f(\zeta)=(1-\zeta\overline{z})^{-2}$ for fixed z in the disk, we see that

$$\frac{1}{\pi}\int_{\mathbb{D}}\frac{1}{|1-\overline{\zeta}z|^4}\,dA(\zeta)=\frac{1}{(1-|z|^2)^2}\,,\qquad z\in\mathbb{D}\,.$$

§1.3. Conformal invariance.

Knowing the kernel function for the unit disk, we can calculate it for an arbitrary simply connected domain in terms of the Riemann mapping function. In fact, the kernel function is conformally invariant in the sense of the following theorem.

THEOREM 3. *Let $w=\varphi(z)$ map a domain Ω conformally onto a domain D, and let $J(w,\omega)$ be the kernel function of D. Then the kernel function of Ω is*

$$K(z,\zeta)=J(\varphi(z),\varphi(\zeta))\,\varphi'(z)\,\overline{\varphi'(\zeta)}\,.$$

PROOF. First note that

$$\int_D |f(w)|^2 \, dA = \int_\Omega |f(\varphi(z))|^2 |\varphi'(z)|^2 \, dA$$

for each function $f \in A^2(D)$, since $|\varphi'(z)|^2$ is the Jacobian of the conformal mapping φ. In other words, the linear mapping $T(f) = g$ defined by

$$g(z) = f(\varphi(z))\varphi'(z)$$

is an *isometry* of $A^2(D)$ into $A^2(\Omega)$, since the norm is preserved. In fact, it easy to see that T maps $A^2(D)$ *onto* $A^2(\Omega)$. Indeed, if $g \in A^2(\Omega)$ and $z = \psi(w)$ is the inverse mapping, then

$$f(w) = g(\psi(w))\psi'(w)$$

is in $A^2(D)$ and $T(f) = g$.

Since $J(w, \omega)$ is the kernel function of D, the reproducing formula

$$f(w) = \int_D J(w, \omega) \, f(\omega) \, dA(\omega) \, , \qquad w \in D \, ,$$

holds for every function $f \in A^2(D)$. Changing variables, we find that

$$f(\varphi(z)) = \int_\Omega J(\varphi(z), \varphi(\zeta)) \, f(\varphi(\zeta)) \, |\varphi'(\zeta)|^2 \, dA(\zeta) \, , \qquad z \in \Omega \, ,$$

which reduces to

$$g(z) = \int_\Omega K(z, \zeta) \, g(\zeta) \, dA(\zeta) \, , \qquad z \in \Omega \, ,$$

where $g(z)$ is defined as above. Thus $K(z, \zeta)$ has the reproducing property, and is therefore the kernel function of Ω. □

As a consequence, it is possible to compute the Riemann mapping function in terms of the Bergman kernel function.

COROLLARY. *Let $\Omega \subset \mathbb{C}$ be a simply connected domain, not the whole plane, and let $K(z, \zeta)$ be its kernel function. Let $w = \varphi(z)$ map Ω conformally onto the unit disk, with $\varphi(\zeta) = 0$ and $\varphi'(\zeta) > 0$. Then*

$$\varphi'(z) = \sqrt{\frac{\pi}{K(\zeta, \zeta)}} \, K(z, \zeta) \, .$$

PROOF. The kernel function of $\mathbb{D}$ is

$$J(w,\omega)=\frac{1}{\pi}\frac{1}{(1-w\overline{\omega})^2},$$

so the theorem provides the relation

$$K(z,\zeta)=\frac{1}{\pi}\frac{\varphi'(z)\,\overline{\varphi'(\zeta)}}{(1-\varphi(z)\overline{\varphi(\zeta)})^2}=\frac{1}{\pi}\varphi'(z)\varphi'(\zeta),$$

since $\varphi(\zeta)=0$ and $\varphi'(\zeta)>0$. Consequently, $\pi\,K(\zeta,\zeta)=\varphi'(\zeta)^2$, which gives the desired result. □

The corollary is of practical importance in numerical calculation of the Riemann mapping function. If Ω is a Jordan domain, for instance, then the polynomials are dense in $A^2(\Omega)$ and an orthonormal basis can be computed by the Gram-Schmidt procedure. By Theorem 2, this allows the kernel function and hence the Riemann mapping function to be calculated.

§1.4. An extremal problem.

In Section 1.1 it was observed that $\|f\|_2 \geq K(\zeta,\zeta)^{-1/2}$ for all $f\in A^2$ with $f(\zeta)=1$, and equality occurs for the function $f(z)=K(z,\zeta)/K(\zeta,\zeta)$. This simple result suggests a more general extremal problem. Suppose for simplicity that Ω is a bounded domain, so that $A^2(\Omega)$ contains all polynomials. Given distinct points $\zeta_1,\zeta_2,\dots,\zeta_n$ in Ω and arbitrary complex numbers $\alpha_1,\alpha_2,\dots,\alpha_n$, the problem is to find the function $f\in A^2(\Omega)$ of smallest norm such that $f(\zeta_j)=\alpha_j$ for $j=1,2,\dots,n$. The existence of a unique extremal function is guaranteed by a theorem of functional analysis (*cf.* Section 2.2), since the family

$$\mathcal{F}=\{f\in A^2(\Omega)\ :\ f(\zeta_j)=\alpha_j\,,\qquad j=1,2,\dots,n\}$$

can be viewed as a closed convex subset of a Hilbert space. Note that $\mathcal{F}\neq\emptyset$ because it contains suitable polynomials. Alternatively, the existence of an extremal function is easily inferred by observing that competition may as well be restricted to the normal family of functions $f\in\mathcal{F}$ with $\|f\|_2\leq\|P\|_2$, where P is some polynomial in $\mathcal{F}$. A variational argument now shows that an extremal function F must be orthogonal to every function in the family

$$\mathcal{F}_0=\{f\in A^2(\Omega)\ :\ f(\zeta_j)=0\,,\qquad j=1,2,\dots,n\}.$$

Indeed, choose an arbitrary function $g\in\mathcal{F}_0$ and note that $(F+\lambda g)\in\mathcal{F}$ for each $\lambda\in\mathbb{C}$. Thus by the norm-minimizing property of F, it is clear that

$$\|F\|^2\leq\|F+\lambda g\|^2=\|F\|^2+2\,\mathrm{Re}\{\lambda\,\langle g,F\rangle\}+|\lambda|^2\,\|g\|^2\,,$$

so that $2\,\mathrm{Re}\{\lambda\,\langle g, F\rangle\} + |\lambda|^2\,\|g\|^2 \ge 0$. Now let λ tend to zero along the ray $\lambda = re^{i\theta}$ to conclude that $\mathrm{Re}\{e^{i\theta}\langle g, F\rangle\} = 0$ for every $e^{i\theta}$. Thus $\langle g, F\rangle = 0$ for each $g \in \mathcal{F}_0$, as asserted.

Observe now that by virtue of its reproducing property, each of the kernel functions $K(\cdot, \zeta_j)$ is orthogonal to $\mathcal{F}_0$. We claim that F is actually a linear combination of these kernel functions. To see this, let M be the set of all such linear combinations

$$h(z) = \sum_{j=1}^{n} c_j\, K(z, \zeta_j)\,, \qquad c_j \in \mathbb{C}\,.$$

Since M is a subspace of $A^2(\Omega)$, our extremal function has a unique decomposition as $F = h + g$, where $h \in M$ and $g \in M^{\perp}$. Here $M^{\perp}$ denotes the orthogonal complement of M, the set of all functions in $A^2(\Omega)$ that are orthogonal to every function in M. But it is clear that $M^{\perp} = \mathcal{F}_0$. Since the extremal function F is orthogonal to $\mathcal{F}_0$, it follows that $F = h \in M$, as asserted. Indeed,

$$\|g\|^2 = \langle g, g+h\rangle = \langle g, F\rangle = 0\,.$$

The uniqueness of the extremal function now follows by linear algebra, since it need only be shown that the interpolation conditions $F(\zeta_j) = \alpha_j$ uniquely determine the coefficients c_j. This in turn follows from the nonvanishing of the Gram determinant

$$\begin{vmatrix} K(\zeta_1,\zeta_1) & K(\zeta_1,\zeta_2) & \dots & K(\zeta_1,\zeta_n) \\ K(\zeta_2,\zeta_1) & K(\zeta_2,\zeta_2) & \dots & K(\zeta_2,\zeta_n) \\ \vdots & \vdots & & \vdots \\ K(\zeta_n,\zeta_1) & K(\zeta_n,\zeta_2) & \dots & K(\zeta_n,\zeta_n) \end{vmatrix},$$

a consequence of the fact that the associated Hermitian form

$$\sum_{j=1}^{n}\sum_{k=1}^{n} K(\zeta_k,\zeta_j)\beta_j\overline{\beta_k} = \left\|\sum_{j=1}^{n} \beta_j K(\cdot,\zeta_j)\right\|^2$$

is positive definite. The uniqueness also follows from a general theorem of functional analysis (Section 2.2, Theorem 1).

When $n = 1$ and the interpolation condition is simply $f(\zeta) = 1$, the above argument can be adapted to prove directly (without appeal to the Riesz representation theorem) the existence of a kernel function $k_\zeta(z) = K(z,\zeta)$ with the reproducing property $f(\zeta) = \langle f, k_\zeta\rangle$ for every $f \in A^2(\Omega)$.

For an elementary treatment of kernel functions and their properties in a more general context, the reader is referred to Epstein [1]. A more comprehensive source is Saitoh [1].

§1.5. Connection with Green's function.

Suppose now for the sake of simplicity that Ω is a finitely connected domain in the plane, bounded by analytic Jordan curves. For $\zeta \in \Omega$, *Green's function* $G(z,\zeta)$ is the function harmonic in Ω except at ζ, where it has a logarithmic singularity, and continuous in the closure $\overline{\Omega}$, with boundary values $G(z,\zeta) = 0$ for all $z \in \partial\Omega$. To say it has a logarithmic singularity at ζ means that $G(z,\zeta) - \log \frac{1}{|z-\zeta|}$ is harmonic in a neighborhood of ζ. The symmetry relation $G(z,\zeta) = G(\zeta,z)$ is easily established with the help of *Green's formula*

$$\int_\Omega (u\,\Delta v - v\,\Delta u)\,dA = \int_{\partial\Omega} \left(u\frac{\partial v}{\partial n} - v\frac{\partial u}{\partial n} \right) ds\,, \qquad u,v \in C^2(\overline{\Omega})\,,$$

where ds is the element of arclength, $\partial/\partial n$ is the outer normal derivative, and

$$\Delta u = \frac{\partial^2 u}{\partial x^2} + \frac{\partial^2 u}{\partial y^2} = 4\frac{\partial^2 u}{\partial z \partial \overline{z}}\,, \qquad z = x + iy\,,$$

denotes the Laplacian.

Green's function is conformally invariant. In other words, if $w = \varphi(z)$ maps a domain D conformally onto Ω, and if $G(w,\omega)$ is Green's function of Ω, then $H(z,\zeta) = G(\varphi(z),\varphi(\zeta))$ is Green's function of D.

It is well known that the normal derivative of Green's function provides a resolvent kernel for the Dirichlet problem. Specifically, a function $u(z)$ harmonic in Ω and continuous in $\overline{\Omega}$ can be recovered from its boundary values by the formula

$$u(z) = -\frac{1}{2\pi}\int_{\partial\Omega} \frac{\partial G}{\partial n}(z,\zeta)\,u(\zeta)\,|d\zeta|\,,$$

Green's function can also be used to recover an arbitrary function $u \in C^2(\overline{\Omega})$ from its Laplacian under the assumption that $u(z) = 0$ on $\partial\Omega$. The formula is

$$u(z) = -\frac{1}{2\pi}\int_\Omega G(z,\zeta)\,\Delta u(\zeta)\,dA(\zeta)\,, \qquad z \in \Omega\,.$$

(See for instance Nehari [1] or Duren [6] for this elementary potential theory.)

It is less well known that Green's function is closely connected with the Bergman kernel function. The following result is due to Bergman and Schiffer [1].

THEOREM 4. *Let Ω be a finitely connected domain bounded by analytic Jordan curves, and let $G(z,\zeta)$ be Green's function of Ω. Then the Bergman kernel function is*

$$K(z,\zeta) = -\frac{2}{\pi}\frac{\partial^2 G}{\partial z \partial \overline{\zeta}}\,, \qquad z \neq \zeta\,.$$

Before passing to the general case, let us outline a proof for simply connected domains. Green's function of a simply connected domain Ω has the form $G(z,\zeta) = -\log|\varphi(z)|$, where φ maps Ω conformally onto $\mathbb{D}$ and $\varphi(\zeta) = 0$. In particular, Green's function of $\mathbb{D}$ is

$$G(z,\zeta) = -\log\left|\frac{z-\zeta}{1-\overline{\zeta}z}\right|,$$

and it is easy to verify the formula in this very special case. To deduce it for a general simply connected domain $\Omega \neq \mathbb{C}$, one has only to observe that both sides of the formula transform in the same manner under conformal mapping.

PROOF OF THEOREM. First observe that the singularity of Green's function disappears under the indicated differentiation. Indeed, Green's function has the form

$$G(z,\zeta) = \log\frac{1}{|z-\zeta|} + h(z,\zeta)$$

in some neighborhood of ζ, where $h(z,\zeta)$ is a harmonic function of z. Thus

$$\frac{\partial G}{\partial z}(z,\zeta) = -\frac{1}{2}\frac{1}{z-\zeta} + \frac{\partial h}{\partial z}(z,\zeta),$$

and so

$$\frac{\partial^2 G}{\partial z \partial\overline{\zeta}}(z,\zeta) = \frac{\partial^2 h}{\partial z \partial\overline{\zeta}}(z,\zeta), \qquad z \neq \zeta.$$

Since the boundary curves are analytic and Green's function vanishes on the boundary, it has a harmonic extension across the boundary. Also note that $\frac{\partial G}{\partial z}(z,\zeta)$ is an analytic function of z, so the same is true of $\frac{\partial^2 G}{\partial z\partial\overline{\zeta}}(z,\zeta)$. We will show that it reproduces each function $f \in A^2(\Omega)$ up to a multiplicative constant.

The proof makes use of the *Cauchy–Green theorem*

$$\int_{\partial\Omega} F(z)\,dz = 2i\int_{\Omega}\frac{\partial F}{\partial\overline{z}}\,dA, \qquad F \in C^1(\overline{\Omega}).$$

Suppose first that f is analytic in Ω and continuous in $\overline{\Omega}$. Let Ω_ε be the domain Ω with a small disk $|z-\zeta| \leq \varepsilon$ excised, and let Γ_ε denote the boundary of this disk. Since G and therefore also $\frac{\partial G}{\partial z}$ vanishes for ζ on $\partial\Omega$, the Cauchy–Green theorem gives

$$\frac{1}{2i}\int_{\Gamma_\varepsilon}\frac{\partial G}{\partial z}(z,\zeta)\,f(\zeta)\,d\zeta = -\int_{\Omega_\varepsilon}\frac{\partial^2 G}{\partial z\partial\overline{\zeta}}(z,\zeta)\,f(\zeta)\,dA(\zeta),$$

where the circle Γ_ε is described in the counterclockwise direction. But by the Cauchy formula,

$$\int_{\Gamma_\varepsilon}\frac{\partial G}{\partial z}(z,\zeta)\,f(\zeta)\,d\zeta = \int_{\Gamma_\varepsilon}\left(-\frac{1}{2}\frac{1}{z-\zeta} + \frac{\partial h}{\partial z}(z,\zeta)\right) f(\zeta)\,d\zeta \to \pi i\,f(z)$$

as $\varepsilon \to 0$. This shows that

$$-\frac{2}{\pi}\int_\Omega \frac{\partial^2 G}{\partial z \partial \overline{\zeta}}(z,\zeta)\, f(\zeta)\, dA(\zeta) = f(z)$$

for functions f analytic in Ω and continuous in $\overline{\Omega}$.

Finally, because Ω has analytic boundary, it can be shown that the kernel function of Ω has an analytic continuation across the boundary. Hence the function $f(z) = K(z,\eta)$ qualifies, and the above formula gives

$$-\frac{2}{\pi}\int_\Omega \frac{\partial^2 G}{\partial z \partial \overline{\zeta}}(z,\zeta)\, K(\zeta,\eta)\, dA(\zeta) = K(z,\eta)\,.$$

But by the reproducing property of the kernel function, the integral is equal to

$$-\frac{2}{\pi}\frac{\partial^2 G}{\partial z \partial \overline{\zeta}}(z,\eta)\,.$$

This completes the proof. □

It may be remarked that the restriction to analytic boundaries can be removed by conformal mapping, since both sides of the formula transform in the same manner.

§1.6. The biharmonic Green function.

A function u is said to be *biharmonic* if its Laplacian Δu is harmonic, or equivalently if it has a vanishing *bilaplacian*: $\Delta^2 u = \Delta(\Delta u) = 0$. An important example is $u(z) = |z|^2 \log |z|$, for which

$$\Delta u = 4(1 + \log |z|)\,, \qquad z \neq 0\,,$$

so that $\Delta^2 u = 0$ for $z \neq 0$.

To prepare the ground for applications to the theory of contractive divisors in Chapter 5, we shall now discuss the biharmonic Green function in some detail. Postponing a formal definition, we begin with some preliminary remarks. The biharmonic Green function of a smoothly bounded domain $\Omega \subset \mathbb{C}$ is a real-valued function $\Gamma(z,\zeta)$ defined for $z, \zeta \in \Omega$ with $z \neq \zeta$. For each fixed $\zeta \in \Omega$, it is a biharmonic function of z in $\Omega \setminus \{\zeta\}$, and it vanishes together with its normal derivative on the boundary $\partial\Omega$. Just as the ordinary Green function $G(z,\zeta)$ can be used to recover a function from its Laplacian, the biharmonic Green function recovers a function from its bilaplacian according to the formula

$$u(z) = \frac{1}{\pi}\int_\Omega \Gamma(z,\zeta)\, \Delta^2 u(\zeta)\, dA(\zeta)\,,$$

where $u \in C^4(\overline{\Omega})$ and $u(z) = \frac{\partial u}{\partial n}(z) = 0$ on $\partial\Omega$.

The biharmonic Green function has physical significance as an "influence function". If a thin metal plate in the shape of the domain Ω is clamped at constant height on the boundary and is subjected to a unit load at the point ζ, then $\Gamma(z,\zeta)$ is its resulting deflection at the point z. This physical interpretation led Hadamard to conjecture that $\Gamma(z,\zeta) > 0$ for every convex domain Ω, but Duffin [1] found that the infinite strip is a counterexample. Garabedian [1,2] then showed that $\Gamma(z,\zeta)$ changes sign when Ω is a sufficiently eccentric ellipse. Shapiro and Tegmark [1] have devised an elementary proof of Garabedian's result. On the other hand, $\Gamma(z,\zeta)$ is positive when Ω is a circular disk. It is an open problem to determine the domains for which $\Gamma(z,\zeta) > 0$.

To give a more formal definition, let Ω be a domain whose boundary consists of finitely many analytic Jordan curves. Recall that the ordinary Green function $G(z,\zeta)$ is taken to be harmonic except for a logarithmic singularity at ζ. Analogously, the *biharmonic Green function* $\Gamma(z,\zeta)$ of Ω is defined by the following requirements.

(*i*) $\Gamma(z,\zeta)$ is a real-valued biharmonic function of z in $\Omega\setminus\{\zeta\}$ for each $\zeta \in \Omega$.

(*ii*) For each fixed $\zeta \in \Omega$, the function $\Gamma(z,\zeta) - \frac{1}{8}|z-\zeta|^2 \log|z-\zeta|$ is biharmonic in a neighborhood of ζ.

(*iii*) $\Gamma(z,\zeta)$ is of class $C^4(\overline{\Omega}\setminus\{\zeta\})$, and $\Gamma(z,\zeta) = \frac{\partial\Gamma}{\partial n}(z,\zeta) = 0$ for all $z \in \partial\Omega$.

Here $\frac{\partial}{\partial n}$ denotes the outer normal derivative.

Note that the condition (*ii*) does not impose a discontinuity on $\Gamma(z,\zeta)$ or on its first partial derivatives at the point ζ, since the function $|z|^2\log|z|$ is of class C^1 at the origin. This shows in particular that $\Gamma(z,\zeta)$ is of class $C^1(\overline{\Omega})$. However, in view of the formula

$$\Delta\{|z|^2\log|z|\} = 4\log|z| + 4\,,$$

the condition (*ii*) implies that the Laplacian of $\Gamma(z,\zeta)$ has a logarithmic singularity at ζ. Specifically, it implies that

$$\Delta\Gamma(z,\zeta) - \frac{1}{2}\log|z-\zeta|$$

is harmonic in a neighborhood of ζ.

The existence and uniqueness of the biharmonic Green function are not at all obvious. The uniqueness will follow from the reproducing formula, whose proof is developed below. Our calculations will also establish the symmetry property $\Gamma(\zeta,z) = \Gamma(z,\zeta)$. Existence is more difficult to prove, but for each fixed $\zeta \in \Omega$ it reduces to finding a function

$$u(z) = \Gamma(z,\zeta) - \frac{1}{8}|z-\zeta|^2\log|z-\zeta| \tag{1}$$

biharmonic in Ω with boundary data

$$\begin{aligned} u(z) &= -\frac{1}{8}\,|z-\zeta|^2 \log|z-\zeta|\,, \\ \frac{\partial u}{\partial n}(z) &= -\frac{1}{8}\frac{\partial}{\partial n}\{|z-\zeta|^2 \log|z-\zeta|\} \end{aligned} \tag{2}$$

for $z \in \partial\Omega$. This is a generalized Dirichlet problem which has a unique solution smooth up to the boundary. (See for instance Friedman [1], p. 37.) In the case of particular interest where Ω is the unit disk, we will prove the existence of the biharmonic Green function by deriving an explicit formula for it.

We show now that the biharmonic Green function recovers a function u from its bilaplacian according to the integral formula displayed above, provided that $u = \partial u/\partial n = 0$ on $\partial\Omega$. Fix an arbitrary point $\zeta \in \Omega$, and let Ω_ε denote the domain obtained from Ω by removing a small closed disk centered at ζ with radius ε. Let C_ε denote the circle $|z-\zeta| = \varepsilon$ bounding that disk. Then for $u \in C^4(\overline{\Omega})$, Green's formula gives

$$\begin{aligned} \int_{\Omega_\varepsilon} \Gamma(z,\zeta)\Delta^2 u(z)\,dA(z) &= \int_{\Omega_\varepsilon} \Delta\Gamma(z,\zeta)\,\Delta u(z)\,dA(z) \\ &\quad + \int_{\partial\Omega_\varepsilon}\left(\Gamma(z,\zeta)\,\frac{\partial}{\partial n}\Delta u(z) \;-\; \Delta u(z)\,\frac{\partial \Gamma}{\partial n}(z,\zeta)\right)|dz| \\ &= \int_{\Omega_\varepsilon} \Delta\Gamma(z,\zeta)\Delta u(z)\,dA(z) \\ &\quad + \int_{C_\varepsilon}\left(\Gamma(z,\zeta)\,\frac{\partial}{\partial n}\Delta u(z) \;-\; \Delta u(z)\,\frac{\partial \Gamma}{\partial n}(z,\zeta)\right)|dz|\,, \end{aligned} \tag{3}$$

since Γ and $\partial\Gamma/\partial n$ both vanish on $\partial\Omega$. Here and elsewhere, $\Delta\Gamma(z,\zeta)$ denotes the Laplacian with respect to z. But by another application of Green's formula, the first term on the last line of (3) is

$$\begin{aligned} &\int_{\Omega_\varepsilon} \Delta\Gamma(z,\zeta)\Delta u(z)\,dA(z) \\ &= \int_{\Omega_\varepsilon} \Delta^2\Gamma(z,\zeta)u(z)\,dA(z) + \int_{\partial\Omega_\varepsilon}\left(\Delta\Gamma(z,\zeta)\frac{\partial u}{\partial n}(z) \;-\; u(z)\frac{\partial}{\partial n}\Delta\Gamma(z,\zeta)\right)|dz| \\ &= \int_{C_\varepsilon}\left(\Delta\Gamma(z,\zeta)\,\frac{\partial u}{\partial n}(z) \;-\; u(z)\,\frac{\partial}{\partial n}\Delta\Gamma(z,\zeta)\right)|dz|\,, \end{aligned} \tag{4}$$

because Γ is biharmonic in Ω_ε and both u and $\partial u/\partial n$ are assumed to vanish on $\partial\Omega$. Recalling that $\Delta\Gamma(z,\zeta)$ behaves like $\frac{1}{2}\log|z-\zeta|$ near the point ζ, we see that

$$\int_{C_\varepsilon} \Delta\Gamma(z,\zeta)\frac{\partial u}{\partial n}(z)\,|dz| \to 0 \qquad \text{and} \qquad \int_{C_\varepsilon} u(z)\frac{\partial}{\partial n}\Delta\Gamma(z,\zeta)\,|dz| \to -\pi u(\zeta)$$

as $\varepsilon \to 0$. Thus we find

$$\int_\Omega \Delta\Gamma(z,\zeta)\Delta u(z)\, dA(z) = \pi\, u(\zeta)\,,$$

a formula that is also of interest. Finally, since $\Gamma(z,\zeta)$ and $\frac{\partial\Gamma}{\partial n}(z,\zeta)$ are continuous at ζ, we see that the last term in (3) satisfies

$$\lim_{\varepsilon\to 0}\int_{C_\varepsilon}\left(\Gamma(z,\zeta)\,\frac{\partial}{\partial n}\Delta u(z) - \Delta u(z)\,\frac{\partial\Gamma}{\partial n}(z,\zeta)\right)|dz| = 0\,.$$

Thus we conclude that

$$\int_\Omega \Gamma(z,\zeta)\,\Delta^2 u(z)\, dA(z) = \pi\, u(\zeta)\,,$$

which is the desired formula since $\Gamma(\zeta,z) = \Gamma(z,\zeta)$.

Observe now that if u is biharmonic in Ω and has a smooth extension to the closure, the same calculations give the explicit formula

$$u(\zeta) = -\frac{1}{\pi}\int_{\partial\Omega}\left(\Delta\Gamma(z,\zeta)\,\frac{\partial u}{\partial n}(z)\; - u(z)\,\frac{\partial}{\partial n}\Delta\Gamma(z,\zeta)\right)|dz|\,, \qquad \zeta\in\Omega\,, \quad (5)$$

for the recovery of u from its boundary data. Now the integral on the left-hand side of (3) vanishes, since $\Delta^2 u(z) = 0$. In (4), the integral over $\partial\Omega$ is coupled with the integral over C_ε to give (5) as $\varepsilon \to 0$.

For a proof of the symmetry property $\Gamma(\zeta,z) = \Gamma(z,\zeta)$, we choose a pair of distinct points ζ and η in Ω and apply the formula (5), or rather its derivation, first to $\Gamma(z,\zeta)$ with $u(z) = \Gamma(z,\eta)$, then to $\Gamma(z,\eta)$ with $u(z) = \Gamma(z,\zeta)$. We then arrive at the symmetry relation $\Gamma(\zeta,\eta) = \Gamma(\eta,\zeta)$.

The reproducing formula (5) shows in particular that if $u \in C^4(\overline{\Omega})$ is a biharmonic function with boundary data $u(z) = \frac{\partial u}{\partial n}(z) = 0$ on $\partial\Omega$, then $u(z) \equiv 0$ in Ω. This proves the uniqueness of the biharmonic Green function, since the difference of any two functions with the defining properties (i), (ii), and (iii) is biharmonic in Ω and has zero boundary data.

Our aim is now to derive an explicit formula for $\Gamma(z,\zeta)$ when the domain Ω is the unit disk $\mathbb{D}$. In this case we first make the general observation that each boundary value problem of the form

$$\begin{cases} \Delta^2 u = 0 & \text{in } \mathbb{D}\,, \\ u = \varphi\,, \quad \dfrac{\partial u}{\partial n} = \psi & \text{on } \mathbb{T} = \partial\mathbb{D}\,, \end{cases} \quad (6)$$

reduces to two ordinary Dirichlet problems. It is a known fact, not difficult to verify (*cf.* Garabedian [2], p. 269), that a function u is biharmonic if and only if it has the form

$$u(z) = (|z|^2 - 1)h(z) + H(z)\,, \quad (7)$$

where h and H are harmonic. In fact, we will need only the easy part of this statement, that every such function u is biharmonic. A simple calculation then shows that

$$\frac{\partial u}{\partial n} = 2h + \frac{\partial H}{\partial n} \qquad \text{on } \mathbb{T}.$$

Thus the boundary value problem (6) reduces to finding a pair of functions h and H harmonic in $\mathbb{D}$, satisfying $H = \varphi$ and $h = \frac{1}{2}(\psi - \frac{\partial H}{\partial n})$ on $\mathbb{T}$. Any smoothness properties of φ and ψ will be transferred to H and h, hence to u, by the Poisson formula. In particular, if φ and ψ are real-analytic on the boundary, then $u \in C^\infty(\overline{\mathbb{D}})$.

Returning now to the biharmonic Green function, we recall that the problem is to find a function u biharmonic in $\mathbb{D}$ with the properties (2) on the boundary $\mathbb{T} = \partial\mathbb{D}$, for each fixed $\zeta \in \mathbb{D}$. This is a problem of the form (6) with

$$\varphi(z) = -\frac{1}{16}|z-\zeta|^2 \log|z-\zeta|^2, \qquad z \in \mathbb{T},$$

and $\psi = \partial\varphi/\partial n$. If $u(z) = u(z,\zeta)$ is the solution of (6), then by (1) the biharmonic Green function of $\mathbb{D}$ is

$$\Gamma(z,\zeta) = u(z,\zeta) + \frac{1}{16}|z-\zeta|^2 \log|z-\zeta|^2. \tag{8}$$

We will look for a solution to (6) of the form (7). Our task is then to find the function H harmonic in $\mathbb{D}$ and continuous in $\overline{\mathbb{D}}$, with $H = \varphi$ on $\mathbb{T}$. This we can do by inspection, since

$$|z-\zeta|^2 = (z-\zeta)(1-\overline{\zeta}z)/z, \qquad z \in \mathbb{T}.$$

Thus the function

$$H(z) = -\frac{1}{8}\operatorname{Re}\left\{(z-\zeta)(1-\overline{\zeta}z)\frac{1}{z}\log(1-\overline{\zeta}z)\right\} \tag{9}$$

agrees with $\varphi(z)$ on $\mathbb{T}$ and is harmonic in $\overline{\mathbb{D}}$, since it is the real part of an analytic function.

The next step is to find the function h harmonic in $\mathbb{D}$ and continuous in $\overline{\mathbb{D}}$, with boundary values $h = \frac{1}{2}(\psi - \frac{\partial H}{\partial n})$, where $\psi = \frac{\partial\varphi}{\partial n}$. Note that although $H = \varphi$ on the boundary, it does not follow that $\frac{\partial H}{\partial n} = \psi$. Straightforward but tedious calculations yield the formulas

$$\begin{aligned}\psi(e^{i\theta}) &= -\frac{1}{16}\left[\frac{\partial}{\partial r}\left\{|re^{i\theta}-\zeta|^2 \log|re^{i\theta}-\zeta|^2\right\}\right]_{r=1} \\ &= -\frac{1}{8}\left(1-\operatorname{Re}\{\zeta e^{-i\theta}\}\right)\left(1+\operatorname{Re}\{\log(e^{i\theta}-\zeta)^2\}\right),\end{aligned}$$

and

$$\frac{\partial H}{\partial n}(e^{i\theta}) = -\frac{1}{8}\,\mathrm{Re}\left[\frac{\partial}{\partial r}\left\{\left(re^{i\theta}-\zeta\right)\left(\frac{1}{r}\,e^{-i\theta}-\overline{\zeta}\right)\log\left(1-\overline{\zeta}re^{i\theta}\right)\right\}\right]_{r=1}$$
$$= -\frac{1}{8}\left(\mathrm{Im}\{\overline{\zeta}e^{i\theta}\}\mathrm{Im}\{\log\left(1-\overline{\zeta}e^{i\theta}\right)^2\} - \mathrm{Re}\{\overline{\zeta}e^{i\theta}\} + |\zeta|^2\right).$$

In view of the identity $\mathrm{Re}\{(x+iy)(u+iv)\} = xu - yv$, we therefore arrive at the expression

$$\frac{1}{2}\left(\psi - \frac{\partial H}{\partial n}\right) = \frac{1}{16}\,\mathrm{Re}\{\left(\zeta e^{-i\theta}-1\right)\log\left(1-\overline{\zeta}e^{i\theta}\right)^2 + |\zeta|^2 - 1\}\,.$$

It is now apparent that the function

$$h(z) = \frac{1}{16}\,\mathrm{Re}\left\{(\zeta - z)\frac{1}{z}\log(1-\overline{\zeta}z)^2 + |\zeta|^2 - 1\right\} \tag{10}$$

is harmonic in $\overline{\mathbb{D}}$ and agrees with $\frac{1}{2}(\psi - \frac{\partial H}{\partial n})$ on $\mathbb{T}$.

Putting the expressions (9) and (10) for h and H into (7) and (8), we conclude after a short calculation that

$$\Gamma(z,\zeta) = \frac{1}{16}\left\{|z-\zeta|^2\log\left|\frac{z-\zeta}{1-\overline{\zeta}z}\right|^2 + (1-|\zeta|^2)(1-|z|^2)\right\}. \tag{11}$$

Thus we have derived a formula for the biharmonic Green function of the unit disk. Easy calculations confirm that it has the required properties (i), (ii), and (iii).

For the purpose of later applications, we now observe that $\Gamma(z,\zeta) > 0$ for all pairs of points z and ζ in the disk. In view of the identity

$$\left|\frac{1-\overline{\zeta}z}{z-\zeta}\right|^2 = 1 + \frac{(1-|\zeta|^2)(1-|z|^2)}{|z-\zeta|^2}\,,$$

the positivity of $\Gamma(z,\zeta)$ follows from the simple inequality

$$\log(1+x) < x\,, \qquad x > 0\,.$$

Specifically,

$$|z-\zeta|^2\log\left|\frac{z-\zeta}{1-\overline{\zeta}z}\right|^2 = -\,|z-\zeta|^2\log\left(1+\frac{(1-|\zeta|^2)(1-|z|^2)}{|z-\zeta|^2}\right)$$
$$> -\,(1-|\zeta|^2)(1-|z|^2)\,.$$

Thus $\Gamma(z,\zeta) > 0$ for all $z,\zeta \in \mathbb{D}$.

CHAPTER 2

Linear Space Properties

The study of Bergman spaces involves two rather different types of investigation. One is to explore the analytic or structural properties of individual functions in a Bergman space. The other is to study the special characteristics of A^p as a linear space. This book is concerned more with function-theoretic properties, but the present chapter will discuss basic linear space aspects of the Bergman spaces. Here our treatment is strongly influenced by Sheldon Axler's expository article [3], which may be consulted for further information. To set the stage, we begin with a brief introduction to Hardy spaces. Then follows a basic discussion of convexity properties for Banach spaces, which will prove useful in later chapters. A primary focus of the chapter is the Bergman projection and its applications. There is a long section on Bloch spaces and their connection with duality problems for the Bergman space A^1. Another section introduces the pseudohyperbolic metric, a tool that is used repeatedly throughout the book. In this chapter it is applied to develop a theory of Carleson measures for the Bergman spaces.

§2.1. Hardy spaces.

The theory of Bergman spaces draws much of its inspiration from the well established theory of Hardy spaces. Classical properties of Hardy spaces suggest analogous problems for Bergman spaces. The problems are generally more difficult for Bergman spaces, requiring new concepts and new techniques for their solution, but standard facts about Hardy spaces are still very relevant. We therefore begin this chapter with a brief overview of Hardy spaces, emphasizing their properties as linear spaces. A corresponding discussion of analytic properties will be deferred to the start of Chapter 3. Proofs and further information will be found in books such as Duren [5], Garnett [1], and Koosis [1].

For a function f analytic in the unit disk $\mathbb{D}$, the *integral means* are defined by

$$M_p(r,f) = \left\{ \frac{1}{2\pi} \int_0^{2\pi} |f(re^{i\theta})|^p \, d\theta \right\}^{1/p}, \qquad 0 < p < \infty ,$$

and

$$M_\infty(r,f) = \max_\theta |f(re^{i\theta})| .$$

These means $M_p(r, f)$ are nondecreasing functions of r. If they stay bounded as $r \to 1$, then f is said to belong to the *Hardy space* H^p. Thus H^∞ consists of all bounded analytic functions in $\mathbb{D}$. Each H^p class is a linear space, preserved under addition and scalar multiplication. The *norm* $\|f\|_{H^p}$ of a function $f \in H^p$ is defined as the limit of $M_p(r, f)$ as $r \to 1$. It is a true norm if $p \geq 1$.

Each function $f \in H^p$ has a radial limit

$$f(e^{i\theta}) = \lim_{r\to 1} f(re^{i\theta})$$

for almost every θ. The boundary function has the properties $f(e^{i\theta}) \in L^p = L^p(\partial\mathbb{D})$ and

$$\int_0^{2\pi} \log|f(e^{i\theta})|\, d\theta > -\infty$$

unless $f(z) \equiv 0$. In particular, the radial limit cannot vanish on any set of positive measure. The norm of f can be defined equivalently as the L^p norm of the boundary function $f(e^{i\theta})$. Thus H^p can be identified with a subspace of L^p. It is a *complete* normed linear space, or a Banach space, if $1 \leq p \leq \infty$.

For $0 < p < \infty$, each function in H^p can be approximated in norm by polynomials. Thus H^p is characterized as the closure of polynomials in the space L^p. A proof is based on the fact that $f(re^{i\theta}) \to f(e^{i\theta})$ in norm, meaning that

$$\lim_{r\to 1} \int_0^{2\pi} |f(re^{i\theta}) - f(e^{i\theta})|^p\, d\theta = 0\,.$$

An equivalent statement is that the *dilations* $f_\rho(z) = f(\rho z)$ tend to f in H^p norm as ρ increases to 1.

Every function $f \in H^1$ can be recovered from its boundary function by the Cauchy integral formula:

$$f(z) = \frac{1}{2\pi} \int_0^{2\pi} \frac{f(e^{it})}{e^{it} - z} e^{it}\, dt\,, \qquad z \in \mathbb{D}\,.$$

More generally, the *Szegő transform*

$$T\varphi(z) = \frac{1}{2\pi} \int_0^{2\pi} \frac{\varphi(t)}{1 - e^{-it}z}\, dt$$

maps an arbitrary function $\varphi \in L^1$ to a function analytic in $\mathbb{D}$. It can be shown that for $1 < p < \infty$, the Szegő transform maps L^p boundedly onto H^p. This result is equivalent to a famous theorem of M. Riesz on conjugate harmonic functions.

For $0 < p < \infty$, a real-valued function u harmonic in $\mathbb{D}$ is said to be of class h^p if its integral means $M_p(r, u)$ remain bounded as $r \to 1$. The class of bounded harmonic functions in the disk is denoted by h^∞. A harmonic function v is said to be a *harmonic conjugate* of u if $f = u + iv$ is analytic, or equivalently if u and v satisfy the Cauchy–Riemann equations

$$\frac{\partial u}{\partial x} = \frac{\partial v}{\partial y}, \qquad \frac{\partial u}{\partial y} = -\frac{\partial v}{\partial x}.$$

The harmonic conjugate is uniquely determined up to an arbitrary additive constant. The theorem of M. Riesz asserts that for $1 < p < \infty$, the class h^p is self-conjugate. More precisely, if $u \in h^p$, then $v \in h^p$, and under the normalization $v(0) = 0$ the mapping from u to v is bounded:

$$\int_0^{2\pi} |v(re^{i\theta})|^p \, d\theta \le C \int_0^{2\pi} |u(re^{i\theta})|^p \, d\theta$$

for some constant C depending only on p. Simple examples show that the theorem fails for $p = 1$ and for $p = \infty$.

If $f = u + iv$ is analytic in $\mathbb{D}$ and continuous in $\overline{\mathbb{D}}$, it can be recovered from the boundary values of u by the *Herglotz formula*

$$f(z) = \frac{1}{2\pi} \int_0^{2\pi} \frac{e^{it} + z}{e^{it} - z} u(e^{it}) \, dt + iv(0),$$

the analytic completion of the Poisson formula. Another formulation of the Riesz theorem is that the *Herglotz transform*

$$H\varphi(z) = \frac{1}{2\pi} \int_0^{2\pi} \frac{e^{it} + z}{e^{it} - z} \varphi(t) \, dt$$

maps L^p boundedly onto H^p for $1 < p < \infty$. Since the Herglotz kernel and the Szegő kernel are related by the simple identity

$$\frac{e^{it} + z}{e^{it} - z} = 2\,\frac{e^{it}}{e^{it} - z} - 1,$$

this is clearly equivalent to the boundedness of the Szegő projection.

Yet another variant relates to the partial sum operator S_N that assigns to an arbitrary analytic function

$$f(z) = \sum_{n=0}^{\infty} a_n z^n$$

the Nth partial sum of its Taylor series, namely $\sum_{n=0}^{N} a_n z^n$. It follows from the boundedness of the Szegő projection T that S_N maps H^p boundedly into itself for $1 < p < \infty$, and the norm $\|S_N\|$ is bounded independently of N. This means that for $1 < p < \infty$,

$$\int_0^{2\pi} |S_N f(e^{i\theta})|^p \, d\theta \le C \int_0^{2\pi} |f(e^{i\theta})|^p \, d\theta$$

for all $f \in H^p$ and for some constant C depending only on p and not on N. The proof is only to observe that

$$\overline{S_N f(e^{i\theta})} = e^{-iN\theta} T\big(e^{iN\theta}\, \overline{f(e^{i\theta})}\big).$$

It is enough to verify this relation when f is a polynomial, since the polynomials are dense in H^p.

§2.2. Strict and uniform convexity.

A Banach space X is said to be *strictly convex* if for each pair of elements $x, y \in X$ with $\|x\| = \|y\| = 1$ and $x \neq y$, it is true that $\left\|\frac{1}{2}(x+y)\right\| < 1$. Thus the midpoint of the segment joining two distinct points on the unit sphere lies strictly inside the sphere. The space X is said to be *uniformly convex* if to each $\varepsilon > 0$ there corresponds a $\delta > 0$ such that $\|x\| = \|y\| = 1$ and $\left\|\frac{1}{2}(x+y)\right\| > 1 - \delta$ imply $\|x - y\| < \varepsilon$. An equivalent requirement is that for any pair of sequences $\{x_n\}$ and $\{y_n\}$ in X such that $\|x_n\| = \|y_n\| = 1$ and $\|x_n + y_n\| \to 2$ as $n \to \infty$, it is necessarily true that $\|x_n - y_n\| \to 0$. It is clear that every uniformly convex space is strictly convex, but examples show the converse to be false.

A subset $E \subset X$ is *convex* if for each pair of points $x, y \in E$, the entire linear segment

$$\{ax + (1-a)y \; : \; 0 < a < 1\}$$

joining them lies in E. The following theorems demonstrate the importance of strict and uniform convexity.

THEOREM 1. *In a convex subset of a strictly convex Banach space there is at most one element of minimal norm.*

THEOREM 2. *Each closed convex subset of a uniformly convex Banach space contains a unique element of minimal norm.*

Both results have important applications to approximation theory. They will be useful as we develop the theory of Bergman spaces.

PROOF OF THEOREM 1. Let E be the convex set, and suppose there are distinct elements x and y of minimal norm: $\|x\| = \|y\| = d$. Then

$$d \leq \left\|\tfrac{1}{2}(x+y)\right\| \leq \tfrac{1}{2}(\|x\| + \|y\|) = d\,,$$

so $\|\frac{x+y}{2}\| = d$, violating the strict convexity of the space. Thus a minimal element, if it exists, is unique. □

PROOF OF THEOREM 2. Let E be the closed convex set and let $d = \inf_{x \in E} \|x\|$. Choose a sequence of points $x_n \in E$ with $\|x_n\| \to d$. Since X is a complete space and E is a closed subset, it will suffice to show that $\{x_n\}$ is a Cauchy sequence. This is clearly true if $d = 0$, so we may assume that $d > 0$. Then in view of the identity

$$x_n - x_m = \|x_n\| \left(\frac{x_n}{\|x_n\|} - \frac{x_m}{\|x_m\|} \right) + \frac{x_m}{\|x_m\|} \Big(\|x_n\| - \|x_m\| \Big)\,,$$

it will be enough to show that $\{\frac{x_n}{\|x_n\|}\}$ is a Cauchy sequence. By the definition of uniform convexity, this will follow if we can show that

$$\left\| \frac{x_n}{\|x_n\|} + \frac{x_m}{\|x_m\|} \right\| \to 2 \qquad \text{as } n, m \to \infty\,.$$

To this end, note first that, since E is convex,

$$d \leq \left\|\tfrac{1}{2}(x_n + x_m)\right\| \leq \tfrac{1}{2}\left(\|x_n\| + \|x_m\|\right) \to d\,,$$

so that $\|x_n + x_m\| \to 2d$. Next observe that

$$\frac{x_n}{\|x_n\|} + \frac{x_m}{\|x_m\|} = \frac{1}{\|x_n\|\|x_m\|}\Big\{(x_n + x_m)\|x_m\| + x_m\big(\|x_n\| - \|x_m\|\big)\Big\}\,,$$

which leads to the conclusion that

$$2 \geq \left\|\frac{x_n}{\|x_n\|} + \frac{x_m}{\|x_m\|}\right\| \geq \frac{1}{\|x_n\|}\Big\{\|x_n + x_m\| - \big|\|x_n\| - \|x_m\|\big|\Big\} \to 2$$

as $n, m \to \infty$. This concludes the proof that E has an element of smallest norm. Uniqueness follows from Theorem 1. □

It is an easy exercise, based on the parallelogram law, to show that every Hilbert space is uniformly convex. It is known that the spaces L^p and ℓ^p are uniformly convex for $1 < p < \infty$, but not for $p = 1$ or $p = \infty$. (See Clarkson [1], who introduced the concept; or Köthe [1].) Every uniformly convex space is reflexive, but the converse is false.

Since every subspace of a uniformly convex space must have the same property, each of the Bergman spaces $A^p = A^p(\mathbb{D})$ with $1 < p < \infty$ is uniformly convex. The space A^1 is not uniformly convex, as can be seen from the fact that A^1 is not reflexive (*cf.* Section 2.6). However, it is strictly convex. This last assertion can be verified as follows. Suppose $f, g \in A^1$, $\|f\|_1 = \|g\|_1 = 1$, and $\|f + g\|_1 = 2$. Then $\|f + g\|_1 = \|f\|_1 + \|g\|_1$, or

$$\int_{\mathbb{D}} |f(z) + g(z)|\, d\sigma = \int_{\mathbb{D}} |f(z)|\, d\sigma + \int_{\mathbb{D}} |g(z)|\, d\sigma$$

where $d\sigma = \frac{1}{\pi} dA$ denotes normalized area measure, with $\int_{\mathbb{D}} d\sigma = 1$. This is possible only if $f(z) = \lambda(z) g(z)$, where $\lambda(z) > 0$ at every point z where $f(z)g(z) \neq 0$. But $\lambda = f/g$ is a meromorphic function, so the requirement that $\lambda(z) > 0$ forces it to be constant. Thus $f(z) = c\, g(z)$ for some constant $c > 0$. But now it follows that $c = 1$, since $\|f\|_1 = \|g\|_1 = 1$. Thus $f = g$, which proves that A^1 is strictly convex.

On the other hand, it is easy to see that the space $L^1(\mathbb{D})$, with norm $\|f\|_1 = \int_{\mathbb{D}} |f(z)|\, d\sigma$, is not strictly convex. For this purpose we need only consider the functions

$$f(z) = \begin{cases} 2\,, & z \in \mathbb{D}\,,\ \operatorname{Re}\{z\} > 0 \\ 0\,, & z \in \mathbb{D}\,,\ \operatorname{Re}\{z\} \leq 0 \end{cases}$$

and $g(z) = 2 - f(z)$. Then $\|f\|_1 = \|g\|_1 = 1$ and $\|f + g\| = 2$, yet $f \neq g$.

§2.3. The Bergman projection.

In Chapter 1 we discussed the kernel function in the Bergman space $A^2(\Omega)$ over a general domain Ω in the plane. Taking Ω to be the unit disk $\mathbb{D}$, we showed that the polynomials are dense in the space $A^2 = A^2(\mathbb{D})$, and we used the result to calculate the Bergman kernel function for the disk. We shall now develop some generalizations to the Bergman space A^p over the disk, where $0 < p < \infty$. Henceforth σ will always denote normalized area measure, so that $\int_{\mathbb{D}} d\sigma = 1$. With slight change of convention from Chapter 1, the *norm* of a function $f \in A^p$ is now defined by

$$\|f\|_p = \left\{ \int_{\mathbb{D}} |f(z)|^p \, d\sigma(z) \right\}^{1/p}, \qquad 0 < p < \infty .$$

It is a true norm for $1 \le p < \infty$. The triangle inequality fails when $0 < p < 1$, but the inequality

$$\|f + g\|_p^p \le \|f\|_p^p + \|g\|_p^p$$

is often an adequate substitute. (See Duren [5], p. 37.)

We begin with a proof that the polynomials are dense in A^p.

THEOREM 3. *The polynomials form a dense subset of the Bergman space A^p, where $0 < p < \infty$. For each function $f \in A^p$ and each number $\varepsilon > 0$, there is a polynomial Q with $\|f - Q\|_p < \varepsilon$.*

PROOF. Given a function $f \in A^p$, let $f_\rho(z) = f(\rho z)$ be its dilations, where $0 < \rho < 1$. Each function f_ρ is analytic in a larger disk, so it can be approximated uniformly in $\mathbb{D}$ by polynomials, the partial sums of its Taylor series. Thus it will be enough to prove that f can be approximated in A^p norm by its dilations: $\|f - f_\rho\|_p \to 0$ as $\rho \to 1$. This result is surprisingly elusive, since one must make estimates that are uniform in the dilation parameter. To address the problem, recall that the integral means

$$M_p(r, f) = \left\{ \frac{1}{2\pi} \int_0^{2\pi} |f(re^{i\theta})|^p \, d\theta \right\}^{1/p}, \qquad 0 \le r < 1 ,$$

increase with r (see Section 2.1), and observe that $M_p(r, f_\rho) = M_p(r\rho, f)$. Therefore,

$$M_p^p(r, f - f_\rho) \le 2^p \{ M_p^p(r, f) + M_p^p(r, f_\rho) \} \le 2^{p+1} M_p^p(r, f).$$

But the hypothesis that $f \in A^p$ is equivalent to saying that $M_p^p(r, f)$ is integrable over the interval $[0, 1)$ with respect to the measure $r\,dr$, and it is clear that $f_\rho(z) \to f(z)$ uniformly on compact subsets of $\mathbb{D}$ as $\rho \to 1$, which implies that $M_p^p(r, f - f_\rho) \to 0$ for each $r \in [0, 1)$. Thus by the Lebesgue dominated convergence theorem, we may conclude that

$$\|f - f_\rho\|_p^p = 2 \int_0^1 M_p^p(r, f - f_\rho) r \, dr \to 0$$

as $\rho \to 1$, which completes the proof. $\square$

Theorem 3 states that given $f \in A^p$, one can find a sequence of polynomials that approach f in norm. With a little bit of effort, we can improve this result for $p > 1$ by showing that f can be approximated by the most natural choice of polynomials, the partial sums of its Taylor series.

For any analytic function f, define the partial sum operator S_N by $S_N f(z) = \sum_{n=0}^{N} a_n z^n$, where $f(z) = \sum_{n=0}^{\infty} a_n z^n$ is the Taylor series of f.

LEMMA 1. *Let X be an arbitrary Banach space of functions analytic in the unit disk, in which the polynomials are dense. Then $\|S_N f - f\| \to 0$ for every $f \in X$ if and only if $\sup_{N \geq 1} \|S_N\| < \infty$.*

Observe that the lemma may apply to the spaces $X = H^p$ and A^p for $1 \leq p < \infty$, since these are Banach spaces in which polynomials are dense.

PROOF OF LEMMA. If $\|S_N f - f\| \to 0$ for every $f \in X$, then it follows that $\sup_N \|S_N f\| < \infty$ for every f, which implies by the uniform boundedness principle that $\sup_N \|S_N\| < \infty$. Conversely, suppose $\|S_N\| \leq C < \infty$ for all $N \geq 1$. Given $f \in X$ and $\varepsilon > 0$, choose a polynomial Q such that $\|f - Q\| < \varepsilon$. Then

$$\|S_N f - f\| \leq \|S_N f - S_N Q\| + \|S_N Q - f\| \leq (C+1)\varepsilon$$

when N is larger than the degree of Q, because in that case $S_N Q = Q$. $\square$

It follows from the boundedness of the Szegő projection on the Hardy space H^p for $1 < p < \infty$ (see Section 2.1) that $\sup_{N \geq 1} \|S_N\| < \infty$. In other words, there is a constant C such that

$$\int_0^{2\pi} |S_N f(e^{i\theta})|^p \, d\theta \leq C \int_0^{2\pi} |f(e^{i\theta})|^p \, d\theta \,, \qquad f \in H^p \,.$$

By Lemma 1, this implies that for any $f \in H^p$, the partial sums of the Taylor series converge to f in H^p norm. An easy consequence is the analogous result for Bergman spaces.

THEOREM 4. *Let $1 < p < \infty$. If $f \in A^p$, then the partial sums of its Taylor series converge in norm to f.*

PROOF. The above inequality can be applied to the dilates $f_r(z) = f(rz)$, since f_r is analytic in the closed disk and is therefore a member of H^p. Thus we have

$$\begin{aligned} \|S_N f\|_p^p &= \frac{1}{\pi} \int_0^1 \int_0^{2\pi} |S_N f(re^{i\theta})|^p d\theta \, r \, dr \\ &\leq C \frac{1}{\pi} \int_0^1 \int_0^{2\pi} |f(re^{i\theta})|^p \, d\theta \, r \, dr = C \|f\|_p^p \,. \end{aligned}$$

The result then follows from Lemma 1. □

We remark that the above theorem is false when $p = 1$. This can be shown by considering functions of the form $f(z) = (1 - |\alpha|^2)/(1 - \overline{\alpha}z)^3$, $\alpha \in \mathbb{D}$. For details, see Zhu [2].

In terms of normalized area measure, the reproducing formula of Section 1.2 takes the form $f(z) = \langle f, k_z \rangle$ for $f \in A^2$, where $k_z(\zeta) = (1 - \overline{z}\zeta)^{-2}$. Alternatively, it can be written as

$$f(z) = \int_{\mathbb{D}} K(z,\zeta) f(\zeta)\, d\sigma(\zeta), \qquad \text{where } K(z,\zeta) = \frac{1}{(1-\overline{\zeta}z)^2}.$$

Since the polynomials are dense in A^1, the last formula remains valid for every function $f \in A^1$.

Now view A^2 as a (closed) subspace of the Hilbert space $L^2 = L^2(\mathbb{D})$, and let P denote the orthogonal projection of L^2 onto A^2. Thus $Pf \in A^2$ if $f \in L^2$, and so

$$Pf(z) = \langle Pf, k_z \rangle = \langle f, k_z \rangle = \int_{\mathbb{D}} \frac{f(\zeta)}{(1-\overline{\zeta}z)^2}\, d\sigma(\zeta),$$

since $(f - Pf)$ is orthogonal to $k_z \in A^2$. The projection operator P is known as the *Bergman projection*. By the formula

$$Pf(z) = \int_{\mathbb{D}} \frac{f(\zeta)}{(1-\overline{\zeta}z)^2}\, d\sigma(\zeta),$$

it is a well-defined linear operator on L^1, mapping each $f \in L^1$ to a function analytic in $\mathbb{D}$ and mapping each $f \in A^1$ to itself.

In fact, it will be shown that the Bergman projection is a bounded map on L^p for $1 < p < \infty$. The proof will make use of the following lemma, which will be applied repeatedly throughout the book.

LEMMA 2. *Let s and t be real numbers satisfying $1 < t < s$. Then there is a constant C, depending only on s and t, such that*

$$\int_{\mathbb{D}} \frac{(1-|\zeta|^2)^{t-2}}{|1-\overline{z}\zeta|^s}\, d\sigma(\zeta) \le C(1-|z|^2)^{t-s}$$

for all $z \in \mathbb{D}$.

PROOF. It is sufficient to consider $z = \rho \ge 0$. If $\rho \le 1/2$, we have

$$\int_{\mathbb{D}} \frac{(1-|\zeta|^2)^{t-2}}{|1-\rho\zeta|^s}\, d\sigma(\zeta) \le C(1-\rho)^{t-s},$$

where $C = 2^t \int_{\mathbb{D}} (1-|\zeta|^2)^{t-2} d\sigma(\zeta)$. This integral converges because $t > 1$.

Suppose now that $\rho > 1/2$, so that $1/(2\rho) < 1$. Note first that

$$\int_{|\zeta|\le\frac{1}{2\rho}} \frac{(1-|\zeta|^2)^{t-2}}{|1-\rho\zeta|^s}\, d\sigma(\zeta) \le 2^s \int_{\mathbb{D}} (1-|\zeta|^2)^{t-2}\, d\sigma(\zeta) \le C(1-\rho)^{t-s},$$

since $t-s<0$.

It remains to estimate

$$\int_{|\zeta|>\frac{1}{2\rho}} \frac{(1-|\zeta|^2)^{t-2}}{|1-\rho\zeta|^s}\, d\sigma(\zeta)$$
$$= \frac{2}{\pi}\int_{\frac{1}{2\rho}}^1 (1-r^2)^{t-2} r \int_0^\pi \frac{d\theta}{(1-2\rho r\cos\theta+\rho^2r^2)^{s/2}}\, dr.$$

Since $\sin x \ge 2x/\pi$ for $0\le x\le \pi/2$, we have

$$1-2\rho r\cos\theta+\rho^2r^2 = (1-\rho r)^2 + 4\rho r\sin^2(\theta/2) \ge (1-\rho r)^2 + 4\rho r\theta^2/\pi^2.$$

Thus for $r\ge 1/(2\rho)$,

$$\begin{aligned}\int_0^\pi \frac{d\theta}{(1-2\rho r\cos\theta+\rho^2r^2)^{s/2}} &\le \frac{1}{(1-\rho r)^s}\int_0^\pi \frac{d\theta}{\left\{1+\frac{2}{\pi^2}\left(\frac{\theta}{1-\rho r}\right)^2\right\}^{s/2}}\\ &\le \frac{1}{(1-\rho r)^{s-1}}\int_0^\infty \frac{du}{(1+\frac{2}{\pi^2}u^2)^{s/2}}\\ &\le C(1-\rho r)^{-s+1},\end{aligned}$$

because $s>1$ and so the last integral converges. Therefore, our integral is bounded by a constant multiple of

$$\int_0^\rho \frac{(1-r^2)^{t-2}}{(1-\rho r)^{s-1}}\, r\,dr + \int_\rho^1 \frac{(1-r^2)^{t-2}}{(1-\rho r)^{s-1}}\, r\,dr = I_1 + I_2.$$

To estimate I_1, observe that $0\le r\le\rho$ implies $1-\rho r\le 1-r^2$. If $t<2$, we make the estimate

$$I_1 \le \int_0^\rho (1-\rho r)^{t-s-1}\, dr \le \frac{2}{s-t}(1-\rho^2)^{t-s} \le C(1-\rho)^{t-s}.$$

If $t\ge 2$, we use the inequality $1-r^2\le 2(1-\rho r)$ to obtain

$$I_1 \le 2^{t-2}\int_0^\rho (1-\rho r)^{t-s-1}\, dr \le C(1-\rho)^{t-s}.$$

On the other hand, $r\le 1$ implies $1-\rho\le 1-\rho r$, and so

$$I_2 \le C(1-\rho)^{-s+1}\int_\rho^1 (1-r)^{t-2}\, dr \le C(1-\rho)^{t-s}.$$

This completes the proof of the lemma. □

We are now ready to prove that the Bergman projection is bounded.

THEOREM 5. *For $1 < p < \infty$, the Bergman projection P is a bounded operator from L^p onto A^p.*

PROOF. Since the restriction of P to A^p is the identity operator on A^p, and Pf is analytic for every $f \in L^p$, it is enough to show that P maps L^p into L^p in bounded fashion. But for $f \in L^p$ and $1/p + 1/q = 1$, Hölder's inequality gives

$$\begin{aligned} |Pf(z)| &\leq \int_{\mathbb{D}} |1-\overline{\zeta}z|^{-2}\,|f(\zeta)|\,d\sigma(\zeta) \\ &= \int_{\mathbb{D}} (1-|\zeta|^2)^{-1/pq}\,|1-\overline{\zeta}z|^{-2}\,|f(\zeta)|(1-|\zeta|^2)^{1/pq}\,d\sigma(\zeta) \\ &\leq J_1(z)^{1/q}\,J_2(z)^{1/p}, \end{aligned}$$

where

$$J_1(z) = \int_{\mathbb{D}} (1-|\zeta|^2)^{-1/p}\,|1-\overline{z}\zeta|^{-2}\,d\sigma(\zeta)$$

and

$$J_2(z) = \int_{\mathbb{D}} |f(\zeta)|^p\,|1-\overline{z}\zeta|^{-2}(1-|\zeta|^2)^{1/q}\,d\sigma(\zeta).$$

We use Lemma 2 with $s = 2$ and $t = 2 - 1/p$ to obtain the bound

$$|J_1(z)| \leq C(1-|z|^2)^{-1/p}.$$

Applying this estimate, we find by Fubini's theorem that

$$\begin{aligned} &\int_{\mathbb{D}} |Pf(z)|^p\,d\sigma(z) \leq \int_{\mathbb{D}} J_1(z)^{p/q}\,J_2(z)\,d\sigma(z) \\ &\leq C\int_{\mathbb{D}} (1-|z|^2)^{-1/q}\int_{\mathbb{D}} |f(\zeta)|^p\,|1-\overline{z}\zeta|^{-2}(1-|\zeta|^2)^{1/q}\,d\sigma(\zeta)\,d\sigma(z) \\ &= C\int_{\mathbb{D}} |f(\zeta)|^p(1-|\zeta|^2)^{1/q}\int_{\mathbb{D}} (1-|z|^2)^{-1/q}\,|1-\overline{\zeta}z|^{-2}\,d\sigma(z)\,d\sigma(\zeta) \\ &\leq C\int_{\mathbb{D}} |f(\zeta)|^p(1-|\zeta|^2)^{1/q}(1-|\zeta|^2)^{-1/q}\,d\sigma(\zeta) = C\int_{\mathbb{D}} |f(\zeta)|^p\,d\sigma(\zeta), \end{aligned}$$

where for the last inequality we have invoked again the same estimate of $J_1(z)$, but with p replaced by q. Here and elsewhere, C denotes a positive constant depending on p, not necessarily the same in each instance. This completes the proof. □

The theorem was first proved by Zaharjuta and Judovič [1], using the theory of singular integral operators. The foregoing proof is a modification by Axler [3] of an argument given by Forelli and Rudin [1]. The proof actually gives a slightly stronger result, recorded here for later reference.

COROLLARY. *If* $1 < p < \infty$, *then* $Tf(z) = \int_{\mathbb{D}} |1 - z\overline{\zeta}|^{-2} f(\zeta)\, d\sigma(\zeta)$ *defines a bounded linear operator* T *of* L^p *into* L^p.

§2.4. Dual spaces.

A *Banach space* is a complete normed linear space. The *dual space* of a Banach space X is the space X^* of bounded linear functionals on X. Each $\phi \in X^*$ has a norm

$$\|\phi\| = \sup_{\|x\|=1} |\phi(x)|\,,$$

under which X^* is itself a Banach space. According to the Riesz representation theorem, the dual space of L^p for $1 < p < \infty$ is isometrically isomorphic to L^q, where q is the *conjugate index*: $1/p + 1/q = 1$. The pairing between $(L^p)^*$ and L^q is given by

$$\phi(f) = \int_{\mathbb{D}} fg\, d\sigma\,, \qquad f \in L^p\,,$$

where $g \in L^q$.

Essentially the same representation holds for functionals $\phi \in (A^p)^*$. Each such functional is uniquely represented in similar fashion by a function $g \in A^q$, but there is an important difference. For $p \neq 2$, the induced isomorphism between $(A^p)^*$ and A^q is no longer an isometry, although the norms of ϕ and g are *equivalent* in the sense that they are bounded by constant multiples of each other. The following theorem was first obtained by Zaharjuta and Judovič [1].

THEOREM 6. *For* $1 < p < \infty$, *the dual space of* A^p *can be identified with* A^q, *where* $1/p + 1/q = 1$. *Each functional* $\phi \in (A^p)^*$ *has a unique representation*

$$\phi(f) = \phi_g(f) = \int_{\mathbb{D}} f\overline{g}\, d\sigma\,, \qquad f \in A^p\,,$$

for some $g \in A^q$. *The norms of* ϕ *and* g *are equivalent; that is,*

$$C_1\, \|\phi\| \le \|g\|_q \le C_2\, \|\phi\|$$

for some positive constants C_1 *and* C_2. *In fact,* $C_1 = 1$.

Before passing to the proof, let us consider an important example. We know from Chapter 1 that for each fixed $\alpha \in \mathbb{D}$, the point-evaluation functional $\phi(f) = f(\alpha)$ is bounded on A^p for $0 < p < \infty$. For $p \ge 1$ it has the representation

$$\phi(f) = \int_{\mathbb{D}} f\overline{k_\alpha}\, d\sigma\,, \qquad f \in A^p,$$

where $k_\alpha(z) = (1 - \overline{\alpha}z)^{-2}$ is the Bergman kernel function. If $q = \frac{p}{p-1}$ is the conjugate index, Hölder's inequality and Lemma 2 give the norm estimate

$$\|\phi\| \le \|k_\alpha\|_q \le C(1 - |\alpha|^2)^{\frac{2}{q}-2} = C(1 - |\alpha|^2)^{-\frac{2}{p}}.$$

In fact, $\|\phi\| = (1 - |\alpha|^2)^{-\frac{2}{p}}$ for $0 < p < \infty$. To prove this, consider first the case $\alpha = 0$. Since $\log|f|$ is subharmonic, the inequality

$$\log|f(0)| \le \int_{\mathbb{D}} \log|f(z)|\, d\sigma$$

holds, and so the arithmetic–geometric mean inequality gives

$$|f(0)|^p \le \int_{\mathbb{D}} |f(z)|^p\, d\sigma, \qquad f \in A^p,$$

and shows that equality occurs only when f is constant. Thus $|f(0)| \le 1$ when $\|f\|_p = 1$, with equality only for constant functions $f(z) \equiv \gamma$, with $|\gamma| = 1$. For $1 \le p < \infty$, this can be proved more directly by applying Hölder's inequality to the mean-value identity

$$f(0) = \int_{\mathbb{D}} f(z)\, d\sigma,$$

but the more circuitous approach identifies the extremal functions when $0 < p < 1$.

For an arbitrary point $\alpha \in \mathbb{D}$, the disk automorphism

$$\varphi_\alpha(z) = \frac{\alpha - z}{1 - \overline{\alpha}z}$$

induces an isometry $f \mapsto g$ of A^p defined by

$$g(z) = f(\varphi_\alpha(z))\varphi_\alpha'(z)^{\frac{2}{p}}$$

(*cf.* Section 1.3), with $|g(0)| = |f(\alpha)|(1 - |\alpha|^2)^{\frac{2}{p}}$. Thus the inequality $|g(0)| \le 1$ gives $|f(\alpha)| \le (1 - |\alpha|^2)^{-\frac{2}{p}}$ when $\|f\|_p = 1$, with equality only for $g(z) \equiv \gamma$, or $f(\varphi_\alpha(z))\varphi_\alpha'(z)^{\frac{2}{p}} \equiv \gamma$ for $|\gamma| = 1$. But $\varphi_\alpha(\varphi_\alpha(z)) \equiv z$, so equality occurs only for $f(z) = \gamma\, \varphi_\alpha'(z)^{\frac{2}{p}}$, or for

$$f(z) = \gamma \frac{(1 - |\alpha|^2)^{\frac{2}{p}}}{(1 - \overline{\alpha}z)^{\frac{4}{p}}}, \qquad |\gamma| = 1.$$

In particular, $\|\phi\| = (1 - |\alpha|^2)^{-\frac{2}{p}}$ for $0 < p < \infty$. This result is due to Vukotić [1], who gave it in more general form.

PROOF OF THEOREM 6. One part of the proof is easy. It follows immediately from Hölder's inequality that the functional ϕ_g defined above is bounded, and $\|\phi_g\| \leq \|g\|_q$. To show that g is unique, we note simply that $\phi_g(z^n) = \overline{a_n}/(n+1)$, where a_n is the n-th Taylor coefficient of g. Thus, if $\phi_{g_1}(z^n) = \phi_{g_2}(z^n)$ for all n, then $g_1 = g_2$.

Conversely, the Hahn–Banach theorem says that any functional $\phi \in (A^p)^*$ can be extended to a functional $\Phi \in (L^p)^*$ without increasing the norm; thus $\|\Phi\| = \|\phi\|$. By the Riesz representation theorem,

$$\Phi(f) = \int_{\mathbb{D}} f\bar{h}\, d\sigma\,, \qquad f \in L^p\,,$$

for some $h \in L^q$ with $\|h\|_q = \|\Phi\|$. Let $g = Ph$, where P is the Bergman projection. Theorem 5 says that $g \in A^q$ and $\|g\|_q \leq C\|h\|_q$. Now for $f \in A^p$,

$$\begin{aligned}\phi(f) = \Phi(f) &= \int_{\mathbb{D}} f\bar{h}\, d\sigma = \int_{\mathbb{D}}\int_{\mathbb{D}} (1-z\bar{\zeta})^{-2} f(\zeta)\, d\sigma(\zeta)\, \overline{h(z)}\, d\sigma(z) \\ &= \int_{\mathbb{D}} f(\zeta) \int_{\mathbb{D}} (1-z\bar{\zeta})^{-2}\, \overline{h(z)}\, d\sigma(z)\, d\sigma(\zeta) = \int_{\mathbb{D}} f(\zeta)\overline{g(\zeta)}\, d\sigma(\zeta) = \phi_g(f)\,,\end{aligned}$$

and $\|g\|_q \leq C\|h\|_q = C\|\phi\|$.

The interchange in the order of integration is justified by Fubini's theorem and the fact (the Corollary to Theorem 5) that the operator T defined by

$$Th(\zeta) = \int_{\mathbb{D}} |1-\bar{z}\zeta|^{-2} |h(z)|\, d\sigma(z)$$

maps L^q (boundedly) into L^q. □

It is evident from the proof of Theorem 6 that the representation of the dual space of A^p depends heavily on the boundedness of the Bergman projection. The proof does not extend to A^1 because the Bergman projection is not bounded for $p = 1$. In other words, P is not a bounded operator from L^1 to L^1. To see this, recall first that the dual space of L^1 is isometrically isomorphic to L^∞, under the integral pairing

$$\phi(f) = \langle f, g\rangle = \int_{\mathbb{D}} f\bar{g}\, d\sigma\,, \qquad f \in L^1\,,$$

where $\phi \in (L^1)^*$ and $g \in L^\infty$. Observe now that the adjoint of P can be identified with P itself under the above pairing. Thus P is a bounded operator from L^1 to L^1 if and only if it is bounded as an operator from L^∞ to L^∞. But for fixed $\zeta \in \mathbb{D}$, define the function

$$g_\zeta(z) = (1-\bar{z}\zeta)^2 |1-\bar{z}\zeta|^{-2}\,.$$

Then $g_\zeta \in L^\infty$ and $\|g_\zeta\|_\infty = 1$, while

$$(Pg_\zeta)(\zeta) = \int_{\mathbb{D}} |1-\bar{\zeta}z|^{-2}\, d\sigma(z) = 2\int_0^1 \frac{r}{1-r^2|\zeta|^2}\, dr$$

is not bounded for $\zeta \in \mathbb{D}$. This shows that P does not map L^∞ boundedly to L^∞, hence that the Bergman projection is not a bounded operator from L^1 to L^1.

The dual space of A^1 can be represented in terms of the Bloch space, and its predual can be identified with the little Bloch space. These constructions will be carried out in Section 2.6. First, however, we must digress to develop some basic material on the pseudohyperbolic metric.

§2.5. The pseudohyperbolic metric.

The theory of Bergman spaces makes abundant use of the pseudohyperbolic metric, a concept more natural than the Euclidean metric for problems in the unit disk. The *pseudohyperbolic distance* between two points z and α in $\mathbb{D}$ is defined by

$$\rho(z,\alpha) = |\varphi_\alpha(z)|\,, \qquad \text{where} \;\; \varphi_\alpha(z) = \frac{\alpha - z}{1-\overline{\alpha}z}\,.$$

The function φ_α is a Möbius transformation, a conformal automorphism of the disk with $\varphi_\alpha(\alpha) = 0$. Note that $\rho(z,0) = |z|$.

It is not hard to verify that ρ is a true metric on $\mathbb{D}$. It is clear that $0 \le \rho(z,\alpha) < 1$, with $\rho(z,\alpha) = 0$ if and only if $z = \alpha$. The symmetry property $\rho(\alpha,z) = \rho(z,\alpha)$ is immediate. The triangle inequality

$$\rho(z,\zeta) \le \rho(z,\alpha) + \rho(\alpha,\zeta)$$

is less obvious, but will be proved below in stronger form.

Direct calculations show that the function φ_α has the following additional properties.

(*i*) φ_α is an involution: $\varphi_\alpha(\varphi_\alpha(z)) = z$ for all $z \in \mathbb{D}$, or $\varphi_\alpha^{-1} = \varphi_\alpha$.

(*ii*) φ_α is an isometry with respect to the pseudohyperbolic metric:

$$\rho(\varphi_\alpha(z), \varphi_\alpha(\zeta)) = \rho(z,\zeta) \qquad \text{for all} \;\; z, \zeta \in \mathbb{D}\,.$$

(*iii*) The quantities $1-|z|^2$ and $1-|\varphi_\alpha(z)|^2$ are related by the equation

$$1 - |\varphi_\alpha(z)|^2 = \frac{(1-|\alpha|^2)(1-|z|^2)}{|1-\overline{\alpha}z|^2}\,. \tag{1}$$

Property (*ii*) says equivalently that ρ is Möbius invariant. Property (*iii*) is a basic identity that will be invoked repeatedly.

We can now verify the triangle inequality for the pseudohyperbolic metric. In fact, we can show that the sharp inequalities

$$\frac{|\rho(z,\alpha) - \rho(\alpha,\zeta)|}{1-\rho(z,\alpha)\rho(\alpha,\zeta)} \le \rho(z,\zeta) \le \frac{\rho(z,\alpha) + \rho(\alpha,\zeta)}{1+\rho(z,\alpha)\rho(\alpha,\zeta)}$$

hold for any points z, ζ, α in $\mathbb{D}$. Because ρ is Möbius invariant, it suffices to assume that $\alpha = 0$. The inequality then reduces to

$$\frac{||z| - |\zeta||}{1 - |z||\zeta|} \leq \left| \frac{\zeta - z}{1 - \overline{\zeta} z} \right| \leq \frac{|z| + |\zeta|}{1 + |z||\zeta|}$$

By symmetry, we may suppose that ζ is real and positive, and that $|z| = r < \zeta$. Then φ_ζ maps the circle $|z| = r$ onto a circle symmetric with respect to the real axis, intersecting the positive real axis at the two points $\frac{\zeta - r}{1 - r\zeta}$ and $\frac{r+\zeta}{1+r\zeta}$. In particular,

$$\frac{\zeta - r}{1 - r\zeta} \leq |\varphi_\zeta(z)| \leq \frac{r + \zeta}{1 + r\zeta},$$

which proves the inequality.

It follows from the strong form of the triangle inequality that the estimates $\rho(z, \alpha) \leq r$ and $\rho(\alpha, \zeta) \leq s$ imply

$$\rho(z, \zeta) \leq \frac{r + s}{1 + rs}.$$

This is true because the function $f(x, y) = \frac{x+y}{1+xy}$ attains a maximum value in the rectangle $[0, r] \times [0, s]$ at the point (r, s).

Another calculation shows that

$$\varphi_\alpha'(z) = -\frac{1 - |\alpha|^2}{(1 - \overline{\alpha} z)^2},$$

so that

$$|\varphi_\alpha'(z)| = \frac{1 - |\varphi_\alpha(z)|^2}{1 - |z|^2}, \tag{2}$$

by the property (*iii*) above. This last identity says that the hyperbolic density $\frac{|dz|}{1-|z|^2}$ is Möbius invariant. The identity can also be derived from the transformation formula

$$\frac{\varphi_\alpha'(z)\overline{\varphi_\alpha'(\zeta)}}{\left(1 - \varphi_\alpha(z)\overline{\varphi_\alpha(\zeta)}\right)^2} = \frac{1}{(1 - z\overline{\zeta})^2}$$

for the Bergman kernel function of the disk (see Section 1.3).

The expression

$$|\varphi_\alpha'(z)|^2 = \frac{(1 - |\alpha|^2)^2}{|1 - \overline{\alpha} z|^4}$$

is a Jacobian that arises when changing variables in an integral over the disk. Thus for $f \in L^1(\mathbb{D})$ and Ω a measurable subset of $\mathbb{D}$,

$$\begin{aligned}
\int_\Omega f(w)\, d\sigma &= \int_{\varphi_\alpha(\Omega)} f(\varphi_\alpha(z))|\varphi_\alpha'(z)|^2\, d\sigma \\
&= (1 - |\alpha|^2)^2 \int_{\varphi_\alpha(\Omega)} f(\varphi_\alpha(z))|1 - \overline{\alpha} z|^{-4}\, d\sigma.
\end{aligned}$$

Another variant of this formula is

$$Bf(\zeta) = \int_{\mathbb{D}} f(z) \frac{(1-|\zeta|^2)^2}{|1-\overline{\zeta} z|^4}\, d\sigma(z) = \int_{\mathbb{D}} f(\varphi_\zeta(w))\, d\sigma(w)\,.$$

The operator B is known as the *Berezin transform.* It arises often in the theory of Bergman spaces, for instance in describing their Carleson measures (*cf.* Section 2.11).

The *pseudohyperbolic disk* with *center* α and *radius* r $(0 < r < 1)$ is

$$\Delta(\alpha, r) = \{z : \rho(z, \alpha) < r\}\,.$$

Since φ_α is an automorphism of the disk, it is clear that $\Delta(\alpha, r) \subset \mathbb{D}$. Moreover, by the Möbius invariance of ρ,

$$\Delta(\alpha, r) = \varphi_\alpha(\Delta(0, r)) = \varphi_\alpha(\{z : |z| < r\})\,,$$

and so $\Delta(\alpha, r)$ is a true Euclidean disk, since linear fractional transformations preserve circles. However, its Euclidean center and radius are different from α and r unless $\alpha = 0$. In order to compute its (normalized) area $|\Delta(\alpha, r)|$, we can change variables in the integral to obtain

$$\begin{aligned} |\Delta(\alpha, r)| &= \int_{\Delta(\alpha,r)} d\sigma(z) = \int_{\Delta(0,r)} |\varphi_\alpha'(z)|^2\, d\sigma(z) = \int_{\Delta(0,r)} \frac{(1-|\alpha|^2)^2}{|1-\overline{\alpha} z|^4}\, d\sigma(z) \\ &= r^2(1-|\alpha|^2)^2 \int_{\mathbb{D}} \frac{d\sigma(w)}{|1-\overline{\alpha} r w|^4} = \left(\frac{r(1-|\alpha|^2)}{1-r^2|\alpha|^2}\right)^2, \end{aligned}$$

since

$$\int_{\mathbb{D}} \frac{d\sigma(z)}{|1-\overline{\beta} z|^4} = \|k_\beta\|_2^2 = \langle k_\beta, k_\beta \rangle = k_\beta(\beta) = \frac{1}{(1-|\beta|^2)^2}\,,$$

where $k_\beta(z) = (1-\overline{\beta} z)^{-2}$ is the Bergman kernel function of $\mathbb{D}$.

It is now apparent that the Euclidean radius of the pseudohyperbolic disk $\Delta(\alpha, r)$ is

$$R = \frac{r(1-|\alpha|^2)}{1-r^2|\alpha|^2}\,,$$

because its area is equal to R^2. The Euclidean center of $\Delta(\alpha, r)$ is

$$\gamma = \frac{(1-r^2)\alpha}{1-r^2|\alpha|^2}\,.$$

To see this, take $\alpha = a > 0$ and observe that the disk $\Delta(a, r)$ is symmetric with respect to the real axis, while its boundary circle crosses at the two points x determined by

$$\frac{a-x}{1-ax} = \pm r\,, \qquad \text{or} \quad x = \frac{a+r}{1+ar} \quad \text{and} \quad \frac{a-r}{1-ar}\,.$$

The center is therefore

$$\gamma = \frac{1}{2}\left(\frac{a+r}{1+ar} + \frac{a-r}{1-ar}\right) = \frac{a(1-r^2)}{1-a^2r^2},$$

and the more general formula follows by a rotation. The Euclidean radius can be calculated directly as

$$R = \frac{1}{2}\left(\frac{|\alpha|+r}{1+|\alpha|r} - \frac{|\alpha|-r}{1-|\alpha|r}\right) = \frac{r(1-|\alpha|^2)}{1-r^2|\alpha|^2}.$$

Any pair of pseudohyperbolically concentric circles are at constant pseudohyperbolic distance from each other. More precisely, if two circles C_1 and C_2 are defined by $\rho(\alpha, z) = r_1$ and $\rho(\alpha, z) = r_2$, then each point $z_1 \in C_1$ has pseudohyperbolic distance

$$\min_{z_2 \in C_2} \rho(z_1, z_2) = \rho(r_1, r_2)$$

from the circle C_2. To prove this, it suffices to take $\alpha = 0$, because of the Möbius invariance of the pseudohyperbolic metric. Then the problem reduces to showing by elementary calculus that $\rho(r_1, r_2 e^{i\theta})$ is smallest for $\theta = 0$.

The *hyperbolic area* of a measurable subset $\Omega \subset \mathbb{D}$ is defined by

$$a(\Omega) = \int_\Omega \frac{d\sigma(z)}{(1-|z|^2)^2}.$$

It is easily seen to be Möbius invariant. Indeed, if $w = \varphi_\alpha(z)$, then

$$\int_{\varphi_\alpha(\Omega)} \frac{d\sigma(w)}{(1-|w|^2)^2} = \int_\Omega \frac{|\varphi_\alpha'(z)|^2\, d\sigma(z)}{(1-|\varphi_\alpha(z)|^2)^2} = \int_\Omega \frac{d\sigma(z)}{(1-|z|^2)^2},$$

by (2). As a consequence, a pseudohyperbolic disk $\Delta(\alpha, r)$ has hyperbolic area

$$a(\Delta(\alpha, r)) = a(\Delta(0, r)) = \frac{1}{\pi}\int_0^r \int_0^{2\pi} \frac{s}{(1-s^2)^2}\, d\theta\, ds = \frac{r^2}{1-r^2}.$$

Thus the hyperbolic area of $\Delta(\alpha, r)$ depends only on the radius r and not on the center α.

Finally, we record two lemmas that will be needed later. The first is a basic fact about the pseudohyperbolic metric. The second will play a role in our discussion of Carleson measures later in this chapter.

LEMMA 3. *Any two points z and w in the same pseudohyperbolic disk $\Delta(\alpha, r)$ satisfy an inequality of the form*

$$\frac{1}{C} \le \frac{1-|z|}{1-|w|} \le C,$$

where C is a constant depending only on r.

PROOF. By symmetry, it is enough to prove the upper inequality. Observe first that if z and w belong to the same pseudohyperbolic disk $\Delta(\alpha, r)$, then by the strong form of the triangle inequality,

$$\rho(z,w) \le \frac{\rho(z,\alpha)+\rho(w,\alpha)}{1+\rho(z,\alpha)\rho(w,\alpha)} < \frac{2r}{1+r^2} = s < 1\,.$$

On the other hand, the above property (iii) of the Möbius function φ_α states that

$$1-\rho(z,w)^2 = \frac{(1-|w|^2)(1-|z|^2)}{|1-\overline{w}z|^2}\,,$$

which implies that

$$\frac{1-|z|^2}{1-|w|^2} < \frac{(1-|z|^2)^2}{(1-s^2)|1-\overline{w}z|^2} = \left(\frac{1+r^2}{1-r^2}\right)^2\left(\frac{1-|z|^2}{|1-\overline{w}z|}\right)^2.$$

However, $|1-\overline{w}z| > 1-|z| > \frac{1}{2}(1-|z|^2)$ for every pair of points z and w in $\mathbb{D}$, so the conclusion is that

$$\frac{1-|z|}{1-|w|} < 2\,\frac{1-|z|^2}{1-|w|^2} < 8\left(\frac{1+r^2}{1-r^2}\right)^2. \qquad \square$$

Under the equivalent hypothesis that $\rho(z,w) < r$, the same argument gives the inequality

$$\frac{1-|z|}{1-|w|} < \frac{8}{1-r^2}\,.$$

LEMMA 4. *In each pseudohyperbolic disk* $\Delta(\alpha, r)$, *the kernel function* $k_\alpha(z) = (1-\overline{\alpha}z)^{-2}$ *satisfies the sharp inequalities*

$$\left(\frac{1-r|\alpha|}{1-|\alpha|^2}\right)^2 \le |k_\alpha(z)| \le \left(\frac{1+r|\alpha|}{1-|\alpha|^2}\right)^2, \qquad z \in \Delta(\alpha, r)\,.$$

PROOF. Let $z = \varphi_\alpha(w)$ and observe that $z \in \Delta(\alpha, r)$ if and only if $|w| < r$, by the involutive property of φ_α and the Möbius invariance of the pseudohyperbolic metric. Thus the problem reduces to estimating

$$|k_\alpha(\varphi_\alpha(w))| = \frac{|1-\overline{\alpha}w|^2}{(1-|\alpha|^2)^2}$$

over the disk $|w| < r$, which gives the sharp upper and lower bounds as claimed. $\square$

§2.6. The Bloch space.

Our next objective is to represent the dual space of A^1 in terms of the Bloch space. We shall also find that A^1 is the dual of the little Bloch space. The representation theorems will require some preliminary discussion of the Bloch spaces.

A function f analytic in $\mathbb{D}$ is said to belong to the *Bloch space* $\mathcal{B}$ if the *Bloch norm*

$$\|f\|_{\mathcal{B}} = \sup_{z\in\mathbb{D}} (1-|z|^2)|f'(z)|$$

is finite. Actually, this is not a true norm, since it identifies functions that differ by a constant. One convenient norm on $\mathcal{B}$ is $\|f\| = \|f\|_{\mathcal{B}} + |f(0)|$. This norm will be adopted throughout our discussion of the Bloch space.

An important property of the Bloch space is its invariance under Möbius transformations. In fact, the relation (2) shows that $\|f \circ \varphi_\alpha\|_{\mathcal{B}} = \|f\|_{\mathcal{B}}$ for every $\alpha \in \mathbb{D}$.

PROPOSITION 1. *Let $f \in \mathcal{B}$ and $z, w \in \mathbb{D}$. Then*

$$|f(z) - f(w)| \le \tfrac{1}{2}\,\|f\|_{\mathcal{B}} \log \frac{1+\rho(z,w)}{1-\rho(z,w)}\,,$$

where $\rho(z,w)$ is the pseudohyperbolic distance.

PROOF. Observe first that

$$\begin{aligned} |f(z) - f(0)| &= \left| z \int_0^1 f'(zt)\,dt \right| \le |z|\,\|f\|_{\mathcal{B}} \int_0^1 \frac{dt}{1-|z|^2t^2} \\ &= \tfrac{1}{2}\,\|f\|_{\mathcal{B}} \log \frac{1+|z|}{1-|z|}\,, \end{aligned}$$

which is the desired inequality for $w = 0$. Now replace f by $f \circ \varphi_z$ and z by $\varphi_z(w)$. The general result then follows from the Möbius invariance of $\|\cdot\|_{\mathcal{B}}$, since $|\varphi_z(w)| = \rho(z,w)$. □

This inequality has some immediate consequences. First, letting $w = 0$, we see that

$$|f(z)| \le \|f\| \log \frac{1+|z|}{1-|z|} \tag{3}$$

for all $|z| \ge \frac{1}{2}$, which means that point evaluation is a bounded linear functional on the Bloch space, with a norm that is uniformly bounded on each compact subset of $\mathbb{D}$. This, together with the maximum principle, implies that if a sequence of functions converges in the Bloch norm, then it does so locally uniformly. We now use (3) to prove the completeness of $\mathcal{B}$.

PROPOSITION 2. *The Bloch space $\mathcal{B}$ is a Banach space.*

Proof. Denoting the dilations of f by $f_r(z) = f(rz)$, we have

$$\|f_r\| = \sup_{z\in\mathbb{D}} (1-|z|^2)|f_r'(z)| + |f_r(0)| = r \sup_{z\in\mathbb{D}} (1-|z|^2)|f'(rz)| + |f(0)|\,,$$

which increases to $\|f\|$ as r increases to 1. Using this fact, we now show that $\mathcal{B}$ is complete.

Let $\{f_n\}$ be a Cauchy sequence in $\mathcal{B}$. By (3), this implies that $\{f_n\}$ is a uniform Cauchy sequence on each compact subset of $\mathbb{D}$, and so it converges locally uniformly to some analytic function f. It remains to show that $\|f_n - f\| \to 0$. Given $\varepsilon > 0$, choose N such that $\|f_n - f_m\| < \varepsilon$ when $n, m \ge N$. Then for $r < 1$,

$$\begin{aligned}\|(f_n)_r - f_r\| &\le \|(f_n)_r - (f_m)_r\| + \|(f_m)_r - f_r\| \\ &\le \|f_n - f_m\| + \|(f_m)_r - f_r\| \\ &< \varepsilon + \|(f_m)_r - f_r\|.\end{aligned}$$

Observe now that the last term approaches 0 as $m \to \infty$, since $((f_m)_r)'$ converges to $(f_r)'$ uniformly on $\mathbb{D}$. Thus $\|(f_n)_r - f_r\| \le \varepsilon$ for $n \ge N$ and all $r < 1$. Now let $r \to 1$ to arrive at the desired conclusion. □

The next proposition shows that every bounded analytic function is in $\mathcal{B}$. On the other hand, the function $f(z) = \log(1-z)$ is an example of a Bloch function that is not bounded. However, the estimate (3) shows that Bloch functions belong to every Bergman space A^p for $0 < p < \infty$. The function $g(z) = (\log(1-z))^2$ is not a member of the Bloch space, while it is in A^p for every $p < \infty$.

Proposition 3. *The space H^∞ is contained in the Bloch space. In fact, $\|f\|_{\mathcal{B}} \le \|f\|_\infty$ for every $f \in H^\infty$.*

Proof. Suppose without loss of generality that $\|f\|_\infty = 1$, and apply the Schwarz–Pick lemma (or the maximum modulus theorem) to obtain

$$\left|\frac{f(z)-f(\alpha)}{1-\overline{f(\alpha)}f(z)}\right| \le \left|\frac{z-\alpha}{1-\overline{\alpha}z}\right|, \qquad \alpha \in \mathbb{D}\,.$$

From this it follows that

$$(1-|\alpha|^2)|f'(\alpha)| \le 1 - |f(\alpha)|^2 \le 1\,, \qquad \alpha \in \mathbb{D}\,,$$

so that $\|f\|_{\mathcal{B}} \le 1$. □

The *little Bloch space* $\mathcal{B}_0$ is the subspace of the Bloch space consisting of functions with the property that

$$\lim_{|z|\to 1^-} (1-|z|^2)|f'(z)| = 0.$$

More precisely, the requirement is that for each $\varepsilon > 0$ there exists a $\delta > 0$ such that $(1-|z|^2)|f'(z)| < \varepsilon$ whenever $0 < 1 - |z| < \delta$.

We begin by characterizing the convergence of little Bloch space functions in the Bloch norm. One consequence will be that $\mathcal{B}_0$ is a *closed* subspace of $\mathcal{B}$, so it is itself a Banach space.

LEMMA 5. *If* $f_n \in \mathcal{B}_0$ *and* $f \in \mathcal{B}$, *then* $\|f_n - f\| \to 0$ *as* $n \to \infty$ *if and only if*

(*i*) $f_n(z) \to f(z)$ *uniformly on compact subsets of* $\mathbb{D}$ *; and*
(*ii*) $(1 - |z|^2)|f_n'(z)| \to 0$ *as* $|z| \to 1$, *uniformly in* n.

Furthermore, these conditions imply that $f \in \mathcal{B}_0$.

PROOF. Suppose first that $\|f_n - f\| \to 0$. Then (*i*) follows from (3). To prove (*ii*), note that the hypothesis implies that $\{(1 - |z|^2)|f_n'(z)|\}$ is uniformly convergent, hence is a uniform Cauchy sequence in $\mathbb{D}$. Thus to each $\varepsilon > 0$ there corresponds a number N such that

$$(1 - |z|^2)|f_n'(z)| \le (1 - |z|^2)|f_m'(z)| + \varepsilon\,, \qquad z \in \mathbb{D}\,,$$

whenever $n, m > N$. Now hold $m > N$ fixed and let $|z| \to 1$. Since $f_m \in \mathcal{B}_0$, we conclude that $(1 - |z|^2)|f_n'(z)| < 2\varepsilon$ for all $z \in \mathbb{D}$ with $0 < 1 - |z| < \delta$, and for all $n > N$. This proves (*ii*).

Conversely, suppose (*i*) and (*ii*) hold for some sequence of functions $f_n \in \mathcal{B}_0$ and some $f \in \mathcal{B}$. For each $\varepsilon > 0$, the condition (*ii*) says that

$$(1 - |z|^2)|f_n'(z)| < \varepsilon \qquad \text{whenever } \rho < |z| < 1\,,$$

for some $\rho > 0$ and all n. But (*i*) implies that $f_n'(z) \to f'(z)$ pointwise in $\mathbb{D}$, so it follows that $f \in \mathcal{B}_0$. Therefore,

$$(1 - |z|^2)|f_n'(z) - f'(z)| < \varepsilon\,, \qquad \rho < |z| < 1\,,$$

for some $\rho > 0$ and all n. But in view of (*i*), this inequality holds also for $|z| \le \rho$ and for all n sufficiently large, since $f_n'(z) \to f'(z)$ uniformly in $|z| \le \rho$. Thus $\|f_n - f\|_{\mathcal{B}} \to 0$, and another appeal to (*i*) shows that $\|f_n - f\| \to 0$, since $|f_n(0) - f(0)| \to 0$. □

An important consequence of the lemma is the following characterization of functions in the little Bloch space.

PROPOSITION 4. *Let* $f \in \mathcal{B}$. *Then* $f \in \mathcal{B}_0$ *if and only if* $\|f - f_r\| \to 0$ *as* $r \to 1$.

PROOF. Each dilation f_r is analytic in $\overline{\mathbb{D}}$, so it belongs to $\mathcal{B}_0$. Thus by Lemma 5, the property $\|f - f_r\| \to 0$ implies $f \in \mathcal{B}_0$. For the converse, it is clear that if $f \in \mathcal{B}_0$, then (*i*) and (*ii*) hold with $f_n = f_{r_n}$, where $\{r_n\}$ is any sequence increasing to 1, so $\|f - f_r\| \to 0$ by Lemma 5. □

As a corollary, we see that the little Bloch space is the closure of the polynomials in the Bloch norm. To show that a function $f \in \mathcal{B}_0$ can be approximated by polynomials, apply Proposition 3 to write

$$\|Q - f\|_{\mathcal{B}} \leq \|Q - f_r\|_{\mathcal{B}} + \|f - f_r\|_{\mathcal{B}} \leq \|Q - f_r\|_\infty + \|f - f_r\|_{\mathcal{B}},$$

where Q is any polynomial. But f_r is analytic in $\overline{\mathbb{D}}$, so $\|Q - f_r\|_\infty < \varepsilon$ for some choice of polynomial Q. An appeal to Proposition 4 now completes the proof. Conversely, Lemma 5 shows that if a function $f \in \mathcal{B}$ can be approximated by polynomials, then $f \in \mathcal{B}_0$ because all polynomials are in $\mathcal{B}_0$.

There is a strong connection between the Bergman projection and the Bloch spaces, which we will exhibit shortly. It depends on the following formulas for images of special classes of functions under the projection.

LEMMA 6. *Let P be the Bergman projection, and let m and n be non-negative integers. Then*

$$(a) \qquad P((1 - |z|^2)z^m) = \frac{z^m}{m+2};$$

$$(b) \qquad P(z^m \overline{z}^n) = \begin{cases} \dfrac{m-n+1}{m+1} z^{m-n}, & \text{for} \quad m \geq n \\ 0, & \text{for} \quad m < n; \end{cases}$$

$$(c) \qquad P\left(\frac{1 - |z|^2}{\overline{z}} g'(z)\right) = g(z) \qquad \text{if } g \in A^1 \text{ with } g(0) = 0.$$

PROOF.

(a). Expand the Bergman kernel function into power series and use polar coordinates to obtain

$$\begin{aligned} P((1 - |z|^2)z^m) &= \int_{\mathbb{D}} (1 - |\zeta|^2)\zeta^m (1 - \overline{\zeta} z)^{-2} d\sigma(\zeta) \\ &= \int_{\mathbb{D}} (1 - |\zeta|^2)\zeta^m \sum_{n=0}^\infty (n+1)(\overline{\zeta} z)^n d\sigma(\zeta) \\ &= \sum_{n=0}^\infty (n+1) z^n \int_0^1 (1 - r^2) r^{m+n+1} dr \, \frac{1}{\pi} \int_0^{2\pi} e^{i\theta(m-n)} d\theta \\ &= 2(m+1) z^m \int_0^1 (1 - r^2) r^{2m+1} dr = \frac{z^m}{m+2}. \end{aligned}$$

The change in the order of integration and summation is justified by the fact that the Bergman kernel function is analytic in $\overline{\mathbb{D}}$.

(b). This calculation is similar and is omitted.

(c). Write $g(z)=\sum_{m=1}^{\infty} a_m z^m$ and proceed as above to find

$$
\begin{aligned}
P\Big(\frac{(1-|z|^2)}{\overline{z}}\,g'(z)\Big) &= \int_{\mathbb{D}} \frac{(1-|\zeta|^2)g'(\zeta)}{\overline{\zeta}(1-\overline{\zeta}z)^2}\,d\sigma(\zeta)\\
&= \int_{\mathbb{D}} \frac{(1-|\zeta|^2)g'(\zeta)}{\overline{\zeta}} \sum_{n=0}^{\infty}(n+1)(\overline{\zeta}z)^n\,d\sigma(\zeta)\\
&= \sum_{n=0}^{\infty}(n+1)z^n \lim_{t\to 1}\int_{t\mathbb{D}} (1-|\zeta|^2)g'(\zeta)(\overline{\zeta})^{n-1}d\sigma(\zeta)\\
&= \sum_{n=0}^{\infty}(n+1)z^n \lim_{t\to 1}\int_{t\mathbb{D}} (1-|\zeta|^2)(\overline{\zeta})^{n-1}\sum_{m=1}^{\infty} m a_m \zeta^{m-1}\,d\sigma(\zeta)\\
&= \sum_{n=0}^{\infty}(n+1)z^n \lim_{t\to 1}\sum_{m=1}^{\infty}\frac{m a_m}{\pi}\int_0^t (1-r^2)r^{n+m-1}dr\int_0^{2\pi} e^{i\theta(m-n)}d\theta\\
&= \sum_{n=1}^{\infty} 2(n+1)n a_n z^n \lim_{t\to 1}\int_0^t (1-r^2)r^{2n-1}dr = \sum_{n=1}^{\infty} a_n z^n = g(z)\,. \qquad \square
\end{aligned}
$$

We denote by $C(\overline{\mathbb{D}})$ the set of continuous functions on $\overline{\mathbb{D}}$ (the closure of $\mathbb{D}$) and by $C_0(\mathbb{D})$ the subspace of $C(\overline{\mathbb{D}})$ consisting of functions vanishing on the unit circle. The following theorem expresses a natural relation between the Bergman projection and the Bloch spaces.

THEOREM 7. *The Bergman projection has the properties* $P(L^\infty(\mathbb{D}))=\mathcal{B}$ *and* $P(C(\overline{\mathbb{D}}))=P(C_0(\mathbb{D}))=\mathcal{B}_0$. *In each case,* P *is bounded.*

PROOF. We show first that P maps $L^\infty(\mathbb{D})$ boundedly into $\mathcal{B}$. For $f\in L^\infty(\mathbb{D})$, the Bergman projection is

$$Pf(z)=\int_{\mathbb{D}}\frac{f(\zeta)}{(1-\overline{\zeta}z)^2}\,d\sigma(\zeta)\,, \qquad z\in\mathbb{D}\,.$$

Now take the derivative under the integral sign and multiply by $(1-|z|^2)$ to obtain

$$(1-|z|^2)(Pf)'(z)=2(1-|z|^2)\int_{\mathbb{D}}\frac{\overline{\zeta}f(\zeta)}{(1-\overline{\zeta}z)^3}\,d\sigma(\zeta)\,,$$

which implies by Lemma 2 in Section 2.3 that

$$(1-|z|^2)|(Pf)'(z)|\le 2(1-|z|^2)\|f\|_\infty\int_{\mathbb{D}}\frac{1}{|1-\overline{\zeta}z|^3}\,d\sigma(\zeta)\le C\|f\|_\infty\,.$$

Moreover, $|Pf(0)|=|\int_{\mathbb{D}}f(\zeta)d\sigma(\zeta)|\le\|f\|_\infty$, and so

$$\|Pf\|=\|Pf\|_{\mathcal{B}}+|Pf(0)|\le C\|f\|_\infty\,.$$

This says that the Bergman projection is a bounded map from $L^\infty(\mathbb{D})$ into $\mathcal{B}$.

We next show that P maps $C(\overline{\mathbb{D}})$ boundedly into $\mathcal{B}_0$. By the Stone–Weierstrass theorem (see Rudin [2], p. 162), every $f \in C(\overline{\mathbb{D}})$ can be uniformly approximated by finite linear combinations of functions $z^m \overline{z}^n$, where m and n are nonnegative integers. Since we have already shown that P maps $L^\infty(\mathbb{D})$ boundedly into $\mathcal{B}$, and $\mathcal{B}_0$ is a closed subspace of $\mathcal{B}$, it suffices to observe that $P(z^m \overline{z}^n) \in \mathcal{B}_0$ for all $m, n \geq 0$. But this is clearly true, by formula (b) of Lemma 6. Since $C_0(\mathbb{D}) \subset C(\overline{\mathbb{D}})$, it follows that P also maps $C_0(\mathbb{D})$ boundedly into $\mathcal{B}_0$.

It is not difficult to see that the above maps are onto. To this end, for a given function $g \in \mathcal{B}$, define $f \in L^\infty(\mathbb{D})$ by

$$f(z) = (1 - |z|^2) \left\{ 2g(0) + 3g'(0)z + 2g''(0)z^2 + \frac{g'(z) - g'(0) - g''(0)z}{\overline{z}} \right\}.$$

Applying formulas (a) and (c) of Lemma 6, we find after a straightforward calculation that $Pf = g$. Since $f \in C_0(\mathbb{D})$ if $g \in \mathcal{B}_0$, the proof of the theorem is complete. □

In fact, the construction shows a little more. Lemma 6(c) gives the formula

$$g(z) = g(0) + \int_{\mathbb{D}} \frac{(1 - |\zeta|^2) g'(\zeta)}{\overline{\zeta}(1 - \overline{\zeta} z)^2} \, d\sigma(\zeta), \qquad g \in A^1,$$

and it follows by differentiation under the integral sign that

$$g''(0) = 6 \int_{\mathbb{D}} \overline{\zeta}(1 - |\zeta|^2) g'(\zeta) \, d\sigma(\zeta).$$

Thus $|g''(0)| \leq 6 \|g\|_{\mathcal{B}}$. This shows that the function f constructed above has the property

$$\|f\|_\infty \leq c\|g\| = c(\|g\|_{\mathcal{B}} + |g(0)|), \qquad g \in \mathcal{B},$$

where c is an absolute constant. Thus for each function $g \in \mathcal{B}$ (*resp.*, $\mathcal{B}_0$), it is possible to choose $f \in L^\infty(\mathbb{D})$ (*resp.*, $C_0(\mathbb{D})$) with $Pf = g$ and $\|f\|_\infty \leq c\|g\|$.

The connection between the Bergman space and the Bloch space is evident in the next theorem.

THEOREM 8. *The dual space of A^1 can be identified with the Bloch space $\mathcal{B}$. Each functional $\phi \in (A^1)^*$ has a unique representation*

$$\phi(f) = \phi_g(f) = \lim_{t \to 1} \int_{t\mathbb{D}} f \, \overline{g} \, d\sigma, \qquad f \in A^1,$$

where $g \in \mathcal{B}$. *Furthermore, the norm* $\|\phi\|$ *is equivalent to the norm* $\|g\| = \|g\|_{\mathcal{B}} + |g(0)|$.

It should be noted that the function $f\overline{g}$ need not be in L^1, so it is not clear *a priori* that the integral over $\mathbb{D}$ exists. Here is a specific example. Let

$$f(z) = \frac{1}{(1-z)^2}\left\{\frac{1}{z}\log\frac{1}{1-z}\right\}^{-3/2}, \qquad g(z) = \frac{1}{z}\log\frac{1}{1-z}.$$

Then $g \in \mathcal{B}$, and we claim that $f \in A^1$ while $f\overline{g} \notin L^1$. Indeed, a calculation to be found for instance in Littlewood [1] (pp. 93–96) shows that the function

$$h(z) = \frac{1}{(1-z)^2}\left\{\frac{1}{z}\log\frac{1}{1-z}\right\}^{a}, \qquad a \in \mathbb{R},$$

has an integral mean with growth

$$M_1(r,h) \sim C\,\frac{1}{1-r}\left\{\log\frac{1}{1-r}\right\}^{a}, \qquad r \to 1.$$

Thus the integral $\int_0^1 M_1(r,h)\,dr$ converges if $a < -1$ and diverges if $a \geq -1$. For the pair of functions f and g given above, this shows that $f \in A^1$ while $fg \notin A^1$, hence $f\overline{g} \notin L^1$.

An important part of the proof of Theorem 8 is to show that the limit $\phi_g(f)$ exists for $f \in A^1$ and $g \in \mathcal{B}$. This is done via the following lemma.

LEMMA 7. *If* f *and* g *are analytic in* $\mathbb{D}$, *and* $0 < t < 1$, *then*

$$\int_{t\mathbb{D}} f(z)\overline{g(z)}\,d\sigma(z) = \int_{t\mathbb{D}} (Vf)(z)\overline{g'(z)}(t^2 - |z|^2)\,d\sigma(z) + t^2 f(0)\overline{g(0)},$$

where Vf *is the analytic function defined by* $(Vf)(z) = (f(z) - f(0))/z$.

PROOF. One can compute both integrals by using the power series representations of all involved functions and integrating with polar coordinates. The details are left to the reader. □

PROOF OF THEOREM 8. It is clear that $Vf \in A^1$ whenever $f \in A^1$, so by the closed graph theorem, we have a positive constant C such that $\|Vf\|_1 \leq C\|f\|_1$ for all $f \in A^1$. Together with the Lebesgue dominated convergence theorem and Lemma 7, this shows that the integral defining ϕ_g converges for $f \in A^1$ and $g \in \mathcal{B}$. Moreover,

$$\phi_g(f) = \int_{\mathbb{D}} Vf(z)\overline{g'(z)}(1 - |z|^2)\,d\sigma(z) + f(0)\overline{g(0)},$$

and it follows that

$$|\phi_g(f)| \leq \|Vf\|_1\|g\|_{\mathcal{B}} + |g(0)|\,\|f\|_1 \leq C\|f\|_1(\|g\|_{\mathcal{B}} + |g(0)|).$$

Thus $\phi_g \in (A^1)^*$ and $\|\phi_g\| \le C(\|g\|_{\mathcal{B}} + |g(0)|)$.

Conversely, let ϕ be a bounded linear functional on A^1. By the Hahn–Banach theorem and the Riesz representation theorem, there is a function $h \in L^\infty(\mathbb{D})$ such that $\|h\|_\infty = \|\phi\|$ and

$$\phi(f) = \int_{\mathbb{D}} f(z)\overline{h(z)}\, d\sigma(z)\,, \qquad f \in A^1\,.$$

By Theorem 7, the Bergman projection sends h to a function $g = Ph$ in the Bloch space, and $\|g\|_{\mathcal{B}} + |g(0)| \le C\|h\|_\infty$. As we have already proved, this function $g \in \mathcal{B}$ defines a functional $\phi_g \in (A^1)^*$ according to the formula displayed in the theorem. We can complete the proof by showing that $\phi(f) = \phi_g(f)$ for all $f \in A^1$. It will be enough to take f to be a polynomial, since the polynomials are dense in A^1. But then we can write

$$\begin{aligned}
\phi_g(f) = \int_{\mathbb{D}} f(z)\overline{g(z)}\, d\sigma(z) &= \int_{\mathbb{D}} f(z)\overline{\int_{\mathbb{D}} h(\zeta)(1-\overline{\zeta}z)^{-2} d\sigma(\zeta)}\, d\sigma(z) \\
&= \int_{\mathbb{D}} \overline{h(\zeta)} \int_{\mathbb{D}} f(z)(1-\overline{z}\zeta)^{-2} d\sigma(z)\, d\sigma(\zeta) \\
&= \int_{\mathbb{D}} f(\zeta)\overline{h(\zeta)}\, d\sigma(\zeta) = \phi(f)\,,
\end{aligned}$$

by the reproducing property of the Bergman kernel function. The interchange in the order of integration is justified by Fubini's theorem (since f is a polynomial) and the corollary to Theorem 5. Moreover,

$$\|g\|_{\mathcal{B}} + |g(0)| \le C\|h\|_\infty = C\|\phi\|\,.$$

Thus each $\phi \in (A^1)^*$ can be represented in the form ϕ_g for some $g \in \mathcal{B}$, and the functional norm $\|\phi\|$ is equivalent to the norm $\|g\| = \|g\|_{\mathcal{B}} + |g(0)|$. The uniqueness of the representation is proved by the same argument as in Theorem 6, by showing that if two functionals ϕ_{g_1} and ϕ_{g_2} agree on all monomials z^n, then $g_1 = g_2$. □

It may be remarked that the proof of Theorem 8 shows again that the Bergman projection P maps $L^\infty(\mathbb{D})$ *onto* $\mathcal{B}$, given that it maps $L^\infty(\mathbb{D})$ boundedly *into* $\mathcal{B}$. For if $g \in \mathcal{B}$, then $\phi_g \in (A^1)^*$, and the proof of Theorem 8 shows that $\phi_g = \phi_{Ph}$ for some $h \in L^\infty(\mathbb{D})$. It then follows that $g = Ph$, by uniqueness of the representation.

We next define a linear mapping Q that will play an important part in identifying the little Bloch space as the predual of A^1. For $f \in L^1(\mathbb{D})$, define Qf by the formula

$$Qf(z) = 3(1-|z|^2)^2 \int_{\mathbb{D}} f(\zeta)(1-\overline{\zeta}z)^{-4}\, d\sigma(\zeta)\,.$$

The connection between Q and the Bergman projection P is given by the following lemma.

LEMMA 8. *If $f \in L^1(\mathbb{D})$, then $PQf = Pf$ and $QPf = Qf$.*

PROOF. Let g be any analytic function satisfying

$$\int_{\mathbb{D}} |g(w)|(1-|w|^2)^2 \, d\sigma(w) < \infty .$$

Since $(1-|w|^2)^2 d\sigma(w)$ is a radial measure of norm $\frac{1}{3}$,

$$g(0) = 3 \int_{\mathbb{D}} g(w)(1-|w|^2)^2 \, d\sigma(w) .$$

Apply this to the function $g \circ \varphi_z$ and use formulas (2) and (1) from Section 2.5 to obtain

$$\begin{aligned} g(z) &= 3 \int_{\mathbb{D}} g(\varphi_z(w))(1-|w|^2)^2 \, d\sigma(w) \\ &= 3 \int_{\mathbb{D}} g(\zeta)(1-|\varphi_z(\zeta)|^2)^4 (1-|\zeta|^2)^{-2} \, d\sigma(\zeta) \\ &= 3(1-|z|^2)^4 \int_{\mathbb{D}} g(\zeta)(1-|\zeta|^2)^2 |1-\overline{z}\zeta|^{-8} \, d\sigma(\zeta) . \end{aligned}$$

Now replace g by the function $(1-\overline{z}\zeta)^4 g(\zeta)$ to arrive at the formula

$$g(z) = 3 \int_{\mathbb{D}} g(\zeta)(1-\overline{\zeta}z)^{-4}(1-|\zeta|^2)^2 \, d\sigma(\zeta) .$$

With the special choice of the Bergman kernel function $g(w) = (1-\overline{z}w)^{-2}$, this formula yields the curious identity

$$(1-\overline{z}w)^{-2} = 3 \int_{\mathbb{D}} (1-\overline{z}\zeta)^{-2}(1-\overline{\zeta}w)^{-4}(1-|\zeta|^2)^2 \, d\sigma(\zeta) .$$

For any function $f \in L^1(\mathbb{D})$, an application of Fubini's theorem therefore gives

$$\begin{aligned} PQf(z) &= 3 \int_{\mathbb{D}} \int_{\mathbb{D}} (1-|\zeta|^2)^2 f(w)(1-\overline{\zeta}z)^{-2}(1-\overline{w}\zeta)^{-4} \, d\sigma(w) \, d\sigma(\zeta) \\ &= 3 \int_{\mathbb{D}} f(w) \int_{\mathbb{D}} (1-|\zeta|^2)^2 (1-\overline{\zeta}z)^{-2}(1-\overline{w}\zeta)^{-4} \, d\sigma(\zeta) \, d\sigma(w) \\ &= \int_{\mathbb{D}} f(w)(1-\overline{w}z)^{-2} \, d\sigma(w) = Pf(z) . \end{aligned}$$

In a similar way, Fubini's theorem and the standard reproducing formula combine to give

$$\begin{aligned} QPf(z) &= 3(1-|z|^2)^2 \int_{\mathbb{D}} \int_{\mathbb{D}} f(w)(1-\overline{w}\zeta)^{-2} \, d\sigma(w) \, (1-\overline{\zeta}z)^{-4} \, d\sigma(\zeta) \\ &= 3(1-|z|^2)^2 \int_{\mathbb{D}} f(w) \int_{\mathbb{D}} (1-\overline{w}\zeta)^{-2}(1-\overline{\zeta}z)^{-4} \, d\sigma(\zeta) \, d\sigma(w) \\ &= 3(1-|z|^2)^2 \int_{\mathbb{D}} f(w)(1-\overline{w}z)^{-4} \, d\sigma(w) = Qf(z) . \qquad \square \end{aligned}$$

LEMMA 9. *The operator Q is an embedding from $\mathcal{B}_0$ into $C_0(\mathbb{D})$. In other words, it maps $\mathcal{B}_0$ linearly into $C_0(\mathbb{D})$ and is bounded above and below. In particular, Q gives a one-to-one mapping of $\mathcal{B}_0$ onto a subspace of $C_0(\mathbb{D})$.*

PROOF. If $f \in \mathcal{B}_0$, then $f = Pg$ for some $g \in C_0(\mathbb{D})$, by Theorem 7. By the remarks following the proof of that theorem, we can choose g to satisfy $\|g\|_\infty \le c(\|f\|_{\mathcal{B}} + |f(0)|)$, where c is an absolute constant. Noting that the definition of Q gives

$$Qh(0) = 3\int_{\mathbb{D}} h(\zeta)\,d\sigma(\zeta)\,, \qquad h \in L^1(\mathbb{D})\,,$$

we can take $h = g \circ \varphi_z$ to infer that

$$Qg(z) = 3\int_{\mathbb{D}} g(\varphi_z(\zeta))\,d\sigma(\zeta)\,.$$

But $g(\varphi_z(\zeta)) \to 0$ as $|z| \to 1$, for each point $\zeta \in \mathbb{D}$, so we may conclude from the Lebesgue bounded convergence theorem that $Qg(z) \to 0$ as $|z| \to 1$. In other words, $Qg \in C_0(\mathbb{D})$. But Lemma 8 tells us that $Qf = QPg = Qg$, so Q maps $\mathcal{B}_0$ into $C_0(\mathbb{D})$.

To prove that Q is an embedding, observe first that its definition gives

$$|Qg(z)| \le 3(1-|z|^2)^2\,\|g\|_\infty \int_{\mathbb{D}} |1-\overline{\zeta}z|^{-4}\,d\sigma(\zeta) = 3\,\|g\|_\infty\,, \qquad z \in \mathbb{D}\,,$$

by the reproducing property of the Bergman kernel. Therefore,

$$\|Qf\|_\infty = \|Qg\|_\infty \le 3\,\|g\|_\infty \le 3c\,(\|f\|_{\mathcal{B}} + |f(0)|)\,,$$

showing that Q is bounded. On the other hand, Lemma 8 gives $f = Pf = PQf$ for every $f \in \mathcal{B}_0$. By Theorem 7, then,

$$\|f\|_{\mathcal{B}} + |f(0)| \le \|P\|\|Qf\|_\infty\,,$$

showing that Q is bounded below. □

We are now ready to prove that the little Bloch space is the predual of the Bergman space A^1.

THEOREM 9. *The dual space of $\mathcal{B}_0$ can be identified with A^1. Each functional $\phi \in (\mathcal{B}_0)^*$ has a unique representation*

$$\phi(f) = \phi_g(f) = \lim_{t\to 1}\int_{t\mathbb{D}} f\,\overline{g}\,d\sigma\,, \qquad f \in \mathcal{B}_0\,,$$

where $g \in A^1$. Furthermore, the norm $\|\phi\|$ is equivalent to the norm $\|g\|_1$.

PROOF. It follows from the proof of Theorem 8 that $\phi_g(f)$ is well defined for every pair of functions $g \in A^1$ and $f \in \mathcal{B}$. The formula

$$\phi_g(f) = \int_{\mathbb{D}} f'(z)\overline{Vg(z)}(1-|z|^2)\,d\sigma(z) + f(0)\overline{g(0)}\,,$$

which follows from Lemma 7, shows again that

$$|\phi_g(f)| \le C\|g\|_1(\|f\|_{\mathcal{B}} + |f(0)|)\,,$$

so ϕ_g defines a bounded linear functional on $\mathcal{B}$ (hence on $\mathcal{B}_0$) with norm $\|\phi_g\| \le C\|g\|_1$.

Conversely, let an arbitrary functional $\phi \in (\mathcal{B}_0)^*$ be given. By Lemma 9, $Q\mathcal{B}_0$ is a closed subspace of $C_0(\mathbb{D})$, and $\phi \circ Q^{-1}$ is a bounded linear functional on $Q\mathcal{B}_0$. By the Hahn–Banach theorem and the Riesz representation theorem, there is a finite measure μ on $\mathbb{D}$ such that

$$\phi \circ Q^{-1}(h) = \int_{\mathbb{D}} h(z)\,d\mu(z)\,, \qquad h \in C_0(\mathbb{D})\,,$$

and $\|\mu\| = \|\phi \circ Q^{-1}\|$. Now define the analytic function

$$g(z) = 3\int_{\mathbb{D}} (1-|\zeta|^2)^2(1-\overline{\zeta}z)^{-4}\,d\overline{\mu}(\zeta)\,.$$

By Fubini's theorem and the reproducing property of the Bergman kernel function,

$$\begin{aligned}\int_{\mathbb{D}} |g(z)|d\sigma(z) &\le 3\int_{\mathbb{D}} (1-|\zeta|^2)^2 \int_{\mathbb{D}} |1-\overline{z}\zeta|^{-4} d\sigma(z)\,d|\mu|(\zeta) \\ &= 3\int_{\mathbb{D}} d|\mu|(\zeta) = 3\,\|\mu\| = 3\,\|\phi\circ Q^{-1}\| \le C\,\|\phi\|\,.\end{aligned}$$

This shows that $g \in A^1$ and that the norm of g is bounded by a constant multiple of the norm of ϕ. To complete the proof of the theorem, we need to show that $\phi(f) = \phi_g(f)$ for all $f \in \mathcal{B}_0$. Since polynomials are dense in the little Bloch space, it will be sufficient to assume that f is a polynomial. Then by Fubini's theorem,

$$\begin{aligned}\phi(f) = \phi\circ Q^{-1}(Qf) &= \int_{\mathbb{D}} Qf(z)\,d\mu(z) \\ &= 3\int_{\mathbb{D}} (1-|z|^2)^2 \int_{\mathbb{D}} f(\zeta)(1-\overline{\zeta}z)^{-4} d\sigma(\zeta)\,d\mu(z) \\ &= 3\int_{\mathbb{D}} f(\zeta) \int_{\mathbb{D}} (1-|z|^2)^2(1-\overline{\zeta}z)^{-4} d\mu(z)\,d\sigma(\zeta) \\ &= \int_{\mathbb{D}} f(\zeta)\overline{g(\zeta)}\,d\sigma(\zeta) = \phi_g(f)\,.\end{aligned}$$

The uniqueness of g is proved in the same way as in Theorem 6. $\square$

The dual space of A^p for $0 < p < 1$ has a similar description. Here the requirement is that the fractional derivative $g^{[\frac{2}{p}-1]}(z)$ be dominated by $\frac{C}{1-|z|}$. This result is due to Joel Shapiro [1] and is similar to the description of $(H^p)^*$ for $0 < p < 1$ by Duren, Romberg, and Shields [1]. An extension to higher dimensions appears in Djrbashian and Shamoian [1].

Philosophically, the Bloch space may be considered to have the same relation to Bergman spaces as the space of functions of bounded mean oscillation (BMO) has to Hardy spaces. Speaking very loosely, $\mathcal{B}$ is the limit of A^p as $p \to \infty$, in the same sense that BMO is the limit of H^p. For instance, a theorem of Charles Fefferman says that the dual of H^1 is BMO. Similarly, the little Bloch space $\mathcal{B}_0$ is the Bergman space analogue of VMO, the space of functions of vanishing mean oscillation.

The modern theory of Bloch spaces was initiated in papers of Pommerenke [1] and Anderson, Clunie, and Pommerenke [1]. The survey articles of Anderson [1] and Cima [1] give further information.

§2.7. Harmonic conjugates.

The *harmonic Bergman space* a^p consists of all real-valued functions u harmonic in $\mathbb{D}$ for which the integral $\int_{\mathbb{D}} |u|^p \, d\sigma$ is finite. The norm $\|u\|_p$ is defined in an obvious way. Similarly, the *harmonic Hardy space* h^p is the collection of all harmonic functions whose integral means $M_p(r,u)$ are bounded. It is clear that $h^p \subset a^p$.

In discussing harmonic conjugates, it will be convenient to adopt a normalization. For a real-valued function u harmonic in $\mathbb{D}$, we now define the *harmonic conjugate* as the function v with $v(0) = 0$ such that $u + iv$ is analytic in $\mathbb{D}$. By a classical theorem of M. Riesz, if $1 < p < \infty$ and $u \in h^p$, then $v \in h^p$ and $M_p(r,v) \le C_p \, M_p(r,u)$ for all $u \in h^p$, where C_p is a constant depending only on p. Simple geometric considerations show that the theorem fails for $p = \infty$. It fails also for $p = 1$, but at least $u \in h^1$ implies $v \in h^p$ for every $p < 1$. For $0 < p < 1$, the theorem collapses altogether. A function u may belong to h^p for every $p < 1$, yet its conjugate may fail to be in h^p for any $p > 0$. Proofs and related information may be found in Duren [5].

It follows immediately from the theorem of M. Riesz that the class a^p is also self-conjugate for every p in the range $1 < p < \infty$, and $\|v\|_p \le C_p \, \|u\|_p$ for all $u \in a^p$. It comes as a surprise, however, that in the Bergman norm the theorem remains true for $0 < p \le 1$. This is an old result, due to Hardy and Littlewood [1]. It is one of the few instances where Bergman spaces behave *better* than Hardy spaces.

THEOREM 10. *For $0 < p < \infty$, the class a^p is self-conjugate. If u is a harmonic function of class a^p, then its harmonic conjugate v is also in a^p, and $\|v\|_p \le C_p \, \|u\|_p$, where C_p is a constant depending only on p.*

We shall carry out a direct proof, without appeal to the Riesz theorem, but only for $1 \le p < \infty$. Hardy and Littlewood [1] stated the result and indicated a proof for $0 < p \le 1$. Details were supplied later by Watanabe [1]. The argument for $0 < p < 1$ is quite technical, however, and will not be given here.

In preparation for the proof, we begin with a simple property of the Bergman projection.

LEMMA 10. *If* $f \in A^1$, *then* $P\overline{f} = \overline{f(0)}$.

PROOF. By the mean-value property of analytic functions,

$$\overline{(P\overline{f})(z)} = \int_{\mathbb{D}} f(\zeta)(1 - \overline{z}\zeta)^{-2}\, d\sigma(\zeta) = f(0)\,. \qquad \square$$

PROOF OF THEOREM. Suppose first that $1 < p < \infty$. If $f = u + iv \in A^p$ and $v(0) = 0$, then by the lemma and Theorem 5,

$$\begin{aligned} \|v\|_p \le \|(u + u(0)) + iv\|_p &= \|f + f(0)\|_p = \|P(f + \overline{f})\|_p \\ &= 2\,\|Pu\|_p \le C_p\,\|u\|_p\,, \end{aligned}$$

since $u = \frac{1}{2}(f + \overline{f}) \in L^p$. Now let u be an arbitrary function in a^p, and let $f = u + iv$, where v is the (normalized) harmonic conjugate of u. For $0 < r < 1$, consider the dilation $f_r(z) = f(rz)$; and let u_r and v_r denote the corresponding dilations of u and v. Then $f_r \in A^p$, and it follows from what we have just proved that $\|v_r\|_p \le C_p\,\|u_r\|_p$. Finally, let $r \to 1$ to conclude as in the proof of Theorem 3 that $\|v\|_p \le C_p\,\|u\|_p$. $\square$

The above proof fails when $p = 1$, because the Bergman projection is no longer bounded in that case. As a substitute, we consider the linear operator R defined for $f \in L^1(\mathbb{D})$ by

$$Rf(z) = 3\int_{\mathbb{D}} f(\zeta)(1 - |\zeta|^2)^2(1 - \overline{\zeta}z)^{-4}\, d\sigma(\zeta)\,.$$

By Fubini's theorem and the reproducing property of the kernel function,

$$\begin{aligned} \int_{\mathbb{D}} |Rf(z)|\,d\sigma(z) &\le 3\int_{\mathbb{D}}\int_{\mathbb{D}} |f(\zeta)|(1 - |\zeta|^2)^2|1 - \overline{\zeta}z|^{-4}\, d\sigma(\zeta)\, d\sigma(z) \\ &= 3\int_{\mathbb{D}} |f(\zeta)|(1 - |\zeta|^2)^2 \int_{\mathbb{D}} |1 - \overline{\zeta}z|^{-4}\, d\sigma(z)\, d\sigma(\zeta) \\ &= 3\int_{\mathbb{D}} |f(\zeta)|\,d\sigma(\zeta)\,, \end{aligned}$$

and so R is a bounded mapping from $L^1(\mathbb{D})$ to A^1. The identity derived in the proof of Lemma 8 in Section 2.6 shows that $Rf = f$ whenever $f \in A^1$, thus showing that R is a bounded projection from $L^1(\mathbb{D})$ onto A^1.

The analogue of Lemma 10 holds when $p = 1$ as well. Indeed, for $f \in A^1$ and $z \in \mathbb{D}$, we have by the mean-value property for analytic functions,

$$\begin{aligned}(R\overline{f})(z) &= \frac{3}{\pi}\int_0^1 (1-r^2)^2 r \overline{\int_0^{2\pi} f(re^{i\theta})(1-\overline{z}re^{i\theta})^{-4}\,d\theta}\,dr \\ &= 6\,\overline{f(0)}\int_0^1 (1-r^2)^2 r\,dr = \overline{f(0)}\,.\end{aligned}$$

The proof of Theorem 10 for $p = 1$ is now identical to that for $p > 1$, with R playing the role of P and the above equation the role of Lemma 10.

§2.8. Linear isometries.

In this section we present a theorem of Kolaski [1] that characterizes the surjective linear isometries of the Bergman space. One such isometry is generated by an arbitrary analytic automorphism

$$\varphi(z) = \gamma\,\frac{\alpha - z}{1 - \overline{\alpha}z}$$

of the unit disk, where $\alpha \in \mathbb{D}$ and γ is a unimodular constant. An application of the change of variables formula for integrals shows that the linear map $T : A^p \to A^p$ defined by $Tf(z) = f(\varphi(z))(\varphi'(z))^{\frac{2}{p}}$ is indeed a surjective isometry, since $|\varphi'|^2$ is the Jacobian of φ.

The Bergman space A^2 is a Hilbert space, and consequently it has many surjective isometries. They can be constructed, for instance, by permuting elements of an orthonormal basis. On the other hand, these isometries are surprisingly scarce in A^p for $p \neq 2$. Kolaski showed that they are all generated as above by automorphisms of the disk.

THEOREM 11. *Suppose $0 < p < \infty$, $p \neq 2$, and let $T : A^p \to A^p$ be a surjective linear isometry. Then T has the form*

$$Tf(z) = \beta\, f(\varphi(z))(\varphi'(z))^{\frac{2}{p}}\,,$$

where φ is an automorphism of $\mathbb{D}$ and β is a constant of modulus 1.

A major ingredient in the proof is the following theorem essentially due to Forelli [1], who used it to describe the isometries of H^p for $p \neq 2$. The theorem in present form may be found in Rudin [4,5].

THEOREM A. *Suppose $0 < p < \infty$, $p \neq 2$, and let μ and ν be finite positive measures defined on sets X and Y, respectively. Suppose $\mathcal{M} \subset L^p(\mu)$ is an algebra with $1 \in \mathcal{M}$, and let $S : \mathcal{M} \to L^p(\nu)$ be a linear map satisfying $S(1) = 1$ and*

$$\int_Y |Sf|^p\,d\nu = \int_X |f|^p\,d\mu$$

for all $f \in \mathcal{M}$*. Then* $\|Sf\|_{L^\infty(\nu)} = \|f\|_{L^\infty(\mu)}$ *and* S *has the multiplicative property* $S(fg) = S(f)S(g)$ *for all* $f, g \in \mathcal{M}$*. Moreover, if* $\mathcal{M} \subset L^\infty(\mu)$*, then*

$$\mu(f^{-1}(\Omega)) = \nu((Sf)^{-1}(\Omega))$$

for every function $f \in \mathcal{M}$ *and every measurable set* $\Omega \subset \mathbb{C}$.

For a proof of Theorem A, which falls outside the scope of this book, the reader is encouraged to consult the literature cited above. An account also appears in the book of Fisher [1]. Here we propose to apply Theorem A to give a proof of Theorem 11. We will also have occasion to use the following lemma.

LEMMA 11. *If* $f \neq 0$ *is analytic in* $\mathbb{D}$ *and* h *is a complex-valued function such that* fh^k *is analytic for every* $k = 1, 2, \dots$*, then* h *is analytic in* $\mathbb{D}$.

PROOF OF LEMMA. Since f and fh are analytic, it suffices to show that wherever f vanishes, fh has a zero of at least the same order, so that $h = (fh)/f$ is analytic. To this end, suppose f has a zero of order m at some point $\alpha \in \mathbb{D}$, and let n be the order of α as a zero of fh. We need to show that $n \geq m$. But since

$$(fh)^k = f^{k-1} fh^k, \qquad k = 1, 2, \dots,$$

and fh^k is analytic, we see that $kn \geq (k-1)m$. Because this inequality holds for all $k \geq 1$, we may infer that $n \geq m$. □

PROOF OF THEOREM 11. Let T be a linear isometry of A^p onto A^p, and define $g = T(1)$. Then $g \in A^p$ and $\|g\|_p = \|1\|_p = 1$, so $g(z) \neq 0$ almost everywhere. Define the measure ν on $\mathbb{D}$ by $d\nu = |g|^p d\sigma$, and consider the linear map $S : H^\infty \to L^p(\nu)$ defined by $S(f) = \frac{T(f)}{g}$. Then $S(1) = 1$ and

$$\int_{\mathbb{D}} |Sf|^p \, d\nu = \int_{\mathbb{D}} \left|\frac{Tf}{g}\right|^p |g|^p \, d\sigma = \int_{\mathbb{D}} |Tf|^p \, d\sigma = \int_{\mathbb{D}} |f|^p \, d\sigma,$$

so the hypotheses of Theorem A are satisfied, where H^∞ is the algebra $\mathcal{M}$. Therefore, $\|Sf\|_{L^\infty(\nu)} = \|f\|_{L^\infty(\sigma)}$ and $S(fg) = S(f)S(g)$ for all $f, g \in H^\infty$.
Moreover, since

$$g(Sf)^k = gS(f^k) = T(f^k) \in A^p, \qquad k = 1, 2, \dots,$$

Lemma 11 implies that Sf is analytic. Therefore, S is a multiplicative linear isometry from H^∞ to itself. If T is surjective, the analyticity of Sf also implies that $g(z) \neq 0$ in $\mathbb{D}$.

We now define the analytic function $\varphi = S(h)$, where $h(z) = z$. Note that $S(1) = 1 \circ S(h)$ and $S(h) = h \circ S(h)$. If $S(h^n) = h^n \circ S(h)$ for some $n \geq 1$, then by the multiplicative property of S,

$$S(h^{n+1}) = S(h)S(h^n) = (h \circ S(h))(h^n \circ S(h)) = h^{n+1} \circ S(h).$$

Thus by induction, we find that $S(f) = f \circ S(h) = f \circ \varphi$ for every polynomial f. In view of the definition of S, we see that

$$Tf(z) = g(z)Sf(z) = g(z)f(\varphi(z))$$

for every polynomial f and for all points $z \in \varphi^{-1}(\mathbb{D})$.

Suppose now that f is an arbitrary function in A^p. By Theorem 3, there is a sequence of polynomials $\{P_n\}$ such that $\|TP_n - Tf\|_p = \|P_n - f\|_p \to 0$ as $n \to \infty$. This implies that $TP_n(z) \to Tf(z)$ for every $z \in \mathbb{D}$, and therefore that $Tf(z) = g(z)f(\varphi(z))$ for every $f \in A^p$ and all $z \in \varphi^{-1}(\mathbb{D})$.

We next show that $\varphi(\mathbb{D}) \subset \mathbb{D}$. By Theorem A, $\nu(\varphi^{-1}(\Omega)) = \sigma(\Omega)$ for every measurable set $\Omega \subset \mathbb{C}$. Therefore, $\nu(\varphi^{-1}(\mathbb{D})) = \sigma(\mathbb{D}) = 1$. But

$$\nu(\mathbb{D}) = \int_{\mathbb{D}} d\nu = \int_{\mathbb{D}} |g|^p \, d\sigma = \int_{\mathbb{D}} |T1|^p \, d\sigma = \int_{\mathbb{D}} d\sigma = 1.$$

Thus $\nu(\mathbb{D} \setminus \varphi^{-1}(\mathbb{D})) = 0$, and $\sigma(\mathbb{D} \setminus \varphi^{-1}(\mathbb{D})) = 0$, since σ is absolutely continuous with respect to ν. This implies that $\varphi^{-1}(\mathbb{D})$ is dense in $\mathbb{D}$ and consequently that $\varphi(\mathbb{D}) \subset \overline{\mathbb{D}}$, since φ is continuous. To see this, let $\alpha \in \mathbb{D}$ and choose a sequence of points $z_n \in \mathbb{D}$ such that $w_n = \varphi(z_n) \in \mathbb{D}$ and $z_n \to \alpha$. Then $w_n \to \varphi(\alpha)$, so $\varphi(\alpha) \in \overline{\mathbb{D}}$. But analytic functions carry open sets to open sets, so from the inclusion $\varphi(\mathbb{D}) \subset \overline{\mathbb{D}}$ we may infer that $\varphi(\alpha) \in \mathbb{D}$. Thus $\varphi(\mathbb{D}) \subset \mathbb{D}$. Furthermore,

$$f(\varphi(\alpha)) = \lim_{n\to\infty} f(\varphi(z_n)) = \lim_{n\to\infty} \frac{Tf(z_n)}{g(z_n)} = \frac{Tf(\alpha)}{g(\alpha)}, \qquad f \in A^p.$$

This shows that

$$Tf(z) = g(z)f(\varphi(z)), \qquad f \in A^p, \ z \in \mathbb{D}.$$

Since T is a surjective isometry, so is its inverse T^{-1}, and we see by the above arguments that $T^{-1}f = G(f \circ \psi)$, where $G = T^{-1}(1)$ and ψ is an analytic function with $\psi(\mathbb{D}) \subset \mathbb{D}$. We also find that

$$f = TT^{-1}f = g((T^{-1}f) \circ \varphi) = g(G \circ \varphi)(f \circ \psi \circ \varphi), \qquad f \in A^p.$$

But $g(G \circ \varphi) = TG = 1$, so $f = f \circ \psi \circ \varphi$ for every function $f \in A^p$. In particular, $\psi(\varphi(z)) = z$ for all $z \in \mathbb{D}$. Reversing the roles of T and T^{-1}, we conclude in a similar manner that $\varphi(\psi(z)) = z$ for all $z \in \mathbb{D}$. This shows that φ is an analytic automorphism of $\mathbb{D}$ with $\psi = \varphi^{-1}$.

It remains to show that $g = \beta(\varphi')^{\frac{2}{p}}$ for some unimodular constant β. Since $\nu(\varphi^{-1}(\Omega)) = \sigma(\Omega)$ for every measurable set $\Omega \subset \mathbb{D}$, we see that

$$\int_{\mathbb{D}} (f \circ \varphi)|g|^p \, d\sigma = \int_{\mathbb{D}} f \circ \varphi \, d\nu = \int_{\mathbb{D}} f \, d\sigma$$

for every function f continuous in $\overline{\mathbb{D}}$. Now set $F = f \circ \varphi$ and apply the change of variables formula to infer that

$$\int_{\mathbb{D}} F|g|^p \, d\sigma = \int_{\mathbb{D}} F \circ \varphi^{-1} \, d\sigma = \int_{\mathbb{D}} F|\varphi'|^2 \, d\sigma \, .$$

Since this holds for all continuous functions F, and since g and φ' are themselves continuous in $\mathbb{D}$, it follows that $|g|^p = |\varphi'|^2$. Thus we conclude that $g = \beta(\varphi')^{\frac{2}{p}}$, where β is a constant of modulus 1. □

§2.9. Function multipliers.

The results of this section are formulated for the standard Bergman spaces A^p over the unit disk, but they extend with little change to Bergman spaces over general domains in the plane. A function φ analytic in $\mathbb{D}$ is said to be a *multiplier* of a space A^p $(0 < p < \infty)$ if $\varphi f \in A^p$ for all $f \in A^p$. It is of interest to describe the multipliers of A^p. An obvious necessary condition is that $\varphi \in A^p$. On the other hand, every bounded function is clearly a multiplier.

In fact, the converse is also true; every multiplier of A^p is bounded. To see this, let $M_\varphi : A^p \to A^p$ be the multiplication operator associated with an arbitrary function $\varphi \in A^p$, with domain of definition

$$\mathcal{D} = \mathcal{D}(M_\varphi) = \{f \in A^p \ : \ \varphi f \in A^p\}$$

and $M_\varphi(f) = \varphi f$ for $f \in \mathcal{D}$. Note that M_φ is densely defined because $\mathcal{D}$ contains all polynomials. It is easy to verify that M_φ is a closed operator. If φ is a multiplier of A^p, then $\mathcal{D} = A^p$ and we conclude from the closed graph theorem that M_φ is bounded: $\|M_\varphi(f)\|_p \le C\|f\|_p$ for all $f \in A^p$. It should be emphasized that the conclusion remains valid for $0 < p < 1$. Then A^p is no longer a Banach space, but it is a complete metric space with translation-invariant metric (sometimes called a *Fréchet space*), and the closed graph theorem extends to such spaces. When φ is a multiplier, the *norm* of the operator M_φ is defined by

$$\|M_\varphi\| = \sup_{\|f\|_p \le 1} \|M_\varphi(f)\|_p \, .$$

THEOREM 12. *Let $0 < p < \infty$. If a function φ is a multiplier of A^p, then $\varphi \in H^\infty$ and $|\varphi(z)| \le \|M_\varphi\|$ for all $z \in \mathbb{D}$.*

PROOF. For fixed $z \in \mathbb{D}$, let λ_z denote the point-evaluation functional defined by $\lambda_z(f) = f(z)$. Recall (*cf.* Section 1.1) that λ_z is bounded on each space A^p. Therefore,

$$\begin{aligned} |\varphi(z)\lambda_z(f)| &= |\varphi(z)f(z)| = |\lambda_z(M_\varphi(f))| \\ &\le \|\lambda_z\|\|M_\varphi(f)\|_p \le \|\lambda_z\|\|M_\varphi\|\|f\|_p \, . \end{aligned}$$

Now take the supremum over the unit ball of A^p to conclude that

$$|\varphi(z)|\|\lambda_z\| \le \|\lambda_z\|\|M_\varphi\|, \qquad z \in \mathbb{D},$$

or $|\varphi(z)| \le \|M_\varphi\|$. □

Each function $\varphi \in A^p$ is said to *generate* a subspace $[\varphi]$, the closure of the set of multiples of φ by polynomials. When $\varphi \in H^\infty$, it is important to observe that $[\varphi]$ does not necessarily coincide with

$$\varphi A^p = \{\varphi f \ : \ f \in A^p\},$$

since the latter set need not be closed. The crucial requirement is that the operator of multiplication by φ be *bounded below*; in other words, that there exist a constant $c > 0$ such that $\|\varphi f\|_p \ge c\|f\|_p$ for all $f \in A^p$.

THEOREM 13. *For a nonzero function $\varphi \in H^\infty$ and $0 < p < \infty$, the set φA^p is closed if and only if the operator of multiplication by φ is bounded below on A^p.*

PROOF. Suppose first that the multiplication operator is bounded below. Then if $f_n \in A^p$ and $\{\varphi f_n\}$ converges to some function $g \in A^p$, it follows that $f_n \to f$ for some $f \in A^p$. Thus $\varphi f_n \to \varphi f$ since $\varphi \in H^\infty$, and so $g \in \varphi A^p$. This shows that φA^p is closed.

The converse is a simple consequence of the closed graph theorem. We define the linear operator $T(g) = g/\varphi$ with domain

$$\mathcal{D}(T) = \{g \in [\varphi] \ : \ g/\varphi \in A^p\}.$$

It is easy to check that T is a closed operator. Observe first that φA^p is closed if and only if $\varphi A^p = [\varphi]$, since the polynomials are dense in A^p. If φA^p is closed, then by the closed graph theorem T is a bounded operator. This says that $\|f\|_p \le C\|\varphi f\|_p$ for some constant $C > 0$ and for all $f \in A^p$. In other words, multiplication by φ is bounded below. □

Essentially the same proof can be applied to establish a more general result. For $0 < p < \infty$, suppose a function $\varphi \in H^\infty$ is a multiplier of a subspace S of A^p, in the sense that $\varphi S \subset S$; that is, if $\varphi f \in S$ whenever $f \in S$. Then the set φS is closed if and only if the operator of multiplication by φ is bounded below on S.

To give an explicit nontrivial example, we shall now show that the operator of multiplication by z is bounded below on each space A^p, for $0 < p < \infty$. The key to the proof is the inequality

$$|f(z_0)|^p \le \frac{1}{|\Omega|}\int_\Omega |f(\zeta)|^p \, d\sigma, \qquad z_0 \in \mathbb{D},$$

where $\Omega \subset \mathbb{D}$ is an arbitrary annulus centered at z_0, with normalized area $|\Omega|$. This follows easily from the subharmonic property of $|f|^p$. As a consequence, we can see that

$$|f(z)|^p \le C \int_{r\le|\zeta|<1} |f(\zeta)|^p \, d\sigma$$

throughout the disk $|z| < \varepsilon = \frac{r}{2}$, where $0 < r < \frac{1}{2}$ and C is a constant depending only on r. Hence

$$\int_{|z|\le\varepsilon} |f(z)|^p \, d\sigma \le C \int_{r\le|\zeta|<1} |f(\zeta)|^p \, d\sigma \le r^{-p} C \int_{\mathbb{D}} |zf(z)|^p \, d\sigma .$$

On the other hand, it is immediate that

$$\int_{\mathbb{D}} |zf(z)|^p \, d\sigma \ge \varepsilon^p \int_{\varepsilon\le|z|<1} |f(z)|^p \, d\sigma .$$

The two estimates together give the desired result.

The proof is easily adapted to show that multiplication by any finite Blaschke product is bounded below. In Chapter 4 we will give a description (due to Horowitz [3]) of the *infinite* Blaschke products whose multiplication operators are bounded below on A^p.

§2.10. Carleson measures.

The problem here is to characterize those finite Borel measures μ on $\mathbb{D}$ with the property that $\int_{\mathbb{D}} |f(z)|^p \, d\mu < \infty$ for every function f in the Bergman space A^p, where $0 < p < \infty$. Ordinary area measure is clearly one example. If μ is any such measure, it follows from the closed graph theorem that

$$\int_{\mathbb{D}} |f(z)|^p \, d\mu \le K \, \|f\|_p^p , \qquad f \in A^p ,$$

for some constant K depending only on p. A measure μ with this property will be called a *Carleson measure for the Bergman space* A^p.

Around 1960, Lennart Carleson [2,3] solved the corresponding problem for the Hardy spaces H^p, applying the result to interpolation problems and to his proof of the corona theorem. Thus a measure μ on $\mathbb{D}$ is called a *Carleson measure for the Hardy space* H^p if

$$\int_{\mathbb{D}} |f(z)|^p \, d\mu \le C \, \|f\|_{H^p}^p , \qquad f \in H^p ,$$

for some constant C depending only on p. Dividing out the zeros by a Blaschke product, one sees that if the inequality holds for *nonvanishing* functions in H^p, then it holds for all H^p functions. Thus it becomes apparent that the property is actually independent of p. Carleson showed that a necessary and sufficient condition for μ to be a Carleson measure is that $\mu(S) \leq Ch$ for some constant C and all *Carleson squares*

$$S = \{re^{i\theta} \in \mathbb{D} : 1 - h \leq r < 1, \, |\theta - \theta_0| \leq h\}$$

of "sidelength" h at the boundary of the disk. A proof of Carleson's theorem can be found in Duren [5].

It is a result of Hastings [1] that the Carleson measures for the Bergman space A^p have the analogous description $\mu(S) \leq Ch^2$ for all Carleson squares S of sidelength h. In particular, the Carleson measures for A^p are again independent of p, a fact that is not so evident. Oleinik [1] obtained the same description in a broader context at about the same time as Hastings. We shall present a more streamlined version of the result, due to Luecking [1,2], that brings into play the pseudohyperbolic metric. Our treatment follows that of Axler [3].

Here is the main theorem.

THEOREM 14. *Let μ be a positive Borel measure on $\mathbb{D}$. Then for each fixed p with $0 < p < \infty$, the following three statements are equivalent.*

(*i*) *The inequality $\int_{\mathbb{D}} |f(z)|^p \, d\mu \leq K\|f\|_p^p$ holds for some constant $K > 0$ and all $f \in A^p$.*

(*ii*) *An inequality $\mu(\Delta(\alpha, r)) \leq C\,|\Delta(\alpha, r)|$ holds for all r $(0 < r < 1)$, for some constant $C > 0$ depending only on r, and for all pseudohyperbolic disks $\Delta(\alpha, r)$, $\alpha \in \mathbb{D}$.*

(*iii*) *An inequality $\mu(\Delta(\alpha, r)) \leq C\,|\Delta(\alpha, r)|$ holds for some r $(0 < r < 1)$, for some constant C, and for all pseudohyperbolic disks $\Delta(\alpha, r)$.*

COROLLARY. *The Carleson measures for the Bergman space A^p are the same for every index p in the interval $0 < p < \infty$.*

Before passing to the proof, let us remark that Hastings' condition is equivalent to the condition (*ii*), where Carleson squares are replaced by pseudohyperbolic disks. The equivalence is not obvious, but it can be verified by a direct geometric argument. Since the details are rather involved, they will be deferred to the end of this section.

The proof of the theorem will require two further lemmas. One is a technical covering lemma, and the other is a kind of uniform subharmonic property with respect to pseudohyperbolic disks.

LEMMA 12. *For each pseudohyperbolic radius r $(0 < r < 1)$, there exist a sequence $\{\alpha_k\}$ of points in $\mathbb{D}$ and an integer N such that*

$$\bigcup_{k=1}^{\infty} \Delta(\alpha_k, r) = \mathbb{D}$$

and no point $z \in \mathbb{D}$ belongs to more than N of the dilated disks $\Delta(\alpha_k, R)$, where $R = \frac{1}{2}(1 + r)$.

PROOF. Let $\{B_k\}$ be a sequence of disks in $\mathbb{D}$ of pseudohyperbolic radius $r/3$ for which $\bigcup_{k=1}^{\infty} B_k = \mathbb{D}$. Select a subsequence $\{D_k\}$ of $\{B_k\}$ inductively as follows. Begin with $D_1 = B_1$. Having chosen $D_1, D_2, \ldots, D_n$, let D_{n+1} be the first disk of the sequence $\{B_k\}$ that is disjoint from $D_1 \cup D_2 \cup \cdots \cup D_n$. This defines a subsequence of disjoint disks $D_k \subset \mathbb{D}$ with common radius $r/3$. Let α_k denote the pseudohyperbolic center of D_k.

We claim now that $\bigcup_{k=1}^{\infty} \Delta(\alpha_k, r) = \mathbb{D}$. Indeed, if any point $\alpha \in \mathbb{D}$ belongs to none of the disks $\Delta(\alpha_k, r)$, then by the triangle inequality the disk $\Delta(\alpha, 2r/3)$ is disjoint from all of the disks $D_k = \Delta(\alpha_k, r/3)$. But by construction the point α must belong to some disk B_j, and another application of the triangle inequality then shows that $B_j \subset \Delta(\alpha, 2r/3)$. In particular, B_j is disjoint from all of the disks D_k, which is impossible because of the way the sequence $\{D_k\}$ was constructed. This contradiction proves that $\bigcup_{k=1}^{\infty} \Delta(\alpha_k, r) = \mathbb{D}$.

To complete the proof, we must now show that each given point $z \in \mathbb{D}$ belongs to at most N of the dilated disks $\Delta(\alpha_k, R)$. For this purpose we consider the set of points

$$E = \left\{\varphi_z(\alpha_k) \,:\, z \in \Delta(\alpha_k, R)\right\}.$$

By definition, each point $w \in E$ lies in the disk $|w| < R$. On the other hand, by the Möbius invariance of the pseudohyperbolic metric, any two distinct elements $w_j = \varphi_z(\alpha_j)$ and $w_k = \varphi_z(\alpha_k)$ in E must satisfy

$$\rho(w_j, w_k) = \rho(\alpha_j, \alpha_k) \geq \tfrac{1}{3} r\,,$$

since the disks D_j and D_k are disjoint. But

$$\rho(w_j, w_k) = \left| \frac{w_k - w_j}{1 - \overline{w_j} w_k} \right| \leq \frac{|w_k - w_j|}{1 - R^2}\,,$$

so $|w_k - w_j| \geq \frac{1}{3} r(1 - R^2)$. In other words, the set E consists of points in the disk $|w| < R$ that are at least a distance $\frac{1}{3} r(1 - R^2)$ apart in the Euclidean metric. This shows that E can contain no more than N points, where N is some integer depending only on r. □

LEMMA 13. *If $0 < p < \infty$, $0 < r < 1$, and f is analytic in $\mathbb{D}$, then for arbitrary $\alpha \in \mathbb{D}$ and for all $z \in \Delta(\alpha, r)$,*

$$|f(z)|^p \leq \frac{4(1-R)^{-4}}{|\Delta(\alpha, R)|} \int_{\Delta(\alpha, R)} |f(\zeta)|^p \, d\sigma\,, \qquad R = \tfrac{1}{2}(1 + r).$$

PROOF. Fix $\alpha \in \mathbb{D}$ and $z \in \Delta(\alpha, r)$. Denote by B the Euclidean disk with center $\varphi_\alpha(z)$ and radius $\frac{1}{2}(1-r)$. Then $B \subset \Delta(0, R)$, since $|\varphi_\alpha(z)| < r$.

Because φ_α is involutive and the function $|f\circ\varphi_\alpha|^p$ is subharmonic in $\mathbb{D}$, the sub-mean-value property gives

$$|f(z)|^p = |(f\circ\varphi_\alpha)(\varphi_\alpha(z))|^p \le \frac{1}{|B|}\int_B |(f\circ\varphi_\alpha)(\zeta)|^p\,d\sigma\,,$$

so that

$$|f(z)|^p \le \frac{4}{(1-r)^2}\int_{\Delta(0,R)} |(f\circ\varphi_\alpha)(\zeta)|^p\,d\sigma\,.$$

It follows that

$$|f(z)|^p \le \frac{(1+R)^4}{(1-R)^2(1-|\alpha|^2)^2}\int_{\Delta(0,R)} |f(\varphi_\alpha(\zeta))|^p|\varphi_\alpha'(\zeta)|^2\,d\sigma\,,$$

since

$$|\varphi_\alpha'(\zeta)|^2 = \frac{(1-|\alpha|^2)^2}{|1-\overline{\alpha}\zeta|^4} \ge \frac{(1-|\alpha|^2)^2}{(1+R)^4} \qquad \text{for } \zeta\in\Delta(0,R)\,.$$

Now make the substitution $w=\varphi_\alpha(\zeta)$, or $\zeta=\varphi_\alpha(w)$, to conclude that

$$|f(z)|^p \le \frac{(1+R)^4}{(1-R)^2(1-|\alpha|^2)^2}\int_{\Delta(\alpha,R)} |f(w)|^p\,d\sigma\,,$$

which leads to the desired result because (*cf.* Section 2.5)

$$|\Delta(\alpha,R)| = \left(\frac{R(1-|\alpha|^2)}{1-R^2|\alpha|^2}\right)^2 \le \left(\frac{1-|\alpha|^2}{1-R^2}\right)^2. \qquad \square$$

PROOF OF THEOREM 14.

$(i)\implies(ii)$. First suppose that μ is a Carleson measure for A^p. In other words, suppose μ has the property (i) for some index p with $0<p<\infty$. Choose the function $f(z)=k_\alpha(z)^{2/p}$, where $k_\alpha(z)=(1-\overline{\alpha}z)^{-2}$ is the Bergman kernel function of $\mathbb{D}$. Then

$$\|f\|_p^p = \|k_\alpha\|_2^2 = (1-|\alpha|^2)^{-2}\,,$$

while

$$\begin{aligned}\int_{\mathbb{D}} |f(z)|^p\,d\mu &= \int_{\mathbb{D}} |k_\alpha(z)|^2\,d\mu \ge \int_{\Delta(\alpha,r)} |k_\alpha(z)|^2\,d\mu \\ &\ge \mu(\Delta(\alpha,r))\min_{z\in\Delta(\alpha,r)} |k_\alpha(z)|^2 = \mu(\Delta(\alpha,r))\left(\frac{1-r|\alpha|}{1-|\alpha|^2}\right)^4,\end{aligned}$$

by Lemma 4 in Section 2.5. Thus in view of (i),

$$\begin{aligned}\mu(\Delta(\alpha,r)) &\le \left(\frac{1-|\alpha|^2}{1-r|\alpha|}\right)^4 K\,\|f\|_p^p = K\frac{(1-|\alpha|^2)^2}{(1-r|\alpha|)^4} \\ &= K\left(\frac{1+r|\alpha|}{r(1-r|\alpha|)}\right)^2 |\Delta(\alpha,r)| \le \frac{4K}{r^2(1-r)^2}\,|\Delta(\alpha,r)|\,.\end{aligned}$$

$(ii)\implies(iii)$. If the inequality holds for *all* r, then it holds for *some* r.

$(iii) \implies (i)$. Suppose now that μ has the property (iii) for *some* pseudohyperbolic radius r. Let $\{\Delta(\alpha_k, r)\}$ be the sequence of pseudohyperbolic disks in Lemma 12. Then

$$\int_{\mathbb{D}} |f(z)|^p \, d\mu \leq \sum_{k=1}^{\infty} \int_{\Delta(\alpha_k,r)} |f(z)|^p \, d\mu \, .$$

But Lemma 13 and the hypothesis (iii) imply

$$\begin{aligned}\int_{\Delta(\alpha_k,r)} |f(z)|^p \, d\mu &\leq \frac{4}{(1-R)^4} \frac{\mu(\Delta(\alpha_k, r))}{|\Delta(\alpha_k, R)|} \int_{\Delta(\alpha_k,R)} |f(z)|^p \, d\sigma \\ &\leq \frac{4C}{(1-R)^4} \frac{|\Delta(\alpha_k, r)|}{|\Delta(\alpha_k, R)|} \int_{\Delta(\alpha_k,R)} |f(z)|^p \, d\sigma \\ &\leq \frac{4C}{(1-R)^6} \int_{\Delta(\alpha_k,R)} |f(z)|^p \, d\sigma \, ,\end{aligned}$$

where $R = \frac{1}{2}(1+r)$. Now invoke Lemma 12 to conclude that

$$\int_{\mathbb{D}} |f(z)|^p \, d\mu \leq \frac{4CN}{(1-R)^6} \int_{\mathbb{D}} |f(z)|^p \, d\sigma = K \, \|f\|_p^p \, ,$$

where the constant K depends only on r. This completes the proof. □

Finally, let us show that the condition (ii), stating that $\mu(\Delta(\alpha, r)) \leq C\,|\Delta(\alpha, r)|$ for all r and all α, is equivalent to the requirement that $\mu(S) \leq Ch^2$ for Carleson squares S of arbitrary sidelength h. We first show that the hypothesis $\mu(S) \leq Ch^2$ implies $\mu(\Delta(\alpha, r)) \leq C|\Delta(\alpha, r)|$. It is sufficient to take α near the boundary of the disk, and to assume that $\alpha = a > r$. Then (see Section 2.5) the pseudohyperbolic disk $\Delta(a, r)$ lies in the annulus $\frac{a-r}{1-ar} < |z| < 1$, and it has Euclidean center γ and Euclidean radius R given by

$$\gamma = \frac{(1-r^2)a}{1-r^2a^2}, \qquad R = \frac{r(1-a^2)}{1-r^2a^2} \, .$$

Thus $\Delta(a, r) \subset S$, where S is the Carleson square centered on the real axis with sidelength

$$h = \max\left\{1 - \frac{a-r}{1-ar}, \; \sin^{-1}\left(\frac{R}{\gamma}\right)\right\} .$$

But since $\frac{2}{\pi}\theta \leq \sin\theta \leq \theta$, the angle θ is comparable to

$$\sin\theta = \frac{R}{\gamma} = \frac{r(1-a^2)}{(1-r^2)a} \, ,$$

while

$$1 - \frac{a-r}{1-ar} = \frac{(1+r)(1-a)}{1-ar} \, .$$

Thus h is comparable to $1-a$. On the other hand, the area of the pseudohyperbolic disk is

$$|\Delta(a,r)| = \left(\frac{r(1-a^2)}{1-r^2a^2}\right)^2,$$

which is comparable to $(1-a)^2$. Thus

$$\mu(\Delta(a,r)) \le \mu(S) \le Ch^2 \le C|\Delta(a,r)|,$$

showing that the property $\mu(S) \le Ch^2$ for Carleson squares implies the corresponding property (ii) for pseudohyperbolic disks.

The converse is more difficult. Now we must cover a given Carleson square

$$S = \{z \in \mathbb{D} : 1-h \le |z| < 1, \quad |\arg z - \theta_0| \le h\}$$

in an economical way by pseudohyperbolic disks of equal radius. We claim first that for sufficiently small h the "inner half"

$$\widetilde{S} = \{z \in \mathbb{D} : 1-h \le |z| < 1-\tfrac{1}{2}h, \quad |\arg z - \theta_0| \le h\}$$

of the square S can be covered by a disk $\Delta(\alpha, r)$ with $|\alpha| = 1-h$ and $r = \frac{3}{4}$. To see this, assume for convenience that $\theta_0 = 0$ and take $\alpha = 1-h$, then calculate the pseudohyperbolic distance

$$\rho(\alpha, (1-\tfrac{h}{2})e^{ih}) = \sqrt{\tfrac{5}{13}} + O(h), \qquad h \to 0,$$

between α and the farthest point in the half-square $\widetilde{S}$. (Note that $\sqrt{\frac{5}{13}} = 0.620\cdots < \frac{3}{4}$.)

Now divide the "inner half" of $S \setminus \widetilde{S}$ into two equal parts and apply the same process, covering each part by a pseudohyperbolic disk $\Delta(\alpha_j, r)$ with $r = \frac{3}{4}$ and $|\alpha_j| = 1 - \frac{h}{2}$ for $j = 1, 2$. At the kth stage of the process, 2^k congruent "inner halves" of Carleson squares with sidelengths $2^{-k}h$ are covered by pseudohyperbolic disks $\Delta(\alpha_{jk}, r)$ with $|\alpha_{jk}| = 1 - 2^{-k}h$ for $j = 1, 2, \ldots, 2^k$; and $r = \frac{3}{4}$. Since

$$S \subset \bigcup_{k=0}^{\infty} \bigcup_{j=1}^{2^k} \Delta(\alpha_{jk}, r),$$

we conclude that

$$\begin{aligned}
\mu(S) &\le \sum_{k=0}^{\infty} \sum_{j=1}^{2^k} \mu(\Delta(\alpha_{jk}, r)) \le C \sum_{k=0}^{\infty} \sum_{j=1}^{2^k} |\Delta(\alpha_{jk}, r)| \\
&\le 16\, C \sum_{k=0}^{\infty} 2^k (1 - |\alpha_{jk}|^2)^2 \le 64\, C \sum_{k=0}^{\infty} 2^k (2^{-k}h)^2 = 128\, Ch^2 .
\end{aligned}$$

Thus the pseudohyperbolic version of the property (condition (iii) with $r = \frac{3}{4}$) implies the Carleson square version.

§2.11. Uniformly discrete sequences.

We now turn to a basic concept that will lead us to important examples of Carleson measures for Bergman spaces. A sequence $\{z_k\}$ of distinct points in the unit disk $\mathbb{D}$ is said to be *uniformly discrete* if there is a constant δ with $0 < \delta < 1$ such that $\rho(z_j, z_k) \geq \delta$ for all pairs of indices j, k with $j \neq k$. Observe that the uniformly discrete sequences are Möbius invariant, since the same is true of the pseudohyperbolic metric. In other words, if $\rho(z_j, z_k) \geq \delta$ and $w_k = \varphi_\alpha(z_k)$, then $\rho(w_j, w_k) \geq \delta$.

Uniformly discrete sequences sometimes play the same role for Bergman spaces that uniformly separated sequences play for Hardy spaces. A sequence $\{z_k\}$ in $\mathbb{D}$ is called a *Blaschke sequence* if $\sum (1 - |z_k|) < \infty$. A Blaschke sequence is said to be *uniformly separated* if

$$\prod_{j \neq k} \left| \frac{z_j - z_k}{1 - \overline{z_j} z_k} \right| \geq \delta \,, \qquad k = 1, 2, \dots \,,$$

for some constant $\delta > 0$ independent of k. It is easy to see that every uniformly separated sequence is uniformly discrete, but the converse is not true. One property of a uniformly separated sequence, to be proved in Chapter 4, is that

$$\sum_{k=1}^{\infty} (1 - |z_k|^2) |f(z_k)|^p \leq C \, \|f\|_{H^p}^p \qquad \text{for every } f \in H^p \,.$$

Equivalently, the discrete measure μ that places mass $(1 - |z_k|^2)$ at z_k is a Carleson measure for H^p. The following lemma makes a corresponding statement for Bergman spaces.

LEMMA 14. *If $\{z_k\}$ is uniformly discrete, then*

$$\sum_{k=1}^{\infty} (1 - |z_k|^2)^2 |f(z_k)|^p \leq C \, \|f\|_p^p \qquad \textit{for every } f \in A^p \,,$$

where $0 < p < \infty$. Equivalently, the discrete measure μ that places mass $(1 - |z_k|^2)^2$ at z_k is a Carleson measure for the Bergman space A^p.

PROOF. If $\rho(z_j, z_k) \geq \delta$, then the pseudohyperbolic disks $\Delta(z_j, \frac{\delta}{2})$ and $\Delta(z_k, \frac{\delta}{2})$ are disjoint for $j \neq k$, in view of the triangle inequality. By the submean-value property of $|g|^p$ for any analytic function g, we have

$$|g(0)|^p \leq \frac{4}{\delta^2} \int_{\Delta(0, \frac{\delta}{2})} |g(\zeta)|^p \, d\sigma \,.$$

Now let $g(\zeta) = f(\varphi_{z_k}(\zeta))\varphi_{z_k}'(\zeta)^{2/p}$ to deduce that

$$\begin{aligned}(1-|z_k|^2)^2|f(z_k)|^p &\le \frac{4}{\delta^2}\int_{\Delta(0,\frac{\delta}{2})}|f(\varphi_{z_k}(\zeta))|^p|\varphi_{z_k}'(\zeta)|^2\,d\sigma\\ &= \frac{4}{\delta^2}\int_{\Delta(z_k,\frac{\delta}{2})}|f(z)|^p\,d\sigma\,,\end{aligned}$$

by a change of variables. Since the disks $\Delta(z_k,\frac{\delta}{2})$ are disjoint, we conclude that

$$\sum_{k=1}^{\infty}(1-|z_k|^2)^2|f(z_k)|^p \le \frac{4}{\delta^2}\int_{\mathbb{D}}|f(z)|^p\,d\sigma\,. \qquad \square$$

COROLLARY 1. *Let f be analytic in $\mathbb{D}$. Then*

$$(1-|\alpha|^2)^2|f(\alpha)|^p \le \frac{1}{r^2}\int_{\Delta(\alpha,r)}|f(z)|^p\,d\sigma\,, \qquad 0<p<\infty,\quad 0<r<1\,,$$

for each point $\alpha\in\mathbb{D}$.

This is a corollary of the proof. The inequality can be viewed as a pseudohyperbolic version of the submean-value property of the subharmonic function $|f|^p$.

COROLLARY 2. *If $\{z_k\}$ is uniformly discrete and $\rho(z_j,z_k)\ge\delta>0$ for $j\ne k$, then*

$$\sum_{k=1}^{\infty}(1-|z_k|^2)^2 \le \frac{4}{\delta^2}\,.$$

For a proof, take $f(z)\equiv 1$. In fact, it can be shown (*cf.* Duren, Schuster, and Vukotić [1]) that every uniformly discrete sequence has the property

$$\sum_{k=1}^{\infty}(1-|z_k|)\left(\log\frac{1}{1-|z_k|}\right)^{-1-\varepsilon} < \infty$$

for each $\varepsilon>0$, and that the series may diverge if $\varepsilon=0$. We will see in Chapter 4 that zero-sets of A^p functions have exactly the same property.

We now formulate a density property of uniformly discrete sequences.

LEMMA 15. *Suppose $\rho(z_j,z_k)\ge\delta>0$ for $j\ne k$, and let $n(\Gamma,\alpha,r)$ denote the number of points in the sequence $\Gamma=\{z_k\}$ that lie in the pseudohyperbolic disk $\Delta(\alpha,r)$. Then*

$$n(\Gamma,\alpha,r) \le \left(\tfrac{2}{\delta}+1\right)^2\frac{1}{1-r^2}$$

for every point $\alpha\in\mathbb{D}$ and $0<r<1$. In particular, $n(\Gamma,\alpha,r)=O\left(\frac{1}{1-r}\right)$ as $r\to 1$.

PROOF. In view of the Möbius invariance of uniformly discrete sequences, there is no loss of generality in taking $\alpha = 0$. The uniform discreteness assumption implies that the disks $\Delta(z_k, \frac{\delta}{2})$ are pairwise disjoint. It follows from the strong form of the triangle inequality for the pseudohyperbolic metric (*cf.* Section 2.5) that for each point $\zeta \in \Delta(0, r)$,

$$\Delta(\zeta, \tfrac{\delta}{2}) \subset \Delta(0, R), \qquad \text{where } R = \frac{r + \frac{\delta}{2}}{1 + \frac{\delta r}{2}}.$$

Indeed, if $\rho(0, \zeta) < r$ and $\rho(z, \zeta) < \frac{\delta}{2}$, then

$$\rho(0, z) \leq \frac{\rho(0, \zeta) + \rho(\zeta, z)}{1 + \rho(0, \zeta)\rho(\zeta, z)} < \frac{r + \frac{\delta}{2}}{1 + \frac{\delta r}{2}} = R.$$

Recall now that a pseudohyperbolic disk $\Delta(\zeta, s)$ has hyperbolic area

$$a(\Delta(\zeta, s)) = \frac{s^2}{1 - s^2}, \qquad \zeta \in \mathbb{D},\ 0 < s < 1.$$

Because the disks $\Delta(z_k, \frac{\delta}{2})$ are disjoint, hyperbolic areas can be compared to give

$$\sum_{z_k \in \Delta(0,r)} a(\Delta(z_k, \tfrac{\delta}{2})) \leq a(\Delta(0, R)),$$

or

$$n(\Gamma, 0, r) \frac{(\frac{\delta}{2})^2}{1 - (\frac{\delta}{2})^2} \leq \frac{R^2}{1 - R^2},$$

which reduces to the inequality

$$n(\Gamma, 0, r) \leq \left(\frac{2r + \delta}{\delta}\right)^2 \frac{1}{1 - r^2} \leq \left(\tfrac{2}{\delta} + 1\right)^2 \frac{1}{1 - r^2}. \qquad \square$$

The next result is a partial converse.

LEMMA 16. *Let $\Gamma = \{z_k\}$ be a sequence of distinct points in the unit disk such that for some fixed radius $r > 0$, each pseudohyperbolic disk $\Delta(\alpha, r)$ contains at most N points z_k. Then Γ is the disjoint union of at most N uniformly discrete sequences.*

PROOF. Consider first the disk $\Delta(z_1, r)$. By hypothesis, it contains at most N points in the sequence Γ, including z_1. Let those points be assigned to n different subsets, $\Gamma_1, \Gamma_2, \ldots, \Gamma_n$, where $n \leq N$. Let z_{n_1} be the first point of Γ not already assigned. Then $\rho(z_{n_1}, z_1) \geq r$, so z_{n_1} is placed into the set Γ_j containing z_1. Now proceed inductively. Suppose a finite number of points have been assigned to subsets $\Gamma_1, \Gamma_2, \ldots, \Gamma_m$ with $m \leq N$, and that $\rho(z, \zeta) \geq r$ for all points $z, \zeta \in \Gamma_j$, where $j = 1, \ldots, m$. Let z^* be the first point of Γ not already assigned to a subset Γ_j. By hypothesis, the disk $\Delta(z^*, r)$ contains at most $N-1$ points of Γ that have already been assigned, and they represent at most $N-1$ different subsets Γ_j, so the point z^* can be assigned to some subset Γ_k not represented in this list. It is clear by construction that $\rho(z^*, \zeta) \geq r$ for all points $\zeta \in \Gamma$ already assigned to Γ_k. This inductive process therefore divides the given set Γ into disjoint subsets $\Gamma_1, \ldots, \Gamma_m$ with $m \leq N$ and $\rho(z, \zeta) \geq r$ for all $z, \zeta \in \Gamma_j$, $j = 1, \ldots, m$. □

We can now state a theorem that combines these lemmas with one further condition. Again let $n(\Gamma, \alpha, r)$ denote the number of points of a given sequence $\Gamma = \{z_k\}$ that lie in the disk $\Delta(\alpha, r)$.

THEOREM 15. *For a sequence Γ of distinct points $z_k \in \mathbb{D}$, the following six statements are equivalent.*

(*i*) *Γ is a finite union of uniformly discrete sequences.*

(*ii*) $\sup_{\alpha \in \mathbb{D}} n(\Gamma, \alpha, r) < \infty$ *for some* $r \in (0, 1)$.

(*iii*) $\sup_{\alpha \in \mathbb{D}} n(\Gamma, \alpha, r) < \infty$ *for every* $r \in (0, 1)$.

(*iv*) *For some* $p \in (0, \infty)$, *there exists a constant* C *such that*

$$\sum_{k=1}^{\infty} (1 - |z_k|^2)^2 |f(z_k)|^p \leq C \, \|f\|_p^p \qquad \text{for every } f \in A^p .$$

(*v*) *For each* $p \in (0, \infty)$, *there exists a constant* C *such that*

$$\sum_{k=1}^{\infty} (1 - |z_k|^2)^2 |f(z_k)|^p \leq C \, \|f\|_p^p \qquad \text{for every } f \in A^p .$$

(*vi*)

$$\sup_{\alpha \in \mathbb{D}} \sum_{k=1}^{\infty} (1 - |\varphi_\alpha(z_k)|^2)^2 < \infty .$$

PROOF. Combining Lemmas 15 and 16, we see that (*i*) $\iff$ (*ii*) $\iff$ (*iii*). Theorem 14 shows that (*iv*) $\iff$ (*v*), since the Carleson measures for A^p are independent of p. Lemma 14 asserts that (*i*) $\implies$ (*v*). Thus it will be enough to show that (*v*) $\implies$ (*vi*) $\implies$ (*iii*).

(*v*) $\implies$ (*vi*). For an arbitrary point $\alpha \in \mathbb{D}$, let

$$f_\alpha(z) = \left\{ \frac{1 - |\alpha|^2}{(1 - \overline{\alpha} z)^2} \right\}^{2/p} .$$

Then $f_\alpha \in A^p$ and $\|f_\alpha\|_p = 1$, so we conclude from (*v*) that

$$\sum_{k=1}^{\infty} (1 - |\varphi_\alpha(z_k)|^2)^2 = \sum_{k=1}^{\infty} \frac{(1 - |z_k|^2)^2 (1 - |\alpha|^2)^2}{|1 - \overline{\alpha} z_k|^4}$$

$$= \sum_{k=1}^{\infty} (1 - |z_k|^2)^2 |f_\alpha(z_k)|^p \leq C \, \|f_\alpha\|_p^p = C .$$

$(vi) \implies (iii)$. Since $|\varphi_\alpha(z_k)| < r$ for $n(\Gamma,\alpha,r)$ points z_k, it follows from (vi) that

$$n(\Gamma,\alpha,r)(1-r^2)^2 \le \sum_{k=1}^{\infty}(1-|\varphi_\alpha(z_k)|^2)^2 \le C\,, \qquad \alpha \in \mathbb{D}\,,$$

which shows that $n(\Gamma,\alpha,r)$ is bounded for each fixed r. □

The equivalence of properties (v) and (vi) is actually a special case of a much more general result about measures in the disk. The Berezin transform of a function $f \in L^1(\mathbb{D})$ was defined in Section 2.5 by

$$Bf(\alpha) = \int_{\mathbb{D}} \frac{(1-|\alpha|^2)^2}{|1-\overline{\alpha}z|^4} f(z)\, d\sigma(z)\,.$$

More generally, the Berezin transform of a finite measure μ is

$$B\mu(\alpha) = \int_{\mathbb{D}} \frac{(1-|\alpha|^2)^2}{|1-\overline{\alpha}z|^4}\, d\mu(z)\,.$$

The Berezin transform of the discrete measure $\mu = \sum(1-|z_k|^2)^2\delta_{z_k}$ is found to be

$$B\mu(\alpha) = \sum_{k=1}^{\infty}(1-|\varphi_\alpha(z_k)|^2)^2\,.$$

Thus the condition (vi) says that the Berezin transform of this measure μ is bounded in $\mathbb{D}$. Furthermore, the equivalence of (v) and (vi) is a special case of the following theorem.

THEOREM 16. *A finite measure μ in $\mathbb{D}$ is a Carleson measure for the Bergman spaces if and only if its Berezin transform is bounded.*

PROOF. The Berezin transform of μ can be written as

$$B\mu(\alpha) = \int_{\mathbb{D}} |f_\alpha(z)|^p\, d\mu\,,$$

where f_α is the function of unit A^p norm defined in the proof of Theorem 15. If μ is a Carleson measure for A^p, then this integral is bounded uniformly for all points $\alpha \in \mathbb{D}$. Thus $B\mu(\alpha)$ is bounded in $\mathbb{D}$.

To prove the converse, we invoke the geometric description of Carleson measures as given by Theorem 14. Assuming that $B\mu(\alpha)$ is bounded, we need to show that $\mu(\Delta(\alpha,r)) \le C|\Delta(\alpha,r)|$ for some constant C depending only on r. But

$$B\mu(\alpha) \ge \int_{\Delta(\alpha,r)} \frac{(1-|\alpha|^2)^2}{|1-\overline{\alpha}z|^4}\, d\mu(z) \ge \frac{(1-r)^4}{(1-|\alpha|^2)^2}\, \mu(\Delta(\alpha,r))\,,$$

since

$$\frac{1}{|1-\overline{\alpha}z|} \geq \frac{1-r}{1-|\alpha|^2}\,, \qquad z \in \Delta(\alpha, r)\,,$$

by Lemma 4 in Section 2.5. The proof is completed by noting that the pseudohyperbolic disk $\Delta(\alpha, r)$ has area

$$|\Delta(\alpha, r)| = \left(\frac{r(1-|\alpha|^2)}{1-r^2|\alpha|^2}\right)^2 \geq r^2(1-|\alpha|^2)^2\,. \qquad \square$$

CHAPTER 3

Analytic Properties

This chapter will focus on the analytic properties of individual functions in a Bergman space A^p. Topics will include boundary behavior, growth estimates, and Taylor coefficients. Here the theory of Hardy spaces H^p serves as a useful point of reference. As a general rule, functions in H^p spaces are better behaved, but many of their properties have clear analogues for Bergman spaces. In some respects the results for Bergman spaces are actually more satisfactory. We shall begin with a further review of H^p spaces.

§3.1. More on Hardy spaces.

Chapter 2 began with a brief survey of Hardy spaces, with emphasis on linear space aspects, to set the stage for a corresponding study of Bergman spaces. The survey will now be continued, with its focus shifting to structural properties of functions in H^p spaces. For proofs and other information, the reader is again referred to Duren [5] and other books on H^p spaces, such as Garnett [1] and Koosis [1].

Recall that a function f analytic in the unit disk $\mathbb{D}$ is said to belong to the Hardy space H^p if its integral means $M_p(r, f)$ are bounded for $0 \le r < 1$. The class of bounded analytic functions is called H^∞. For $0 < p \le \infty$, each function $f \in H^p$ has a *nontangential limit*

$$f(e^{i\theta}) = \lim_{z \to e^{i\theta}} f(z) \qquad \text{a.e.},$$

where the limit is taken as $z \in \mathbb{D}$ approaches the boundary point $e^{i\theta}$ within a sector defined by $|\arg\{e^{i\theta} - z\}| \le k$ for any constant $k < \pi/2$. The boundary function $f(e^{i\theta})$ is in L^p and $\log|f(e^{i\theta})|$ belongs to L^1. As a consequence, the boundary function cannot vanish on any set of positive measure unless f is the zero function. For $0 < p < \infty$, the *norm* of an H^p function f is defined by

$$\|f\|_{H^p} = \left\{ \frac{1}{2\pi} \int_0^{2\pi} |f(e^{i\theta})|^p \, d\theta \right\}^{1/p} = \lim_{r \to 1} M_p(r, f) .$$

It is a genuine norm only for $1 \le p < \infty$. The norm of a function $f \in H^\infty$ is

$$\|f\|_{H^\infty} = \sup_{z \in \mathbb{D}} |f(z)| = \lim_{r \to 1} M_\infty(r, f) ,$$

where $M_\infty(r,f) = \max_\theta |f(re^{i\theta})|$.

The zeros z_k of a nonnull function $f \in H^p$, repeated according to multiplicity, satisfy the *Blaschke condition* $\sum(1-|z_k|) < \infty$, which ensures the convergence of the *Blaschke product*

$$B(z) = z^m \prod_k \frac{|z_k|}{z_k} \frac{z_k - z}{1 - \overline{z_k} z}\,.$$

This analytic function B has precisely the same zeros as f, taking account of multiplicity. The nonvanishing function f/B again belongs to H^p and has the same norm as f. Blaschke products have the properties $|B(z)| < 1$ and $|B(e^{i\theta})| = 1$ for almost every θ. In particular, a Blaschke product has H^p norm $\|B\|_{H^p} = 1$ for $0 < p \le \infty$.

In the terminology introduced by Beurling [2], an *inner function* is any function $\varphi \in H^\infty$ whose boundary values satisfy $|\varphi(e^{i\theta})| = 1$ almost everywhere. Blaschke products are inner functions. Another type of inner function is

$$S(z) = \exp\left\{-\int_0^{2\pi} \frac{e^{it}+z}{e^{it}-z}\, d\mu(t)\right\},$$

where μ is a finite singular measure. In other words, $\mu(t)$ is a bounded nondecreasing function with $\mu'(t) = 0$ a.e. These functions S are called *singular inner functions.* An important example is the *atomic singular inner function*

$$S(z) = \exp\left\{-\frac{1+z}{1-z}\right\},$$

with its measure μ concentrated at one point. The most general inner function is a product $\varphi(z) = e^{i\gamma}B(z)S(z)$, where $e^{i\gamma}$ is a unimodular constant, B is a Blaschke product, and S is a singular inner function. An *outer function for* H^p is any function of the form

$$F(z) = \exp\left\{\frac{1}{2\pi}\int_0^{2\pi} \frac{e^{it}+z}{e^{it}-z}\, h(t)\, dt\right\}, \qquad z \in \mathbb{D},$$

where $h \in L^1$ and $e^h \in L^p$. It is easily shown that $F \in H^p$ and that its boundary function satisfies $\log|F(e^{i\theta})| = h(\theta)$ a.e.

Each function $f \in H^p$ has a unique factorization $f = \varphi F$, where φ is an inner function and F is the outer function

$$F(z) = \exp\left\{\frac{1}{2\pi}\int_0^{2\pi} \frac{e^{it}+z}{e^{it}-z} \log|f(e^{it})|\, dt\right\}.$$

The representation $f(z) = e^{i\gamma}B(z)S(z)F(z)$ is unique and is known as the *canonical factorization* of an H^p function. Conversely, any such function f belongs to H^p, where B is an arbitrary Blaschke product, S is an arbitrary singular inner function, and F is an arbitrary outer function for H^p. Moreover, $|f(e^{i\theta})| = |F(e^{i\theta})|$ a.e.

It follows trivially from the Schwarz inequality that the product of any pair of H^2 functions belongs to H^1. Conversely, the factorization $f = Bg$ of an H^1 function as a Blaschke product B times a nonvanishing H^1 function g shows that every H^1 function has a representation $f = (B\sqrt{g})\sqrt{g}$ as a product of two H^2 functions, one of them nonvanishing. The same factorization shows that for $0 < p \le \infty$, every function $f \in H^p$ can be expressed as a sum of two nonvanishing H^p functions with comparable norms:

$$f(z) = B(z)g(z) = (B(z) - 1)g(z) + g(z)\,.$$

An analytic function f is said to be in the *Nevanlinna class* N if

$$\sup_{r<1} \int_0^{2\pi} \log^+ |f(re^{i\theta})|\, d\theta < \infty\,,$$

where $\log^+ x = \max\{\log x, 0\}$. Thus $H^p \subset N$ for every $p > 0$. It is known that functions in the Nevanlinna class are quotients of H^∞ functions, so their zero-sets are again Blaschke sequences. Also, each function $f \in N$ has a nontangential limit $f(e^{i\theta})$ at almost every boundary point $e^{i\theta}$.

Functions in Hardy spaces have nice growth properties. For instance, if $f \in H^p$ and $0 < p < q \le \infty$, then

$$M_q(r, f) = o\big((1-r)^{\frac{1}{q}-\frac{1}{p}}\big), \qquad r \to 1\,. \tag{1}$$

To see that the exponent is best possible, one need only consider functions of the form $f(z) = (1-z)^{-\lambda}$, which belong to H^p if and only if $\lambda < \frac{1}{p}$. In fact, the estimate is best possible in a stronger sense. (See Duren and Taylor [1] for details.) However, the result can be slightly sharpened to say that if $f \in H^p$, then

$$\int_0^1 (1-r)^{\lambda\alpha-1}\{M_q(r, f)\}^\lambda\, dr < \infty\,, \qquad \alpha = \tfrac{1}{p} - \tfrac{1}{q}\,, \tag{2}$$

for all $\lambda \ge p$. In the converse direction, no growth condition on integral means weaker than boundedness can place a function in any H^p space. More precisely, for any prescribed continuous function $\psi(r)$ that increases to infinity as $r \to 1$, it is possible to construct a function f analytic in $\mathbb{D}$ such that

$$M_\infty(r, f) = O\big(\psi(r)\big)\,, \qquad r \to 1,$$

yet $f(z)$ has a radial limit almost nowhere. (A proof will be given in the next section. See also Duren [5], Theorem 5.10.)

The derivative of a function $f \in H^p$ need not belong to any H^q space. In fact, there exists a function f analytic in $\mathbb{D}$ and continuous in $\overline{\mathbb{D}}$ whose derivative fails to have a radial limit on any set of positive measure (see Duren [4] for a simple "construction"). On the other hand, the behavior of

H^p functions is improved by integration. If $f' \in H^1$, then f is continuous in $\overline{\mathbb{D}}$ and absolutely continuous on the boundary. In particular, $f' \in H^1$ implies $f \in H^\infty$. If $f' \in H^p$ for some $p < 1$, then $f \in H^q$ for $q = \frac{p}{1-p}$, and the index q is best possible.

Information is available about the Taylor coefficients a_n of functions $f(z) = \sum a_n z^n$ in a Hardy space H^p. The case $p = 2$ is very simple: $f \in H^2$ if and only if $\{a_n\} \in \ell^2$; that is, if and only if $\sum |a_n|^2 < \infty$. In fact, $\|f\|_{H^2} = \|\{a_n\}\|_2$, where

$$\|\{a_n\}\|_p = \left\{\sum_{n=0}^{\infty} |a_n|^p\right\}^{1/p} < \infty, \qquad \{a_n\} \in \ell^p.$$

When $p \neq 2$, the situation is more complicated. Nevertheless, connections between H^p functions and their coefficients can be expressed in several ways.

THEOREM A (HAUSDORFF–YOUNG THEOREM). *For $1 \le p \le \infty$, let q be the conjugate index, with $\frac{1}{p} + \frac{1}{q} = 1$.*

(*i*) *If $1 \le p \le 2$, then $f \in H^p$ implies $\{a_n\} \in \ell^q$, and $\|\{a_n\}\|_q \le \|f\|_{H^p}$.*

(*ii*) *If $2 \le p \le \infty$, then $\{a_n\} \in \ell^q$ implies $f \in H^p$, and $\|f\|_{H^p} \le \|\{a_n\}\|_q$.*

THEOREM B (HARDY–LITTLEWOOD THEOREM).

(*i*) *If $0 < p \le 2$, then $f \in H^p$ implies $\sum n^{p-2}|a_n|^p < \infty$, and*

$$\left\{\sum_{n=0}^{\infty}(n+1)^{p-2}|a_n|^p\right\}^{1/p} \le C_p \|f\|_{H^p},$$

where C_p denotes a constant depending only on p.

(*ii*) *If $2 \le p \le \infty$, then $\sum n^{p-2}|a_n|^p < \infty$ implies $f \in H^p$, and*

$$\|f\|_{H^p} \le C_p \left\{\sum_{n=0}^{\infty}(n+1)^{p-2}|a_n|^p\right\}^{1/p}.$$

THEOREM C. *If $0 < p \le 1$, then $f \in H^p$ implies $a_n = o\left(n^{\frac{1}{p}-1}\right)$ and*

$$|a_n| \le C_p \|f\|_{H^p} n^{\frac{1}{p}-1}.$$

All of these results are best possible in a strong sense. However, none of the four converses of the statements in Theorems A and B are valid. For instance, if $f \in H^p$ for some $p > 2$, we can conclude nothing about the moduli of its Taylor coefficients a_n beyond the obvious fact that the sum $\sum |a_n|^2 < \infty$. On the other hand, no weaker condition on the coefficients can imply that f belongs to any H^p space. These assertions are supported by the following remarkable fact.

THEOREM D (LITTLEWOOD'S THEOREM).

(*i*) *If* $\sum |a_n|^2 < \infty$, *then for almost every choice of signs* $\varepsilon_n = \pm 1$, *the function* $f(z) = \sum \varepsilon_n a_n z^n$ *belongs to* H^p *for every* $p < \infty$.

(*ii*) *If* $\sum |a_n|^2 = \infty$, *then for almost every choice of signs the function* $f(z)$ *has a radial limit almost nowhere; thus* f *does not belong to* H^p *for any* $p > 0$.

In the statement of the theorem, the phrase "almost every choice of signs" refers to a measure on the set of sign sequences $\{\varepsilon_n\}$ that is induced by Lebesgue measure on the interval $[0,1]$ when sequences of signs $\varepsilon_n = \pm 1$ are put in correspondence with dyadic expansions of numbers in that interval.

§3.2. Growth of functions in Bergman spaces.

For $0 < p < \infty$, the Bergman space A^p consists of all functions f analytic in the unit disk for which the normalized area integral $\int_{\mathbb{D}} |f(z)|^p \, d\sigma$ is finite. Note that $A^q \subset A^p$ if $p < q$. It will be seen by simple examples that $A^p \neq A^q$ when $p \neq q$.

In terms of integral means, the condition $f \in A^p$ is equivalent to saying that $\int_0^1 \{M_p(r,f)\}^p \, dr < \infty$, since

$$\pi \|f\|_p^p = \int_0^{2\pi} \int_0^1 |f(re^{i\theta})|^p \, r \, dr \, d\theta = 2\pi \int_0^1 \{M_p(r,f)\}^p \, r \, dr \, .$$

It is then obvious that $H^p \subset A^p$. In fact, it is even true that $H^p \subset A^{2p}$ for $0 < p < \infty$. This is an immediate corollary of the integral means theorem (2) of Hardy and Littlewood, with $\lambda = q = 2p$.

However, there is an *elementary* proof of the inclusion $H^p \subset A^{2p}$, due to Vukotić [4], that gives in addition the sharp inequality $\|f\|_{2p} \leq \|f\|_{H^p}$, where

$$\|f\|_p = \left\{ \int_{\mathbb{D}} |f(z)|^p \, d\sigma \right\}^{1/p}$$

is the Bergman norm and $d\sigma$ denotes normalized area measure. Recall (*cf.* Section 1.2) that the A^2 norm has a simple expression in terms of Taylor coefficients. If $f(z) = \sum a_n z^n$, then

$$\|f\|_2^2 = \sum_{n=0}^{\infty} \frac{|a_n|^2}{n+1} \, .$$

Thus for the A^4 norm, one has by the Cauchy–Schwarz inequality

$$\|f\|_4^4 = \|f^2\|_2^2 = \sum_{n=0}^{\infty} \frac{1}{n+1} \left| \sum_{k=0}^{n} a_{n-k} a_k \right|^2$$

$$\leq \sum_{n=0}^{\infty} \sum_{k=0}^{n} |a_{n-k} a_k|^2 = \left\{ \sum_{n=0}^{\infty} |a_n|^2 \right\}^2 = \|f\|_{H^2}^4 \, .$$

This shows that $H^2 \subset A^4$, and $\|f\|_4 \le \|f\|_{H^2}$. If f is a *nonvanishing* function in H^p, the inequality can be applied to the function $f^{p/2}$ to infer that $\|f\|_{2p} \le \|f\|_{H^p}$. Finally, if f is an arbitrary function in H^p, it can be factored as $f = Bg$, where B is a Blaschke product and g is a nonvanishing function in H^p with the same norm. Thus

$$\|f\|_{2p} = \|Bg\|_{2p} \le \|g\|_{2p} \le \|g\|_{H^p} = \|f\|_{H^p},$$

which completes the proof. Dilations of functions $f(z) = (1-z)^{-2/p}$ show that the inequality is sharp, and they are essentially the only functions for which $\|f\|_{2p} = \|f\|_{H^p}$.

If $M_p(r,f) = O\big((1-r)^{-a}\big)$ for some $a < \frac{1}{p}$, then it is clear by integration over the interval $[0,1)$ that $f \in A^p$. The converse is "almost true", since $M_p(r,f) = o\big((1-r)^{-\frac{1}{p}}\big)$ if $f \in A^p$. For a proof, observe that because the integral means $M_p(r,f)$ are nondecreasing functions of r,

$$(1-r)\{M_p(r,f)\}^p \le \int_r^1 \{M_p(t,f)\}^p\, dt < \varepsilon$$

for r sufficiently near 1. These two results point out a major difference between Hardy spaces and Bergman spaces. As was mentioned earlier, the derivative of a function $f \in H^p$ need not belong to any H^q space. However, if $f \in A^p$, then $M_p(r,f) = O\big((1-r)^{-\frac{1}{p}}\big)$, so it follows from the Cauchy integral formula (see Duren [5] for details) that $M_p(r,f') = O\big((1-r)^{-\frac{1}{p}-1}\big)$ and $f' \in A^q$ for all $q < \frac{p}{p+1}$.

Corresponding to the fact that the function $(1-z)^{-\lambda}$ belongs to H^p for every $\lambda < \frac{1}{p}$, it can be shown that $(1-z)^{-\lambda}$ is in A^p if and only if $\lambda < \frac{2}{p}$. A verification can be based on the following special estimate (see Duren [5], p. 65), implicit in the proof of Lemma 2 in Section 2.3.

LEMMA 1. *For* $p > 1$,

$$\int_0^{2\pi} \frac{d\theta}{|1 - re^{i\theta}|^p} = O\left(\frac{1}{(1-r)^{p-1}}\right), \qquad r \to 1.$$

Alternatively, we can argue as follows. Let $f(z) = (1-z)^{-\lambda}$ for some $\lambda < \frac{2}{p}$. Let Ω denote the disk $|z-1| < 2$, so that $\mathbb{D} \subset \Omega$. Then

$$\int_{\mathbb{D}} |f(z)|^p\, dA \le \int_\Omega |1-z|^{-\lambda p}\, dA = 2\pi \int_0^2 r^{1-\lambda p}\, dr < \infty,$$

with the change of variables $w = z - 1$. Thus $f \in A^p$. Conversely, the function $f(z) = (1-z)^{-\lambda}$ is in A^p only if $\lambda < \frac{2}{p}$. To see this, suppose $\lambda \ge \frac{2}{p}$ and let

$$S = \big\{z \in \mathbb{D} : |\arg\{1-z\}| \le c\big\}$$

for some $c < \pi/2$. Then

$$\int_{\mathbb{D}} |f(z)|^p \, dA \geq \int_S |1-z|^{-\lambda p} \, dA \geq 2c \int_0^{2\cos c} r^{1-\lambda p} \, dr = \infty \,,$$

so that $f \neq A^p$. These examples show that $A^p \neq A^q$ if $p \neq q$.

Since $M_p(r,f) = o\big((1-r)^{-\frac{1}{p}}\big)$ for functions $f \in A^p$, a standard result on integral means (see Duren [5], Theorem 5.9) shows that $M_\infty(r,f) = o\big((1-r)^{-\frac{2}{p}}\big)$. For $1 \leq p < \infty$, the weaker estimate $M_\infty(r,f) = O\big((1-r)^{-\frac{2}{p}}\big)$ is easily obtained by applying Hölder's inequality to the reproducing formula involving the Bergman kernel function. For $0 < p < \infty$, a direct proof of the "o" estimate proceeds as follows. Since $|f|^p$ is subharmonic, the inequality

$$|f(z)|^p \leq \frac{1}{\delta^2} \int_\Delta |f(\zeta)|^p \, d\sigma \,, \qquad z \in \mathbb{D} \,,$$

holds for any disk Δ of radius $\delta < 1 - |z|$ centered at z. In particular,

$$(1-|z|)^2 |f(z)|^p \leq \int_{\Delta_z} |f(\zeta)|^p \, d\sigma \,, \qquad z \in \mathbb{D} \,,$$

where Δ_z is the disk $|\zeta - z| < 1 - |z|$. For $\frac{1}{2} < r < 1$, this implies that

$$(1-r)^2 \{M_\infty(r,f)\}^p \leq \int_{\mathcal{A}_r} |f(\zeta)|^p \, d\sigma \,,$$

where $\mathcal{A}_r$ is the annulus defined by $2r - 1 < |\zeta| < 1$. But for any function f in the space A^p, this last integral tends to zero as $r \to 1$, which proves that $M_\infty(r,f) = o\big((1-r)^{-\frac{2}{p}}\big)$.

The sharp inequality

$$|f(z)| \leq (1-|z|^2)^{-\frac{2}{p}} \, \|f\|_p \,, \qquad z \in \mathbb{D} \,,$$

due to Vukotić [1], was already derived in Section 2.4. The reader is referred to the discussion there, following the statement of Theorem 6. In the converse direction, $f \in A^p$ if $M_\infty(r,f) = O\big((1-r)^{-a}\big)$ for some $a < \frac{1}{p}$.

More generally, if $f \in A^p$ for some p with $0 < p < \infty$, then $M_p(r,f) = o\big((1-r)^{-\frac{1}{p}}\big)$ and it follows from a standard result on integral means (see Duren [5], Theorem 5.9) that

$$M_q(r,f) = o\big((1-r)^{\frac{1}{q}-\frac{2}{p}}\big)$$

for each $q > p$. Furthermore, these estimates are best possible. (Compare with the estimate (1) in Section 3.1 for functions $f \in H^p$.) If $q \leq p$, nothing better than $M_q(r,f) = o\big((1-r)^{-\frac{1}{p}}\big)$ is true for arbitrary functions $f \in A^p$.

For ease of reference, we now collect some of these observations into a formal theorem.

THEOREM 1. *Let $0 < p < \infty$ and let f be a function analytic in $\mathbb{D}$.*

(*i*) *If $M_p(r,f) = O\big((1-r)^{-a}\big)$ for some $a < \frac{1}{p}$, then $f \in A^p$.*

(*ii*) *If $f \in A^p$, then $M_q(r,f) = o\big((1-r)^{\frac{1}{q}-\frac{2}{p}}\big)$ for $p \le q \le \infty$. In particular, $M_p(r,f) = o\big((1-r)^{-\frac{1}{p}}\big)$ and $M_\infty(r,f) = o\big((1-r)^{-\frac{2}{p}}\big)$.*

(*iii*) *If $f \in A^p$, then $f' \in A^q$ for all $q < \frac{p}{p+1}$.*

On the other hand, functions in Bergman spaces need not have radial limits. An estimate of the form $M_p(r,f) = O\big((1-r)^{-a}\big)$ places f in A^p whenever $a < 1/p$, although (as mentioned above) no such estimate for any $a > 0$ can ensure that f has radial limits on any set of positive measure. A typical example is the power series

$$f(z) = \sum_{k=0}^{\infty} z^{2^k} = z + z^2 + z^4 + z^8 + \dots .$$

It will be seen that $M_\infty(r,f) = O\big(\log\frac{1}{1-r}\big)$, so that $f \in A^p$ for every $p < \infty$. However, as a lacunary series whose coefficients are not square-summable, f has a radial limit almost nowhere.

This last example calls for further discussion. A power series

$$f(z) = \sum_{k=0}^{\infty} a_k z^{n_k}$$

is said to be *lacunary* if $n_{k+1}/n_k \ge q$ for some $q > 1$ and for all k. Lacunary power series are nicely behaved if $\sum |a_k|^2 < \infty$ and very badly behaved otherwise. Specifically, if $\sum |a_k|^2 < \infty$, then f belongs not only to H^2 but to H^p for every $p < \infty$; while if $\sum |a_k|^2 = \infty$, then f has a radial limit almost nowhere and so belongs to no H^p space for any $p > 0$. (See for instance Zygmund [1], Ch. V, §6.) Simple estimates based on the Cauchy formula show that the coefficients of any function in the Bloch space $\mathcal{B}$ are bounded, and they tend to zero for functions in $\mathcal{B}_0$. For lacunary series, the converse is also true: $f \in \mathcal{B}$ if $a_k = O(1)$, while $f \in \mathcal{B}_0$ if $a_k = o(1)$. (See Pommerenke [2], p. 188.) Thus in particular, the function $f(z) = \sum z^{2^k}$ is in $\mathcal{B}$ and so $M_\infty(r,f) = O\big(\log\frac{1}{1-r}\big)$, as was shown in Section 2.6. This estimate can also be established directly by adapting the proof of the following lemma (*cf.* Duren [5], p. 87), which gives considerably more information.

LEMMA 2. *Let $\psi(r)$ be continuous and increasing on $0 \le r < 1$, with $\psi(0) = 1$ and $\psi(r) \to \infty$ as $r \to 1$. Then*

$$\sum_{k=1}^{\infty} r^{n_k} \le \psi(r), \qquad 0 \le r < 1,$$

whenever the positive integers n_k increase sufficiently fast.

PROOF. Let $r_0 = 0$ and let $r_1, r_2, \ldots$ be defined by $\psi(r_k) = k + 1$. Then $0 < r_1 < r_2 < \ldots < 1$. Choose n_1 so that ${r_1}^{n_1} \le \frac{1}{2}$, and inductively select $n_k > n_{k-1}$ such that ${r_k}^{n_k} \le 2^{-k}$. For $0 < r < 1$, let s be the index for which $r_{s-1} \le r < r_s$. Then

$$\begin{aligned}\sum_{k=1}^{\infty} r^{n_k} &\le \sum_{k=1}^{\infty} {r_s}^{n_k} = \sum_{k=1}^{s-1} {r_s}^{n_k} + \sum_{k=s}^{\infty} {r_s}^{n_k} \\ &\le (s-1) + \sum_{k=s}^{\infty} {r_k}^{n_k} \le (s-1) + 2^{1-s} \\ &\le s = \psi(r_{s-1}) \le \psi(r)\,. \qquad \square\end{aligned}$$

The proof shows that $\{n_k\}$ may be chosen to be a lacunary sequence. Hence there are functions $f(z) = \sum z^{n_k}$ analytic in $\mathbb{D}$ that grow arbitrarily slowly, yet belong to no H^p space. On the other hand, every lacunary power series with bounded coefficients belongs to $\mathcal{B}$ and hence to A^p for every $p < \infty$, but it has a radial limit almost nowhere if the coefficients are not square-summable.

§3.3. Coefficients of functions in Bergman spaces.

As in the case of Hardy spaces, it is possible to give fairly precise conditions on the Taylor coefficients of a function $f(z) = \sum a_n z^n$ in order that it belong to a Bergman space A^p. In analogy with the fact that $f \in H^2$ if and only if $\{a_n\} \in \ell^2$, we know that $f \in A^2$ if and only if $\sum n^{-1}|a_n|^2 < \infty$, and

$$\|f\|_2 = \left\{\sum_{n=0}^{\infty} \frac{|a_n|^2}{n+1}\right\}^{1/2}.$$

This result can be partially generalized to $1 < p < \infty$. The Hausdorff–Young theorem (Theorem A above) has the following analogue for Bergman spaces, apparently first noted by Horowitz [1].

THEOREM 2. *Let $1 < p < \infty$ and let $q = \frac{p}{p-1}$ be its conjugate exponent.*

(i) If $1 < p \le 2$, then $f \in A^p \implies \sum n^{1-q}|a_n|^q < \infty$; and

$$\left\{\sum_{n=0}^{\infty} \frac{|a_n|^q}{(n+1)^{q-1}}\right\}^{1/q} \le \|f\|_p\,.$$

(ii) If $2 \le p < \infty$, then $\sum n^{1-q}|a_n|^q < \infty \implies f \in A^p$; and

$$\|f\|_p \le \left\{\sum_{n=0}^{\infty} \frac{|a_n|^q}{(n+1)^{q-1}}\right\}^{1/q}.$$

PROOF. (*ii*). The case $2 \le p < \infty$ is a little easier, and will be considered first. Let μ be the discrete measure on the set $\mathbb{N}$ of nonnegative integers which assigns the mass $\mu(n) = n+1$ to the integer $n = 0, 1, 2, \dots$. Consider the linear operator T that maps the sequence $\mathbf{b} = \{b_n\} = \left\{\frac{1}{n+1}a_n\right\}$ to the formal power series $f(z) = \sum_{n=0}^{\infty} a_n z^n$. For $2 \le p \le \infty$, we want to show that T is bounded as an operator from $L^q(\mathbb{N}, d\mu)$ to $L^p(\mathbb{D}, d\sigma)$, with norm $\|T\| \le 1$. But for $p = 2$ this follows from the relation

$$\|f\|_2^2 = \sum_{n=0}^{\infty} \frac{|a_n|^2}{n+1} = \sum_{n=0}^{\infty} (n+1)|b_n|^2 .$$

For $p = \infty$ it is the trivial fact that $|f(z)| \le \sum_{n=0}^{\infty} |a_n|$ for $z \in \mathbb{D}$. Thus we may invoke the Riesz–Thorin interpolation theorem (see Zygmund [1], Ch. *XII*, §1) to draw the conclusion that $\|T(\mathbf{b})\|_p \le \|\mathbf{b}\|_q$ for $2 \le p \le \infty$. More explicitly, we find that

$$\left\{\int_{\mathbb{D}} |f(z)|^p \, d\sigma\right\}^{1/p} \le \left\{\sum_{n=0}^{\infty} (n+1)|b_n|^q\right\}^{1/q} = \left\{\sum_{n=0}^{\infty} (n+1)^{1-q}|a_n|^q\right\}^{1/q}$$

for $2 \le p < \infty$ and conjugate index q, which is the desired result (*ii*).

(*i*). Now consider the case $1 < p \le 2$. For a function $f \in L^p(\mathbb{D}, d\sigma)$, define the linear operator $T(f) = \{b_n\}$, where

$$b_n = \int_{\mathbb{D}} f(z)\, \overline{z^n} \, d\sigma .$$

For a function $f(z) = \sum a_n z^n$ in A^p, a standard calculation shows that $b_n = \frac{1}{n+1}a_n$. With the measure μ defined as before, and for $1 \le p \le 2$, we want to show that T is a bounded operator from $L^p(\mathbb{D}, d\sigma)$ to $L^q(\mathbb{N}, d\mu)$, for $1 \le p \le 2$, with norm $\|T\| \le 1$. For $p = 1$ this is the trivial inequality $|b_n| \le \|f\|_1$ for all $n \in \mathbb{N}$. For $p = 2$ it follows from the relation

$$\begin{aligned}
\sum_{n=0}^{\infty} (n+1)|b_n|^2 &= \sum_{n=0}^{\infty} (n+1)\left|\int_{\mathbb{D}} f(z)\overline{z^n} \, d\sigma\right|^2 \\
&= \int_{\mathbb{D}}\int_{\mathbb{D}} f(z)\overline{f(\zeta)} \sum_{n=0}^{\infty} (n+1)(\overline{z}\zeta)^n \, d\sigma(z)\, d\sigma(\zeta) \\
&= \int_{\mathbb{D}}\int_{\mathbb{D}} \frac{f(z)\overline{f(\zeta)}}{(1-\overline{z}\zeta)^2} \, d\sigma(z)\, d\sigma(\zeta) = \langle Pf, f\rangle ,
\end{aligned}$$

where P denotes the Bergman projection (see Section 2.3). Since

$$|\langle Pf, f\rangle| \le \|Pf\|_2 \, \|f\|_2 \le \|f\|_2^2 ,$$

this gives the desired inequality for $p=2$. The inequality for $1<p<2$ now follows as before from the Riesz–Thorin interpolation theorem. Specifically, one can conclude by interpolation that

$$\left\{\sum_{n=0}^{\infty}(n+1)|b_n|^q\right\}^{1/q} \le \left\{\int_{\mathbb{D}}|f(z)|^p\,d\sigma\right\}^{1/p}, \qquad f\in L^p(\mathbb{D}),$$

where $1<p\le 2$ and q is the conjugate exponent. Specializing to $f\in A^p$ and recalling that $b_n=\frac{1}{n+1}a_n$, we arrive at the desired result (i). □

The Hardy–Littlewood theorem (Theorem B) actually takes a more complete form for Bergman spaces, because weak converses of the two implications are now valid. The following theorem is due to Nakamura, Ohya, and Watanabe [1].

THEOREM 3.

(i) If $0<p\le 2$, then

$$\sum_{n=1}^{\infty}n^{-1}|a_n|^p<\infty \Longrightarrow f\in A^p \Longrightarrow \sum_{n=1}^{\infty}n^{p-3}|a_n|^p<\infty .$$

(ii) If $2\le p<\infty$, then

$$\sum_{n=1}^{\infty}n^{p-3}|a_n|^p<\infty \Longrightarrow f\in A^p \Longrightarrow \sum_{n=1}^{\infty}n^{-1}|a_n|^p<\infty .$$

In all four assertions, the exponents of n are best possible.

PROOF. (i). Suppose first that $0<p\le 2$, and let $f\in A^p$. Applying the Hardy–Littlewood inequality (Theorem B) to an arbitrary dilation $f_r(z)=f(rz)$ for $0<r<1$, we see that

$$\sum_{n=0}^{\infty}(n+1)^{p-2}|a_n r^n|^p \le C_p\{M_p(r,f)\}^p .$$

Multiplying this inequality by r and integrating over the interval $(0,1)$, we conclude that

$$\sum_{n=0}^{\infty}(n+1)^{p-2}|a_n|^p\int_0^1 r^{np+1}\,dr \le C_p\|f\|_p^p ,$$

or that

$$\sum_{n=0}^{\infty}(n+1)^{p-3}|a_n|^p \le C_p\|f\|_p^p .$$

Assume now that $\sum n^{-1}|a_n|^p < \infty$. Then since $p \le 2$,

$$\int_0^1 \{M_p(r,f)\}^p\,dr \le \int_0^1 \{M_2(r,f)\}^p\,dr = \int_0^1 \left\{\sum_{n=0}^\infty |a_n|^2 r^{2n}\right\}^{p/2} dr$$
$$\le \int_0^1 \left\{\sum_{n=0}^\infty |a_n|^p r^{pn}\right\} dr = \sum_{n=0}^\infty (pn+1)^{-1}|a_n|^p < \infty\,.$$

In fact, the above inequality shows that

$$\|f\|_p^p \le C_p \sum_{n=0}^\infty (n+1)^{-1}|a_n|^p\,.$$

(ii). Next suppose that $2 \le p < \infty$. Then the other half of the Hardy–Littlewood theorem can be applied to the dilation f_r to obtain

$$\{M_p(r,f)\}^p \le C_p \sum_{n=0}^\infty (n+1)^{p-2}|a_n r^n|^p\,, \qquad 0 < r < 1\,,$$

so that multiplication by r and integration gives

$$\|f\|_p^p \le C_p \sum_{n=0}^\infty (n+1)^{p-3}|a_n|^p\,.$$

Finally, if $f \in A^p$, then

$$\infty > \int_0^1 \{M_p(r,f)\}^p\,dr \ge \int_0^1 \{M_2(r,f)\}^p\,dr = \int_0^1 \left\{\sum_{n=0}^\infty |a_n|^2 r^{2n}\right\}^{p/2} dr$$
$$\ge \int_0^1 \left\{\sum_{n=0}^\infty |a_n|^p r^{pn}\right\} dr = \sum_{n=0}^\infty (pn+1)^{-1}|a_n|^p\,.$$

Thus $f \in A^p$ implies $\sum n^{-1}|a_n|^p < \infty$, and the argument shows that

$$\sum_{n=0}^\infty (n+1)^{-1}|a_n|^p \le C_p\|f\|_p^p\,.$$

This completes the verification of all four implications stated in the theorem. The proof that the assertions are best possible will be omitted. □

One interesting consequence of Part (i) is that $f' \in A^1$ whenever f has an absolutely convergent power series; that is, whenever $\sum |a_n| < \infty$.

Theorem C also has an analogue for Bergman spaces.

THEOREM 4.

(*i*) *If* $0 < p \le 1$, *then* $f \in A^p \implies a_n = o\big(n^{\frac{2}{p}-1}\big)$.

(*ii*) *If* $1 \le p < \infty$, *then* $f \in A^p \implies a_n = o\big(n^{\frac{1}{p}}\big)$.

In both estimates the exponents are best possible.

PROOF. (*i*). For $p \le 1$, the estimate $M_p(r,f) = o\big((1-r)^{-\frac{1}{p}}\big)$ implies that $M_1(r,f) = o\big((1-r)^{1-\frac{2}{p}}\big)$. On the other hand, the Cauchy formula gives $|a_n| \le r^{-n} M_1(r,f)$ for $0 < r < 1$. Choosing $r = 1 - \frac{1}{n}$, we conclude that $a_n = o\big(n^{\frac{2}{p}-1}\big)$.

(*ii*). If $p \ge 1$, Cauchy's inequality implies that $|a_n| \le r^{-n} M_p(r,f)$, and the result follows similarly from the estimate $M_1(r,f) = o\big((1-r)^{-\frac{1}{p}}\big)$. Functions of the form $(1-z)^{-a}$ show the exponents are best possible. □

In the converse direction, certain conditions on the average growth of coefficients are sufficient to place a function in a Bergman space. In terms of the sums

$$S_n^{(q)} = \sum_{k=0}^{n} |a_k|^q ,$$

Horowitz [2] obtained the following result.

THEOREM 5.

(*i*) *If* $0 < p \le 2$ *and* $\lambda < \frac{2}{p}$, *then*

$$S_n^{(2)} = O(n^\lambda) \implies f \in A^p .$$

(*ii*) *If* $2 \le p < \infty$, $q = \frac{p}{p-1}$ *is the conjugate exponent, and* $\lambda < q - 1$, *then*

$$S_n^{(q)} = O(n^\lambda) \implies f \in A^p .$$

PROOF. (*i*). Since $p \le 2$, we have the inequality

$$M_p^2(r,f) \le M_2^2(r,f) = \sum_{k=0}^{\infty} |a_k|^2 r^{2k} = \sum_{n=0}^{\infty} S_n^{(2)} (r^{2n} - r^{2n+2}) ,$$

after a summation by parts. But $S_n^{(2)} \le C(n+1)^\lambda$ by hypothesis, so we deduce that

$$M_p^2(r,f) \le C\,(1-r^2) \sum_{n=0}^{\infty} (n+1)^\lambda r^{2n} = O\big((1-r)^{-\lambda}\big) ,$$

or $M_p(r,f) = O\big((1-r)^{-\frac{\lambda}{2}}\big)$, which implies that $f \in A^p$, since $\frac{\lambda}{2} < \frac{1}{p}$.

(*ii*). By Theorem 1, it will suffice to show that

$$\sum_{n=0}^{\infty}(n+1)^{1-q}|a_n|^q < \infty .$$

But a summation by parts gives

$$\sum_{n=0}^{\infty}(n+1)^{1-q}|a_n|^q = \sum_{n=0}^{\infty} S_n^{(q)}\{(n+1)^{1-q} - (n+2)^{1-q}\}$$

$$\leq \sum_{n=0}^{\infty} S_n^{(q)}(n+1)^{-q} \leq C\sum_{n=0}^{\infty}(n+1)^{\lambda-q} < \infty ,$$

since $S_n^{(q)} = O(n^\lambda)$ and $\lambda - q < -1$. This completes the proof. □

§3.4. Coefficient multipliers.

Some information about the coefficients of A^p functions can be gleaned from a study of the *coefficient multipliers* of A^p into A^q. These are the complex sequences $\{\lambda_n\}$ with the property that $\sum \lambda_n a_n z^n$ is in A^q whenever $\sum a_n z^n$ is in A^p. The space of coefficient multipliers from A^p to A^q will be denoted by (A^p, A^q) Recent work of Wojtaszczyk [1], Vukotić [2], MacGregor and Zhu [1], and others has led to an extensive theory of coefficient multipliers for the Bergman spaces, parallel to the well-established theory for Hardy spaces (see Duren [5], where further references are given). Once again it turns out that the results for Bergman spaces are in some respects more satisfactory. Here we shall content ourselves with discussing some basic principles and a few sample theorems.

First of all, the closed graph theorem can be applied to obtain a *necessary* condition for coefficient multipliers. Given a sequence $\lambda = \{\lambda_n\}$, let T_λ denote the linear operator that assigns the formal power series $\sum \lambda_n a_n z^n$ to a function $\sum a_n z^n$. For fixed indices p and q in the interval $(0, \infty)$, let $\mathcal{D}(T_\lambda)$ be the vector subspace of functions $f \in A^p$ for which $T_\lambda(f) \in A^q$. Note that $\mathcal{D}(T_\lambda)$ is dense in A^p, since it contains all polynomials; and that $\mathcal{D}(T_\lambda) = A^p$ if and only if $\{\lambda_n\}$ is a multiplier from A^p to A^q. It may also be observed that T_λ is a closed operator from A^p to A^q. Thus if $\lambda \in (A^p, A^q)$, it follows from the closed graph theorem that T_λ is *bounded*: $\|T_\lambda(f)\|_q \leq C\,\|f\|_p$. Suitable choices of the function f now give explicit conditions that a multiplier sequence $\{\lambda_n\}$ must satisfy. The following theorem is due to Vukotić [2].

THEOREM 6. *Let* $\{\lambda_n\} \in (A^p, A^q)$ *for* $0 < p < \infty$.

(*i*) *If* $0 < q \leq 2$, *then*

$$\sum_{n=1}^{N} n^{\frac{2q}{p}-2}|\lambda_n|^q = O(N) \qquad \text{as } N \to \infty .$$

(ii) If $2 \le q < \infty$, *then*

$$\sum_{n=1}^{N} n^{\frac{2q}{p}-q} |\lambda_n|^q = O(N) \qquad \textit{as } N \to \infty .$$

The proof will use an elementary lemma, easily proved via summation by parts.

LEMMA 3. *Let* $a_n \ge 0$, $\alpha > 0$, *and* $\beta > 0$. *Then*

$$\sum_{n=1}^{N} a_n = O(N^{\alpha}) \iff \sum_{n=1}^{N} n^{\beta} a_n = O(N^{\alpha+\beta}) .$$

PROOF OF THEOREM. Choose s with $\frac{1}{p}-1 < s < \frac{1}{p}-1+\frac{1}{q}$, and consider the function

$$h(z) = \frac{1}{(1-z)^{\frac{1}{p}+s+1}} = \sum_{n=0}^{\infty} c_n z^n ,$$

with $c_n \sim \gamma\, n^{\frac{1}{p}+s}$ for some constant $\gamma > 0$. We shall apply the inequality $\|T_\lambda(f)\|_q \le C\,\|f\|_p$ to the function $f = h_R$, where $0 < R < 1$ and $h_R(z) = h(Rz)$ is a dilation. Since $ps + p > 1$, an application of Lemma 1 in Section 3.2 gives the estimate

$$\|h_R\|_p^p = O\left(\frac{1}{(1-R)^{ps+p-1}}\right), \qquad R \to 1 .$$

On the other hand, Theorem 3 (assisted by the closed graph theorem) says that if $g(z) = \sum b_n z^n$ is in A^q, then

$$\sum_{n=1}^{\infty} n^{q-3} |b_n|^q \le C\,\|g\|_q^q , \qquad 0 < q \le 2 ;$$

$$\sum_{n=1}^{\infty} n^{-1} |b_n|^q \le C\,\|g\|_q^q , \qquad 2 \le q < \infty .$$

Suppose first that $0 < q \le 2$. Taking $f = h_R$ and combining our estimates, we find

$$\sum_{n=1}^{\infty} n^{q-3} |\lambda_n c_n R^n|^q \le C\,\|T_\lambda(h_R)\|_q^q \le C\,\|h_R\|_p^q \le C\,(1-R)^{\frac{q}{p}-q-qs} ,$$

where the constant C is not necessarily the same in each instance. Since $c_n \sim \gamma\, n^{\frac{1}{p}+s}$, it follows that

$$R^{Nq} \sum_{n=1}^{N} n^{\frac{q}{p}+qs+q-3} |\lambda_n|^q \le C\,(1-R)^{\frac{q}{p}-q-qs} .$$

Now choose $R = 1 - 1/N$ to deduce that

$$\sum_{n=1}^{N} n^{\frac{q}{p}+qs+q-3} |\lambda_n|^q \le C\, N^{q+qs-\frac{q}{p}}.$$

Therefore, since $\beta = 1 - (q + qs - \frac{q}{p}) > 0$ by our choice of s, we can apply Lemma 3 to arrive at the desired conclusion. The proof in the case (ii), where $2 \le q < \infty$, is essentially the same. □

Although *sufficient* conditions for multipliers are generally more difficult to obtain, Vukotić [2] observed that the following criteria are immediate consequences of Theorems 3 and 4.

THEOREM 7. *If p, q, and $\{\lambda_n\}$ satisfy any of the conditions*

(a) $0 < p \le 1,\ 2 \le q < \infty,\quad \sum_{n=1}^{\infty} n^{\frac{2q}{p}-3} |\lambda_n|^q < \infty$;

(b) $1 \le p < \infty,\ 2 \le q < \infty,\quad \sum_{n=1}^{\infty} n^{\frac{q}{p}+q-3} |\lambda_n|^q < \infty$;

(c) $0 < p \le 1,\ 0 < q \le 2,\quad \sum_{n=1}^{\infty} n^{\frac{2q}{p}-q-1} |\lambda_n|^q < \infty$;

(d) $1 \le p < \infty,\ 0 < q \le 2,\quad \sum_{n=1}^{\infty} n^{\frac{q}{p}-1} |\lambda_n|^q < \infty$,

then $\{\lambda_n\} \in (A^p, A^q)$.

Sufficient conditions can also be expressed in terms of the growth of λ_n as $n \to \infty$. To put the next theorem in perspective, we digress to recall the situation for H^p spaces. In the context of fractional integrals, Hardy and Littlewood [2] proved that if $0 < p < q < \infty$, $\alpha = \frac{1}{p} - \frac{1}{q}$, and

$$\lambda_n = \frac{n!}{\Gamma(n+\alpha+1)} = n^{-\alpha} + O(n^{-\alpha-1}),$$

then $\{\lambda_n\}$ is a multiplier of H^p into H^q. Duren [3] showed that the condition $\lambda_n = O(n^{-\alpha})$ alone implies that $\{\lambda_n\}$ is such a multiplier when $0 < p \le 2 \le q < \infty$ or when $0 < p \le 1$ and $q = \infty$; but the implication is false in all other cases. Vukotić [2] then established the following analogue for Bergman spaces.

THEOREM 8. *If $0 < p \le 2 \le q < \infty$, $\alpha = \frac{1}{p} - \frac{1}{q}$, and $\lambda_n = O(n^{-2\alpha})$, then $\{\lambda_n\} \in (A^p, A^q)$. If $0 < p \le 1$ and $\lambda_n = O(n^{-\frac{2}{p}})$, then $\{\lambda_n\} \in (A^p, H^\infty)$. In both cases the exponent is best possible.*

PROOF. Suppose first that $0 < p \le 1$ and $2 \le q < \infty$. Let $f(z) = \sum a_n z^n$ be a function in A^p, and let $g = T_\lambda(f)$, so that $g(z) = \sum b_n z^n$ with $b_n = \lambda_n a_n$. The hypothesis $\lambda_n = O(n^{-2\alpha})$ ensures that the power series converges in $\mathbb{D}$, since the estimate $a_n = O(n^{\frac{2}{p}-1})$ follows from Theorem 4(i). Let q' be the conjugate index of q, with $\frac{1}{q} + \frac{1}{q'} = 1$. To show that

$g \in A^q$, we appeal to Theorem 2(ii) and observe that

$$\sum_{n=1}^{\infty} n^{1-q'} |b_n|^{q'} = \sum_{n=1}^{\infty} n^{1-q'} |\lambda_n|^{q'} |a_n|^{q'-p} |a_n|^p$$

$$\leq C \sum_{n=1}^{\infty} n^{1-q'} n^{-2\alpha q'} n^{(\frac{2}{p}-1)(q'-p)} |a_n|^p = C \sum_{n=1}^{\infty} n^{p-3} |a_n|^p < \infty ,$$

in view of Theorem 3(i), since $f \in A^p$.

Next suppose that $0 < p \leq 1$ and $q = \infty$. Here we need only show that $\sum |\lambda_n a_n| < \infty$. Applying the hypothesis $\lambda_n = O(n^{-\frac{2}{p}})$ and the estimate $a_n = O(n^{\frac{2}{p}-1})$, we find by Theorem 3$(i)$ that

$$\sum_{n=1}^{\infty} |\lambda_n a_n| \leq C \sum_{n=1}^{\infty} n^{-\frac{2}{p}} n^{(\frac{2}{p}-1)(1-p)} |a_n|^p = C \sum_{n=1}^{\infty} n^{p-3} |a_n|^p < \infty .$$

In the remaining case where $1 \leq p \leq 2 \leq q < \infty$, the proof will appeal to the corresponding result of Duren [3] for H^p spaces, as cited above. To pass between Bergman and Hardy spaces, we use a result of MacGregor and Zhu [1] that $\{n^{-\frac{1}{p}}\} \in (A^p, H^p)$ if $1 \leq p \leq 2$; while $\{n^{\frac{1}{q}}\} \in (H^q, A^q)$ if $2 \leq q < \infty$. Simply write $\lambda_n = n^{\frac{1}{q}} \mu_n n^{-\frac{1}{p}}$ and observe that the hypothesis $\lambda_n = O(n^{-2\alpha})$ implies $\mu_n = O(n^{-\alpha})$. Thus $\{\mu_n\} \in (H^p, H^q)$, and it follows that $\{\lambda_n\} \in (A^p, A^q)$. Functions of the form $(1-z)^a$ show that the exponents are best possible. □

As a final example, the multipliers from A^1 to A^2 will now be described. Hardy and Littlewood [3] proved that $\{\lambda_n\}$ is a multiplier of H^1 into H^2 if

$$\sum_{n=1}^{N} n|\lambda_n|^2 = O(N) , \qquad n \to \infty .$$

Duren and Shields [2] used the closed graph theorem to show that this condition is also necessary. Vukotić [2] obtained the following analogue for Bergman spaces.

THEOREM 9. *A sequence $\{\lambda_n\}$ is a multiplier of A^1 into A^2 if and only if*

$$\sum_{n=1}^{N} n^2 |\lambda_n|^2 = O(N) , \qquad n \to \infty .$$

The proof will appeal to the result of Hardy and Littlewood on the multipliers of H^1 into H^2; their proof may also be found in Duren [5], pp. 103–104. The Hardy–Littlewood argument uses the canonical factorization of H^1 functions to reduce the problem to *nonvanishing* H^1 functions, so it does not work for Bergman spaces. The following lemma, a special case of the MacGregor–Zhu theorem cited above, shifts the problem to Hardy spaces.

LEMMA 4. *The sequence* $\{\frac{1}{n}\}$ *is a multiplier of* A^1 *into* H^1.

PROOF OF LEMMA. An equivalent statement is that $f' \in A^1$ implies $f \in H^1$. But for $0 < r < 1$ a radial integration from 0 to $re^{i\theta}$ gives

$$f(re^{i\theta}) = f(0) + e^{i\theta} \int_0^r f'(\rho e^{i\theta})\, d\rho .$$

A further integration now leads to the inequality

$$\int_0^{2\pi} |f(re^{i\theta})|\, d\theta \leq 2\pi |f(0)| + \int_0^{2\pi} \int_0^r |f'(\rho e^{i\theta})|\, d\rho d\theta ,$$

which shows that $f \in H^1$ if $f' \in A^1$. □

PROOF OF THEOREM 9. The necessity of the given condition is a special case of Theorem 5, with $p = 1$ and $q = 2$. To prove the sufficiency, suppose that $\sum_{n=1}^N n^2 |\lambda_n|^2 = O(N)$ and let $\mu_n = \sqrt{n}\, \lambda_n$. Then $\sum_{n=1}^N n|\mu_n|^2 = O(N)$, and it follows from the Hardy–Littlewood result that $\{\mu_n\} \in (H^1, H^2)$. But it is trivially true that $\{\sqrt{n}\} \in (H^2, A^2)$, so Lemma 4 implies that

$$\{\lambda_n\} = \{\sqrt{n}\, \mu_n \tfrac{1}{n}\} \in (A^1, A^2) . \qquad \square$$

One consequence of Theorem 9 (or of Lemma 4) is that the integral of an A^1 function is in A^2. Thus $f' \in A^1$ implies $f \in A^2$. Another corollary is an analogue of the theorem of Paley for H^1 functions that originally motivated the Hardy–Littlewood theorem (*cf.* Duren [5], p. 104). Specifically, if $\{n_k\}$ is a lacunary sequence in the sense that $n_{k+1}/n_k \geq Q > 1$, then the sequence $\{\lambda_n\}$ defined by $\lambda_n = \sqrt{n_k}$ for $n = n_k$ and $\lambda_n = 0$ for $n \neq n_k$ is a multiplier of A^1 into A^2. Thus for every function $f(z) = \sum a_n z^n$ in A^1, we have

$$\sum_{k=1}^{\infty} n_k^{-2} |a_{n_k}|^2 < \infty .$$

§3.5. Korenblum's maximum principle.

For $0 < c < 1$, let $\mathcal{A}_c$ denote the annulus defined by $c < |z| < 1$. The maximum modulus theorem states that if a function f is analytic in $\mathbb{D}$ and $|f(z)| \leq b$ in $\mathcal{A}_c$ for some constant b and some radius c, then $|f(z)| \leq b$ for all $z \in \mathbb{D}$. Therefore, $\|f\|_p \leq b = \|b\|_p$. A natural question is whether the constant b can be replaced by an arbitrary analytic function. In other words, if f and g are analytic functions with $|f(z)| \leq |g(z)|$ for $z \in \mathcal{A}_c$, does it follow that $\|f\|_p \leq \|g\|_p$? If f/g is analytic, then the maximum modulus theorem implies that $|f(z)| \leq |g(z)|$ in $\mathbb{D}$, and so $\|f\|_p \leq \|g\|_p$. However, a simple example shows that the result does not hold for arbitrary $c < 1$, even when $p = 2$. Specifically, the conclusion fails for each $c > \frac{1}{\sqrt{2}}$. To see this, take $g(z) = z$ and $f(z) \equiv b$, where b is a constant satisfying $\frac{1}{\sqrt{2}} < b < c$. Then $|f(z)| \leq |g(z)|$ for all $z \in \mathcal{A}_c$, but

$$\|f\|_2 = b > \frac{1}{\sqrt{2}} = \|g\|_2 .$$

Korenblum [5] conjectured that the conclusion $\|f\|_p \leq \|g\|_p$ is valid under the assumption that $|f(z)| \leq |g(z)|$ in $\mathcal{A}_c$ for *some* $c > 0$. This he proved for $p = 2$ and $c < \frac{1}{2}e^{-2} = 0.067\ldots$ under the additional hypothesis that g/f is analytic. Then came a succession of partial results by Korenblum, O'Neil, Richards, and Zhu [1], Korenblum and Richards [1], Matero [1], Andersson [1], Schwick [1], and others. Finally, Hayman [1] proved Korenblum's conjecture for $p = 2$ and $c = \frac{1}{25} = 0.04$. Hinkkanen [1] then improved the constant to $c = 0.157$ and generalized the result to $1 \leq p < \infty$. Thus we can state the following theorem, which is known as *Korenblum's maximum principle.*

THEOREM 10. *Let $1 \leq p < \infty$. Then there exists a constant $c > 0$ with the following property. If f and g are functions analytic in $\mathbb{D}$, and $|f(z)| \leq |g(z)|$ for all $z \in \mathcal{A}_c$, then $\|f\|_p \leq \|g\|_p$.*

Korenblum actually conjectured that the theorem holds for $0 < p < \infty$. The case $0 < p < 1$ remains open. Also, the best possible value of c has not been determined.

Theorem 10 plays no role in topics that are discussed in this book. The proof is omitted.

CHAPTER 4

Zero-Sets

A sequence of points $\{z_k\}$ in $\mathbb{D}$ is called an A^p *zero-set* if some function $f \in A^p$ vanishes precisely on this set. If all of the points z_k are distinct, this means that $f(z_k) = 0$ and $f'(z_k) \neq 0$ for $k = 1, 2, \ldots$; and $f(z) \neq 0$ elsewhere in $\mathbb{D}$. If some point α occurs m times in the sequence $\{z_k\}$, the requirement is that f have a zero at α of precise order m. An analytic function f is said to *vanish on* $\{z_k\}$ if $f(z_k) = 0$ for each k, and if the order of its zero at z_k is greater than or equal to the multiplicity of z_k in the sequence. Note that f is allowed to vanish elsewhere in $\mathbb{D}$, and to have a zero of higher order than required at a point z_k. If a nonzero function $f \in A^p$ vanishes on a sequence $\{z_k\}$, it does not follow *a priori* (although it turns out to be true) that $\{z_k\}$ is an A^p zero-set, for this requires that some function $g \in A^p$ vanish *precisely* on $\{z_k\}$. A sequence $\{z_k\}$ is called a *set of uniqueness* for A^p if the only function that vanishes on $\{z_k\}$ is the zero function. Whenever two A^p functions have the same values (respecting multiplicities) on a set of uniqueness, they must agree everywhere in $\mathbb{D}$. It will turn out that a sequence is a set of uniqueness for A^p if and only if it is not an A^p zero-set.

It is an important problem to give a complete description of the zero-sets of functions in the Bergman space A^p. The problem is still open, but a great deal of information is available. This chapter will present some basic properties of A^p zero-sets. Some of the results will be subsumed in deeper theories of contractive divisors and of interpolation and sampling sets, to be developed in later chapters. In order to set the stage for those considerations, however, it is helpful first to study the zero-sets by elementary and more direct methods.

§4.1. Preliminary remarks.

First a word about terminology. We propose to use the terms "zero-set" and "zero-sequence" interchangeably. Neither term is entirely satisfactory. A zero-set of an analytic function is a countable sequence, but the ordering is generally unimportant, and we will need to speak of subsets and unions of zero-sets. On the other hand, a zero-set is more than a set of points, since it must take account of multiplicities. Thus when we say that the zero-set of some function f is a *subset* of the zero-set of another function g, we mean that g vanishes wherever f does, to at least the same multiplicity. The *union* of two zero-sets assigns to any common point a multiplicity equal to the sum of the two multiplicities.

For functions in the Hardy space H^p, the zero-sets are known and are very easy to describe. They are exactly the Blaschke sequences, for which the series $\sum(1-|z_k|)$ is convergent. In particular, the H^p zero-sets are the same for all values of p. Since $H^p \subset A^p$, it is clear that every Blaschke sequence is an A^p zero-set for $0 < p < \infty$. To show that non-Blaschke sequences occur as A^p zero-sets for each p, we shall begin with an explicit example.

Let $\{n_k\}$ be an increasing sequence of positive integers, and consider the infinite product

$$f(z) = \prod_{k=1}^{\infty}(1 - 2z^{n_k}),$$

which converges locally uniformly and therefore represents an analytic function in $\mathbb{D}$. This function f has n_k equally spaced zeros on the circle of radius

$$2^{-1/n_k} = e^{-\frac{\log 2}{n_k}} \sim 1 - \frac{\log 2}{n_k},$$

which contribute a total of

$$n_k(1 - 2^{-1/n_k}) \sim \log 2$$

to the Blaschke sum. Hence the zero-set of f does not satisfy the Blaschke condition for any choice of exponents n_k. However,

$$|f(z)| \le \prod_{k=1}^{\infty}(1 + 2r^{n_k}) \le \exp\left\{2\sum_{k=1}^{\infty} r^{n_k}\right\},$$

where $r = |z|$, in view of the inequality $1+t \le e^t$. By Lemma 2 in Section 3.2, the right-hand expression can be made to grow arbitrarily slowly as $r \to 1$ if the exponents n_k increase sufficiently fast. In particular, the exponents can be chosen to place f in A^p for every $p < \infty$.

On the other hand, every A^p zero-set is "almost" a Blaschke sequence. In fact, this is true for the zeros of any analytic function that satisfies a growth condition of the form $M_\infty(r,f) = O\big((1-r)^{-a}\big)$ as $r \to 1$, for some exponent $a > 0$. Recall from Chapter 3 (Theorem 1) that every function $f \in A^p$ satisfies $M_\infty(r,f) = o\big((1-r)^{-\frac{2}{p}}\big)$, while a growth condition $M_\infty(r,f) = O\big((1-r)^{-a}\big)$ for some $a < 1/p$ implies that $f \in A^p$. This means that the class of analytic functions belonging to *some* Bergman space A^p coincides with the class of functions satisfying some growth condition of the form $M_\infty(r,f) = O\big((1-r)^{-a}\big)$.

Let $z_1, z_2, \ldots$ be the zeros, repeated according to multiplicity, of a function f analytic in $\mathbb{D}$. We will suppose that $f(0) \ne 0$, and that the zeros are ordered by modulus, so that

$$0 < |z_1| \le |z_2| \le |z_3| \le \cdots.$$

Let $n(r)$ be the number of zeros z_k, counted according to multiplicity, with modulus $|z_k| < r$. Observe that

$$(1-r)\,n(r) \le \sum_{|z_k|<r} (1-|z_k|)\,,$$

so that $n(r) = O(1/(1-r))$ for Blaschke sequences. Thus the following proposition says in two ways that every A^p zero-set is almost a Blaschke sequence.

PROPOSITION. *Let f be a function analytic in $\mathbb{D}$, with $f(0) \neq 0$, and suppose that $M_\infty(r,f) = O\big((1-r)^{-a}\big)$ as $r \to 1$, for some exponent $a > 0$. Then the zeros of f have the properties*

$$(i)\quad n(r) = O\left(\frac{1}{1-r}\log\frac{1}{1-r}\right), \qquad r \to 1\,;$$

$$(ii)\quad \sum_k (1-|z_k|)\left(\log\frac{1}{1-|z_k|}\right)^{-1-\varepsilon} < \infty \qquad \textit{for each } \varepsilon > 0\,.$$

The classic tool for expressing the connection between rate of growth of a function and density of zeros is Jensen's formula

$$\frac{1}{2\pi}\int_0^{2\pi} \log|f(re^{i\theta})|\,d\theta = \log|f(0)| + \sum_{k=1}^{n} \log\frac{r}{|z_k|}\,,$$

where $z_1, z_2, \dots, z_n$ are the zeros of f, repeated according to multiplicity, in the disk $|z| < r$. In terms of the counting function $n(r)$, the sum becomes

$$\sum_{k=1}^{n} \log\frac{r}{|z_k|} = \int_0^r \frac{n(x)}{x}\,dx\,.$$

PROOF OF PROPOSITION. Taking $f(0) = 1$ for convenience, we apply Jensen's formula to obtain

$$\int_0^r n(x)\,dx \le \int_0^r \frac{n(x)}{x}\,dx = \frac{1}{2\pi}\int_0^{2\pi} \log|f(re^{i\theta})|\,d\theta = O\left(\log\frac{1}{1-r}\right),$$

by the growth hypothesis on f. Therefore, since $n(x)$ is nondecreasing,

$$(r-r^2)n(r^2) \le \int_{r^2}^r n(x)\,dx = O\left(\log\frac{1}{1-r}\right),$$

which implies (i). To prove (ii), we will apply the stronger estimate

$$N(r) = \int_0^r \frac{n(x)}{x}\,dx = O\left(\log\frac{1}{1-r}\right)$$

that comes from Jensen's formula. Integration by parts and (i) yield

$$\sum_k (1-|z_k|)\left(\log\frac{1}{1-|z_k|}\right)^{-1-\varepsilon} = \int_0^1 (1-r)\left(\log\frac{1}{1-r}\right)^{-1-\varepsilon} dn(r)$$

$$= \int_0^1 r\left(\log\frac{1}{1-r}\right)^{-1-\varepsilon}\left\{1+(1+\varepsilon)\left(\log\frac{1}{1-r}\right)^{-1}\right\}\frac{n(r)}{r}\,dr$$

$$\leq C\int_0^1\left(\log\frac{1}{1-r}\right)^{-1-\varepsilon}\frac{n(r)}{r}\,dr\,.$$

Another integration by parts and the above estimate for $N(r)$ give

$$\int_0^1\left(\log\frac{1}{1-r}\right)^{-1-\varepsilon}\frac{n(r)}{r}\,dr = (1+\varepsilon)\int_0^1\frac{1}{1-r}\left(\log\frac{1}{1-r}\right)^{-2-\varepsilon} N(r)\,dr$$

$$\leq C\int_0^1\frac{1}{1-r}\left(\log\frac{1}{1-r}\right)^{-1-\varepsilon} dr < \infty\,,$$

which proves (ii). □

The proof of (i) is due to Shapiro and Shields [2], who gave an example to show that the estimate is best possible. In other words, the statement is no longer true if "O" is replaced by "o". We will see later that the statement (ii) is also best possible, in the sense that ε cannot be replaced by zero.

§4.2. Density of zero-sets.

Since the H^p zero-sets are characterized by the Blaschke condition, they are independent of p, every subset of an H^p zero-set is an H^p zero-set, and the union of two H^p zero-sets is always an H^p zero-set. What is the situation for the Bergman spaces? Does the collection of A^p zero-sets become smaller as p increases? Is every subset of an A^p zero-set an A^p zero-set? Must the union of two A^p zero-sets be an A^p zero-set?

In his Ph.D. thesis in 1974, Charles Horowitz [1] answered all three of these questions. He showed that the A^p zero-sets vary with p, and every subset of an A^p zero-set is an A^p zero-set, but the union of two A^p zero-sets need not be an A^p zero-set. Much of this chapter is based on the thesis and two related papers by Horowitz [2,3].

The point of departure is again Jensen's formula

$$\log|f(0)| + \sum_{k=1}^n \log\frac{r}{|z_k|} = \frac{1}{2\pi}\int_0^{2\pi}\log|f(re^{i\theta})|\,d\theta\,,$$

where $0<r<1$ and f is an analytic function in $\mathbb{D}$ whose zeros z_k (repeated according to multiplicity) satisfy

$$0<|z_1|\leq|z_2|\leq\cdots\leq|z_n|<r\leq|z_{n+1}|\leq\cdots\,.$$

Multiply the Jensen formula by p, exponentiate, and apply the arithmetic–geometric mean inequality to conclude that

$$|f(0)|^p \prod_{k=1}^{n} \frac{r^p}{|z_k|^p} \le \frac{1}{2\pi} \int_0^{2\pi} |f(re^{i\theta})|^p \, d\theta \, .$$

Now observe that this inequality holds for each $r < 1$ and for *arbitrary* n. Indeed, the product

$$\prod_{k=1}^{m} \frac{r^p}{|z_k|^p}$$

increases to a maximum value as m goes from 1 to n, since the factors are greater than 1 when $|z_k| < r$. It actually decreases for $m > n$, as soon as $|z_k| > r$ and so the extra factors are less than 1. Thus if $f \in A^p$, an integration gives

$$2\,|f(0)|^p \prod_{k=1}^{n} |z_k|^{-p} \int_0^1 r^{np+1} \, dr \le \|f\|_p^p \, ,$$

or

$$\left(\frac{2}{np+2} \right)^{1/p} |f(0)| \prod_{k=1}^{n} |z_k|^{-1} \le \|f\|_p \, .$$

We have therefore arrived at the following theorem.

THEOREM 1. *If $\{z_k\}$ are the ordered zeros of a function $f \in A^p$ with $f(0) \neq 0$, then*

$$\prod_{k=1}^{n} \frac{1}{|z_k|} = O(n^{\frac{1}{p}}) \, , \qquad n \to \infty \, .$$

We shall see that the exponent $1/p$ in the theorem is best possible. However, the condition is far from sufficient for the sequence $\{z_k\}$ to form an A^p zero-set. For instance, if a function $f \in A^p$ has all of its zeros on a single ray, then by a result of Shapiro and Shields [1], they must satisfy the Blaschke condition, which can be expressed in the form $\prod |z_k| > 0$. (This result will be established in Section 4.8.) However, the condition in Theorem 1 admits non-Blaschke sequences, even if they are aligned in a single ray. In view of the Shapiro–Shields result, it is clear that any necessary and sufficient condition must take account of arguments as well as moduli of the points z_k.

On the other hand, the theorem does show that the zeros of functions in Bergman spaces are "almost" Blaschke sequences in the following sense, essentially a restatement of part (*ii*) of the proposition in Section 4.1.

COROLLARY. *Let $\{z_k\}$ be the zero-set of a function f belonging to A^p for some $p > 0$, with $f(0) \neq 0$. Then for each $\varepsilon > 0$,*

$$\sum_k (1 - |z_k|) \left(\log \frac{1}{1 - |z_k|} \right)^{-1-\varepsilon} < \infty .$$

The corollary can be deduced from Theorem 1 by a method of summation by parts. However, the details will be omitted since we have already derived the result directly from Jensen's formula. Recall that f belongs to a Bergman space A^p for some $p > 0$ if and only if $M_\infty(r, f) = O\big((1 - r)^{-a}\big)$ for some $a > 0$. The corollary says in particular that each A^p zero-set $\{z_k\}$ satisfies $\sum (1 - |z_k|)^{1+\delta} < \infty$ for every $\delta > 0$.

§4.3. Dependence on p.

We are now ready to show that the A^p zero-sets vary with p.

THEOREM 2. *If $0 < p < q < \infty$, there exists an A^p zero-set that is not an A^q zero-set.*

PROOF. We shall follow the constructive proof by Horowitz [2]. The strategy is to exhibit a function of the form

$$f(z) = \prod_{\nu=1}^{\infty} (1 - b z^{m^\nu}) , \qquad b > 1, \quad m = 2, 3, \dots ,$$

where the parameters b and m are chosen so that $f \in A^p$ but its zero-set violates the necessary condition of Theorem 1 for index q. For any choice of real number $b > 1$ and integer $m \geq 2$, the infinite product converges uniformly on compact sets of $\mathbb{D}$ and so represents an analytic function f. It is clear by inspection that f has m^ν equally spaced simple zeros on the circle $|z| = b^{-1/m^\nu}$, for each $\nu = 1, 2, \dots$. Thus if these zeros z_k are ordered by nondecreasing modulus, the construction shows that

$$|z_k| = b^{-1/m} , \qquad 1 \leq k \leq m ,$$
$$|z_k| = b^{-1/m^2} , \qquad m + 1 \leq k \leq m + m^2 ,$$

and in general

$$|z_k| = b^{-1/m^\nu} , \qquad N_{\nu-1} < k \leq N_\nu ,$$

where

$$N_\nu = m + m^2 + \cdots + m^\nu = \frac{m(m^\nu - 1)}{m - 1} , \qquad \nu = 1, 2, \dots ,$$

and $N_0 = 0$. If these points z_k were the zeros of a function in A^q, then Theorem 1 would imply that

$$\prod_{k=1}^{N_\nu} \frac{1}{|z_k|} = \prod_{j=1}^{\nu} (b^{1/m^j})^{m^j} = b^\nu = O\big(N_\nu{}^{\frac{1}{q}}\big),$$

or equivalently that $b^\nu = O(m^{\frac{\nu}{q}})$ as $\nu \to \infty$. But $b^\nu = m^{\lambda\nu}$ for $\lambda = \frac{\log b}{\log m}$, so the condition is violated when $\lambda > \frac{1}{q}$. In other words, if the parameters are chosen with $\frac{\log m}{\log b} < q$, the zero-set of f is not an A^q zero-set .

It remains to show that $f \in A^p$ for some choice of parameters b and m satisfying $\frac{\log m}{\log b} < q$. For this purpose we consider the Taylor series expansion

$$f(z) = \sum_{k=0}^{\infty} a_k z^k.$$

and refer to Theorem 4 of Chapter 3, which says that $f \in A^p$ if

$$S_n^{(r)} = \sum_{k=0}^{n} |a_k|^r = O(n^\lambda)$$

for certain values of r and λ. The Taylor coefficients a_k are easily computed because of the lacunarity of the sequence $\{m^\nu\}$; the partial products of the infinite product for f also represent partial sums of its Taylor expansion. In fact, it is easy to see that

$$S_{N_\nu}^{(r)} = (1 + b^r)^\nu, \qquad r > 0, \quad \nu = 1, 2, \dots,$$

where $N_\nu = m + m^2 + \cdots + m^\nu$ as above. Note that the numbers N_ν are comparable to m^ν, since $m^\nu \le N_\nu \le 2m^\nu$. We want to choose the parameters $b > 1$ and $m = 1, 2, \dots$ such that $S_n^{(r)} = O(n^\lambda)$ for given $\lambda > 0$. In particular, we want

$$(1 + b^r)^\nu = S_{N_\nu}^{(r)} = O(N_\nu^\lambda) = O(m^{\lambda\nu}),$$

so we will simply choose b and m to make $1 + b^r = m^\lambda$. Then for $N_{\nu-1} \le n \le N_\nu$, it will follow that

$$S_n^{(r)} \le S_{N_\nu}^{(r)} \le C\, N_\nu^\lambda \le C(2m)^\lambda m^{(\nu-1)\lambda} \le C(2m)^\lambda N_{\nu-1}^\lambda \le C(2m)^\lambda n^\lambda,$$

which will imply that $f \in A^p$.

Suppose first that $0 < p \le 2$. Then $f \in A^p$ if $S_n^{(2)} = O(n^\lambda)$ for some $\lambda < 2/p$, by Theorem 4 in Chapter 3. With $1 + b^2 = m^\lambda$, the requirement $\lambda < 2/p$ then reduces to $p \log(1 + b^2) < 2 \log m$. Thus if

$$p < \frac{2 \log m}{\log(1 + b^2)} \qquad \text{and} \qquad \frac{\log m}{\log b} < q,$$

it will follow that $f \in A^p$ but its zeros z_k do not form an A^q zero-set. For given p and q with $0 < p < q < \infty$, it is possible to realize both of the required inequalities simultaneously with suitable choices of the parameters m and b, since $\log(1+b^2) \sim 2\log b$ as $b \to \infty$.

Next suppose that $p > 2$, and let $p' = \frac{p}{p-1}$ be the conjugate index. Then the criterion is that $f \in A^p$ if $S_n^{(p')} = O(n^\lambda)$ for some $\lambda < p' - 1$. With $1 + b^{p'} = m^\lambda$, the inequality $\lambda < p' - 1$ becomes

$$\frac{\log(1+b^{p'})}{\log m} < p' - 1\,, \qquad \text{or} \qquad p < \frac{\log m}{\frac{p-1}{p}\log\left(1 + b^{\frac{p}{p-1}}\right)}\,.$$

But $\frac{p-1}{p}\log\left(1+b^{\frac{p}{p-1}}\right) \sim \log b$ as $b \to \infty$, so this condition and the inequality $\frac{\log m}{\log b} < q$ can both be satisfied for certain choices of $m = 2, 3, \ldots$ and $b > 1$. Any such choice of parameters produces a function $f \in A^p$ whose zeros do not constitute an A^q zero-set. □

Luecking [4] showed that whenever a set of points is regularly spaced as in Horowitz' construction, the simple condition $\frac{\log m}{\log b} > p$ is in fact sufficient for it to be an A^p zero-set. Duren, Schuster, and Seip [1] obtained sharper results by a different method, a direct calculation of the densities to be described in Chapter 6. These issues are discussed further in Sections 6.6 and 6.7.

Three corollaries can now be noted. The first follows immediately from the theorem, and contrasts once again with the situation for Hardy spaces. The last two corollaries are actually by-products of the construction given in the proof.

COROLLARY 1. *There is a function in A^1 which cannot be factored as the product of two A^2 functions, one of them nonvanishing.*

COROLLARY 2. *The exponent $\frac{1}{p}$ in Theorem 1 is best possible. For each pair of numbers p and s with $0 < p < \infty$ and $0 < s < \frac{1}{p}$, there is a function $f \in A^p$ with $f(0) \neq 0$ whose ordered zeros are distinct and have the property*

$$\prod_{k=1}^{n} \frac{1}{|z_k|} \geq Cn^s, \qquad n = 1, 2, \ldots\,,$$

for some constant $C > 0$ independent of n .

COROLLARY 3. *The corollary to Theorem 1 is best possible. For each p with $0 < p < \infty$, there is a function $f \in A^p$ with zeros $z_k \neq 0$ satisfying*

$$\sum_{k=1}^{\infty}(1 - |z_k|)\left(\log\frac{1}{1 - |z_k|}\right)^{-1} = \infty\,.$$

The divergence of the series in Corollary 3 actually holds for *every* function f given as above by the infinite product construction, regardless of the choice of parameters $b > 1$ and $m = 2, 3, \ldots$. To see this, it will be enough to show that the ordered zeros satisfy $1 - |z_k| \ge \frac{c}{k}$ for some constant $c > 0$ and all k, because the result will then follow from the divergence of the series $\sum \frac{1}{k \log k}$. But $|z_k| = b^{-1/m^\nu}$ for $N_{\nu-1} < k \le N_\nu$, so it is equivalent to show that

$$1 - b^{-1/m^\nu} \ge \frac{c}{m^\nu}, \qquad \nu = 1, 2, \ldots,$$

since $m^\nu \le N_\nu \le 2m^\nu$. For a proof, note first the inequality $-2x \le \log(1-x)$ for $0 \le x \le \frac{1}{2}$. Then choose $c = \frac{1}{2}\log b$ to deduce that

$$\log\big(b^{-1/m^\nu}\big) = -\frac{2c}{m^\nu} \le \log\big(1 - \frac{c}{m^\nu}\big)$$

when $m^\nu \ge 2c$. This shows that $1 - b^{-1/m^\nu} \ge c/m^\nu$ for all ν sufficiently large, and Corollary 3 follows.

§4.4. Unions and subsets of zero-sets.

In further contrast with the situation for Hardy spaces, the A^p zero-sets are not preserved under finite unions. In fact, Horowitz [2] discovered the following more precise result.

THEOREM 3. *Let $0 < p < \infty$. Then the union of each pair of A^p zero-sets is an $A^{\frac{p}{2}}$ zero-set. However, their union need not be an A^q zero-set for any $q > \frac{p}{2}$.*

PROOF. The first assertion is trivial. If f and g are functions in A^p, the Schwarz inequality shows that their product fg belongs to $A^{\frac{p}{2}}$. On the other hand, it is clear that fg vanishes precisely on the union of the respective zero-sets of f and g. Hence the union of any two A^p zero-sets is an $A^{\frac{p}{2}}$ zero-set. In all of these statements, the "union" of zero-sets must be interpreted to take account of multiplicities.

To show that the result is best possible, choose a number s in the interval $\frac{1}{2q} < s < \frac{1}{p}$. Then, according to Corollary 2 of Theorem 2, there is a function $f \in A^p$ with $f(0) \ne 0$ whose ordered zeros z_k are all distinct and satisfy

$$\prod_{k=1}^{n} \frac{1}{|z_k|} \ge C n^s, \qquad n = 1, 2, \ldots.$$

Now rotate the zero-set $\{z_k\}$ through an angle θ chosen so that the A^p zero-set $\{e^{i\theta} z_k\}$ is disjoint from $\{z_k\}$. Let $\{\zeta_k\}$ represent the union of these two sets, with $0 < |\zeta_1| \le |\zeta_2| \le \ldots$. From the construction it is clear that

$$\prod_{k=1}^{2n} \frac{1}{|\zeta_k|} \ge C^2 n^{2s} \ne O(n^{\frac{1}{q}}),$$

which shows by Theorem 1 that $\{\zeta_k\}$ is not an A^q zero-set. □

We shall now conclude with a positive result, also due to Horowitz [2].

THEOREM 4. *For $0 < p < \infty$, every subset of an A^p zero-set is an A^p zero-set.*

The theorem asserts that if a non-null function $f \in A^p$ vanishes on a set $\{z_k\}$, then some function $g \in A^p$ vanishes *precisely* on that set. We shall present Horowitz' elegant proof, although a sharper form of the argument will be given in terms of the "contractive zero-divisors" to be developed in the next chapter. The proof depends on the construction of an infinite product that will "divide out" an arbitrary portion of the zero-set.

For $\alpha \in \mathbb{D}$ and $\alpha \neq 0$, let

$$b_\alpha(z) = \frac{|\alpha|}{\alpha}\frac{\alpha - z}{1 - \overline{\alpha}z}$$

denote the corresponding Blaschke factor. Observe that $b_\alpha(z) \to 1$ for each $z \in \mathbb{D}$ as $|\alpha| \to 1$. Let $b_0(z) = z$. Recall that the Blaschke product

$$B(z) = \prod_{k=1}^{\infty} b_{z_k}(z)$$

converges if and only if the sum $\sum_{k=1}^{\infty}(1-|z_k|)$ converges. However, the zeros of functions in a Bergman space need not satisfy the Blaschke condition, so the Blaschke product may well diverge. To deal with this problem, Horowitz devised the strategy of inserting convergence factors, constructing what is now called a *Horowitz product*

$$H(z) = \prod_{k=1}^{\infty} b_{z_k}(z)\big(2 - b_{z_k}(z)\big).$$

If $f \in A^p$, the corollary to Theorem 1 shows that its zeros $\{z_k\}$ have in particular the property

$$\sum_{k=1}^{\infty}(1 - |z_k|)^2 < \infty.$$

This condition does not imply convergence of the corresponding Blaschke product, but it is enough to ensure the convergence of the Horowitz product, uniformly on compact subsets of the disk.

A proof of convergence uses the estimate

$$\begin{aligned} \big|1 - b_{z_k}(z)\big(2 - b_{z_k}(z)\big)\big| &= \big|1 - b_{z_k}(z)\big|^2 \\ &= \left|\frac{z_k + z|z_k|}{z_k(1 - \overline{z_k}z)}\right|^2 (1 - |z_k|)^2 \le \frac{4(1 - |z_k|)^2}{(1 - r)^2}, \end{aligned}$$

uniformly in any disk $|z| \le r < 1$. Thus the convergence of $\sum(1-|z_k|)^2$ implies that

$$\sum_{k=1}^{\infty} \left|1 - b_{z_k}(z)\big(2 - b_{z_k}(z)\big)\right| < \infty,$$

uniformly on compact subsets of $\mathbb{D}$. The Horowitz product therefore converges absolutely and uniformly on compact sets, so that $H(z)$ is an analytic function whose zero-set is precisely $\{z_k\}$.

The condition $\sum(1-|z_k|)^2 < \infty$ is necessary as well as sufficient for convergence of the Horowitz product. To see this, observe that

$$b_{z_k}(0)\big(2 - b_{z_k}(0)\big) = |z_k|(2-|z_k|) = 1 - (1-|z_k|)^2.$$

We shall now show that the Horowitz product divides out the zeros of a function $f \in A^p$ without essential increase of norm. The following theorem is the main component in the proof of Theorem 4, and it has other applications.

THEOREM 5. *For $0 < p < \infty$, let $f \in A^p$ and let H be the Horowitz product formed with its zero-set. Then $f/H \in A^p$ and $\|f/H\|_p \le C_p \|f\|_p$, where C_p is a constant depending only on p.*

It is clear first of all that $g = f/H$ is analytic in $\mathbb{D}$, since H and f have the same zero-set. To estimate the norm of g, we need a lemma stated in terms of a *weighted Bergman space* A^p_α with norm

$$\|f\|_{p,\alpha} = \left\{\int_{\mathbb{D}} |f(z)|^p (1-|z|^2)^\alpha \, d\sigma\right\}^{1/p}, \qquad \alpha > -1.$$

LEMMA. *Let $f \in A^p_2$ with $f(0) \ne 0$, and let $\{z_k\}$ be its zero-set. Then*

$$|f(0)|\left\{\prod_{k=1}^{\infty} |z_k|(2-|z_k|)\right\}^{-1} \le C_p \, \|f\|_{p,2},$$

where C_p is a constant depending only on p.

PROOF OF LEMMA. First note that

$$M_\infty(r,f) = O\left(\frac{1}{(1-r)^{\frac{2+\alpha}{p}}}\right), \qquad r \to 1,$$

for every function $f \in A^p_\alpha$. This is proved by the same argument that was used in Section 3.2 for the unweighted case, where $\alpha = 0$. Thus $M_\infty(r,f) = O\big((1-r)^{-4/p}\big)$ for $f \in A^p_2$.

Assume for convenience that $f(0) = 1$. Let $n(r)$ be the number of zeros z_k, counted according to multiplicity, in the disk $|z| < r$. Then

$$n(r) = O\left(\frac{1}{1-r}\log\frac{1}{1-r}\right), \qquad r \to 1,$$

by the proposition in Section 4.1. Integration by parts gives

$$-\sum_{k=1}^{\infty}\log\big(|z_k|(2-|z_k|)\big) = -\int_0^1 \log\big(r(2-r)\big)\,dn(r)$$

$$= 2\int_0^1 \frac{1-r}{r(2-r)}n(r)\,dr \;-\; \lim_{r\to 1} n(r)\log\big(r(2-r)\big).$$

But $\log(r(2-r)) = O((1-r)^2)$, so $n(r)\log(r(2-r)) \to 0$ as $r\to 1$. As for the integral, another integration by parts gives

$$\int_0^1 \frac{1-r}{r(2-r)}n(r)\,dr = \int_0^1 \frac{1}{(2-r)^2}N(r)\,dr\,,$$

where

$$N(r) = \int_0^r \frac{n(s)}{s}\,ds = \frac{1}{2\pi}\int_0^{2\pi}\log|f(re^{i\theta})|\,d\theta\,,$$

by Jensen's theorem, since $f(0)=1$. Thus

$$-\sum_{k=1}^{\infty}\log\big(|z_k|(2-|z_k|)\big) = \frac{1}{\pi}\int_0^{2\pi}\int_0^1 \frac{1}{(2-r)^2}\log|f(re^{i\theta})|\,dr\,d\theta$$

$$= \frac{1}{\pi}\int_0^{2\pi}\int_0^1 \frac{1}{(2-r)^2}\log\big\{|f(re^{i\theta})|(1-r^2)^2\big\}\,dr\,d\theta \;-\; 4\int_0^1 \frac{\log(1-r^2)}{(2-r)^2}\,dr\,.$$

Multiplying by p, exponentiating, and applying the arithmetic–geometric mean inequality with respect to the unit measure $\frac{1}{\pi}\frac{1}{(2-r)^2}\,drd\theta$, we obtain the desired estimate. □

PROOF OF THEOREM 5. Now let $f\in A^p$ and let $\{z_k\}$ be its zero-set. For $w\in\mathbb{D}$, define the function

$$F_w(z) = f(\varphi_w(z))\varphi_w'(z)^{\frac{2}{p}}\,,$$

where

$$\varphi_w(z) = \frac{w-z}{1-\overline{w}z}\,.$$

Then $F\in A^p\subset A^p_2$. Since ${\varphi_w}^{-1} = \varphi_w$, the zeros of F_w are the points $\zeta_k = \varphi_w(z_k)$, and $|\zeta_k| = |\varphi_{z_k}(w)|$. Also note that

$$|F_w(0)| = |f(w)|(1-|w|^2)^{\frac{2}{p}}\,.$$

If $f(w)\neq 0$, then in view of the lemma,

$$|g(w)| = |f(w)/H(w)| \le |f(w)|\left\{\prod_{k=1}^{\infty}|\varphi_{z_k}(w)|(2-|\varphi_{z_k}(w)|)\right\}^{-1}$$

$$= |F_w(0)|\,(1-|w|^2)^{-\frac{2}{p}}\left\{\prod_{k=1}^{\infty}|\zeta_k|(2-|\zeta_k|)\right\}^{-1}$$

$$\le C_p\,(1-|w|^2)^{-\frac{2}{p}}\,\|F_w\|_{p,2}\,.$$

The requirement that $f(w) \neq 0$ is now removed by continuity. We have therefore shown that

$$
\begin{aligned}
|g(w)|^p(1-|w|^2)^2 &\leq C_p^p \int_{\mathbb{D}} |f(\varphi_w(\zeta))|^p \, |\varphi_w'(\zeta)|^2 \, (1-|\zeta|^2)^2 \, d\sigma(\zeta) \\
&= C_p^p \int_{\mathbb{D}} |f(z)|^p \, (1-|\varphi_w(z)|^2)^2 \, d\sigma(z)
\end{aligned}
$$

for all $w \in \mathbb{D}$. But

$$
\frac{1-|\varphi_w(z)|^2}{1-|w|^2} = \frac{1-|z|^2}{|1-\overline{z}w|^2} = |\varphi_z'(w)| \,,
$$

so it follows from Fubini's theorem that

$$
\begin{aligned}
\|g\|_p^p &= \int_{\mathbb{D}} |g(w)|^p \, d\sigma(w) \\
&\leq C_p^p \int_{\mathbb{D}} \int_{\mathbb{D}} |f(z)|^p |\varphi_z'(w)|^2 \, d\sigma(w) \, d\sigma(z) = C_p^p \|f\|_p^p \,,
\end{aligned}
$$

which completes the proof of Theorem 5. □

PROOF OF THEOREM 4. The proof of Theorem 5 is easily adapted to show that $f/H \in A^p$ if H is the Horowitz product constructed over an arbitrary subset of the zero-set $\{z_k\}$ of f. But f/H is then an A^p function whose zero-set is precisely the complementary subset of $\{z_k\}$, so the theorem is proved. □

For later reference we now record a result slightly stronger than Theorem 5, actually implicit in the proof.

COROLLARY. *Let $f \in A^p$, let $\{z_k\}$ be its zero-set, and let*

$$
H^*(z) = \prod_{k=1}^{\infty} |\varphi_{z_k}(z)| \big(2 - |\varphi_{z_k}(z)|\big)
$$

Then $f/H^* \in L^p(\mathbb{D})$ and $\|f/H^*\|_p \leq C_p \, \|f\|_p$.

One disadvantage of the Horowitz product is that it need not belong to the space A^p. It is an interesting open problem to determine necessary and sufficient conditions on $\{z_k\}$ in order that $H \in A^p$. Émile LeBlanc [1] gave a probabilistic sufficient condition for H to lie in A^2. Anton Kim [1] exhibited a specific class of Horowitz products generated by A^p zero-sets but not belonging to A^p. The probabilistic approach to A^p zero-sets has been further developed by Bomash [1], Horowitz [4], and by Nowak and Waniurski [1].

As we shall see in the next chapter, there is actually a *contractive zero-divisor* associated with each A^p zero-set $\{z_k\}$. This is a function $G \in A^p$ of unit norm that vanishes precisely on the given zero-set $\{z_k\}$ and has the contractive property $\|f/G\|_p \leq \|f\|_p$ for every function $f \in A^p$ whose zero-set includes $\{z_k\}$. Contractive divisors are natural Bergman-space analogues of Blaschke products, but they are difficult to express analytically, whereas Horowitz products are totally explicit.

§4.5. Blaschke products as generators.

Given a Blaschke product B, let $[B]$ denote the closure in the Bergman space A^p of the set of all multiples of B by polynomials. Let N^p denote the set of functions in A^p that vanish on the zero-set of B, to at least the specified multiplicity. Because point-evaluation is a bounded linear functional on A^p, it is clear that N^p is a (closed) subspace of A^p. Since polynomials are dense in A^p, it is reasonable to expect the subspace $[B]$ to coincide with N^p. In other words, the polynomial multiples of B should generate all of N^p. This turns out to be true, but the result is not trivial. We shall give a proof due to H. S. Shapiro [4], later reproduced in a paper of Duren, Khavinson, and Shapiro [1].

THEOREM 6. *Let p be fixed arbitrarily, $0 < p < \infty$. For any Blaschke product B, the subspace $[B]$ generated in A^p by polynomial multiples of B coincides with the subspace N^p of all functions in A^p that vanish on the zero-set of B.*

PROOF. It is clear that $[B] \subset N^p$, so we have to show that each $f \in N^p$ can be approximated by polynomial multiples of B. Write $B = B_n R_n$, where B_n is the partial product consisting of the first n factors of B. Then $R_n(z) \to 1$ as $n \to \infty$ for each $z \in \mathbb{D}$, while $\|R_n\|_\infty = 1$. By the Lebesgue dominated convergence theorem, this shows that $\|R_n f - f\|_p \to 0$ for each $f \in A^p$. Given $\varepsilon > 0$, choose n so large that $\|R_n f - f\|_p < \frac{\varepsilon}{2}$. Because $f/B_n \in A^p$, there is a polynomial Q such that $\|Q - f/B_n\|_p < \frac{\varepsilon}{2}$. Thus $\|BQ - R_n f\|_p < \frac{\varepsilon}{2}$, since $\|Bg\|_p \leq \|g\|_p$ for every function $g \in A^p$. Now for $1 \leq p < \infty$, the triangle inequality gives $\|BQ - f\|_p < \varepsilon$. In the range $0 < p < 1$ the triangle inequality is not strictly valid but can be replaced by the inequality

$$\|f + g\|_p^p \leq \|f\|_p^p + \|g\|_p^p .$$

This completes the proof. □

It is the content of Theorem 5 that $N^p \subset HA^p$ for any A^p zero-set, where H is the corresponding Horowitz product and HA^p denotes the set of all multiples of H by functions in A^p. The reverse inclusion need not be valid because H is generally not a bounded function and it may not even belong to the space A^p. On the other hand, for a Blaschke product B the inclusion $BA^p \subset N^p$ always holds. It is therefore tempting to carry Theorem 6 a step farther and conjecture that $BA^p = N^p$ for all Blaschke products B. This is not always true, however, because the set BA^p need not be *closed.* Whenever BA^p is closed, the inclusions $[B] \subset BA^p \subset N^p$, in combination with Theorem 6, will show that $[B] = BA^p = N^p$.

We record this last remark for future reference.

COROLLARY. *For any Blaschke product B, the identity $BA^p = N^p$ is valid if and only if BA^p is a closed set.*

It is a theorem of Horowitz, to be proved in the next section, that $N^p = BA^p$ if and only if the zero-set of B is a finite union of uniformly separated sequences.

§4.6. Universal divisors.

We have seen that Horowitz products

$$H(z) = \prod_{k=1}^{\infty} b_{z_k}(z)\big(2 - b_{z_k}(z)\big).$$

act as *universal divisors* of the Bergman space A^p for $0 < p < \infty$. This means specifically that $\|f/H\|_p \le C_p \|f\|_p$ for all functions $f \in A^p$ that vanish on the prescribed zero-set $\{z_k\}$, where C_p denotes a constant depending only on p. The question arises whether Blaschke products play a similar role when $\{z_k\}$ is a Blaschke sequence. In other words, if $\sum (1 - |z_k|) < \infty$ and

$$B(z) = \prod_{k=1}^{\infty} b_{z_k}(z)$$

is the corresponding Blaschke product, is it true that $\|f/B\|_p \le C_p \|f\|_p$ for all functions $f \in A^p$ that vanish on $\{z_k\}$? A Blaschke product with this property for a given A^p space will be called a *universal divisor* for A^p. By the closed graph theorem, it is equivalent to require that $f/B \in A^p$ whenever f vanishes on $\{z_k\}$.

It turns out that a Blaschke product need not be a universal divisor for any Bergman space A^p, but it is a universal divisor for all A^p if its zeros are sufficiently separated. A Blaschke sequence $\{z_k\}$ is said to be *uniformly separated* if

$$\prod_{j \ne k} \left| \frac{z_j - z_k}{1 - \overline{z_j} z_k} \right| \ge \delta\,, \qquad k = 1, 2, \dots\,,$$

for some constant $\delta > 0$ independent of k. The complete description of universal divisors, again due to Horowitz [3], is as follows.

THEOREM 7. *For any Blaschke product B, the following statements are equivalent.*

(i) B is a universal divisor for some space A^p, $0 < p < \infty$.

(ii) B is a universal divisor for every space A^p.

(iii) The zero-set of B is a finite union of uniformly separated sequences.

The most difficult part of the proof is to establish the necessity of the condition *(iii)*. The argument will be embedded in the proof of a broader result (Theorem 10 below).

By way of orientation, we now recall some facts about interpolation in the Hardy spaces H^p. A sequence $\{z_k\}$ in the disk is a *universal interpolation sequence* for H^∞ if for each complex sequence $\{w_k\} \in \ell^\infty$ there exists a function $f \in H^\infty$ with $f(z_k) = w_k$ for $k = 1, 2, \ldots$. Carleson [2] proved that $\{z_k\}$ is a universal interpolation sequence for H^∞ if and only if it is uniformly separated. Shapiro and Shields [1] then generalized the theorem to arbitrary H^p spaces ($1 \le p < \infty$) by showing that the operator

$$T_p(f) = \{(1 - |z_k|^2)^{\frac{1}{p}} f(z_k)\}$$

maps H^p onto ℓ^p if and only if $\{z_k\}$ is uniformly separated. Kabaĭla [1,2] extended the result to $0 < p < 1$. One aspect of the theorem is that

$$\sum_{k=1}^{\infty} (1 - |z_k|^2)|f(z_k)|^p < \infty \qquad \text{for every } f \in H^p$$

if $\{z_k\}$ is uniformly separated.

Observe now that the convergence of this sum for every function $f \in H^p$ is equivalent to saying that the discrete measure

$$\mu = \sum_{k=1}^{\infty} (1 - |z_k|^2)\delta_{z_k}$$

is a *Carleson measure* for H^p, a property independent of p ($0 < p < \infty$), as noted in Section 2.10. McDonald and Sundberg [1] completed this result to the following theorem.

THEOREM 8. *A sequence $\{z_k\}$ of points in $\mathbb{D}$ generates a Carleson measure $\mu = \sum(1 - |z_k|^2)\delta_{z_k}$ for H^p if and only if $\{z_k\}$ is a finite union of uniformly separated sequences.*

On the other hand, we will show that the following theorem of Horowitz [3], used in his proof of Theorem 7, implies and is essentially equivalent to the McDonald–Sundberg theorem.

THEOREM 9. *A sequence $\{z_k\}$ of points in $\mathbb{D}$ has the property*

$$\sup_{z \in \mathbb{D}} \sum_{k=1}^{\infty} (1 - |b_{z_k}(z)|) < \infty$$

if and only if it is a finite union of uniformly separated sequences.

It is convenient to establish all three of these results together (Theorems 7, 8, and 9) by proving the following theorem.

THEOREM 10. *For a Blaschke sequence* $\{z_k\}$ *of points in* $\mathbb{D}$, *the following five statements are equivalent.*

(i) $\{z_k\}$ *is a finite union of uniformly separated sequences.*

(ii) $\sum_{k=1}^{\infty}(1-|z_k|^2)|f(z_k)|^p < \infty$ *for all* $f \in H^p$.

(iii) $\sup_{\zeta\in\mathbb{D}} \sum_{k=1}^{\infty}(1-|\varphi_\zeta(z_k)|) < \infty$, *where* $\varphi_\zeta(z) = \frac{\zeta - z}{1-\bar{\zeta}z}$.

(iv) The Blaschke product B *with zero-set* $\{z_k\}$ *is a universal divisor of* A^p; *that is,* $f/B \in A^p$ *for every function* $f \in A^p$ *which vanishes on* $\{z_k\}$.

(v) The operator of multiplication by the Blaschke product B *is bounded below on* A^p; *that is,* $\|Bf\|_p \geq c\|f\|_p$ *for some constant* $c > 0$ *and every function* $f \in A^p$.

As noted above, the condition (ii) is actually independent of p. However, the last two statements are ambiguous because they do not specify the values of p for which the property is to hold. In fact, it will be clear from the proof that if either property holds for *some* $p\,(0 < p < \infty)$, then it holds for *all* p.

PROOF OF THEOREM 10.

$(i) \implies (ii)$. It is enough to assume that f is a nonvanishing function in H^p, since the presence of a Blaschke product only decreases the sum. But then $f^{p/2} \in H^2$, so it suffices to show that

$$\sum_{k=1}^{\infty}(1-|z_k|^2)|f(z_k)|^2 < \infty$$

for all $f \in H^2$ if $\{z_k\}$ is uniformly separated. This implication is part of the Shapiro–Shields theorem cited above, but we shall give a self-contained and relatively simple proof due to Neville [1]. The idea is to consider the weighted Bergman space A_1^2 with norm defined by

$$\|f\|_{2,1}^2 = \int_{\mathbb{D}} |f(z)|^2(1-|z|^2)\,d\sigma = \sum_{n=0}^{\infty} \frac{|a_n|^2}{(n+1)(n+2)}\,,$$

where

$$f(z) = \sum_{n=0}^{\infty} a_n z^n\,, \qquad z \in \mathbb{D}\,.$$

Using power-series expansions, it is easy to verify that $\|f'\|_{2,1} \leq \|f\|_{H^2}$. If $\{z_k\}$ is uniformly separated, then it is *uniformly discrete*:

$$\rho(z_j, z_k) = \left|\frac{z_j - z_k}{1-\overline{z_j}z_k}\right| \geq \delta > 0\,, \qquad j \neq k.$$

But for arbitrary $\zeta \in \mathbb{D}$ and $0 < d < 1$, it can be checked by straightforward calculation that the Euclidean disk $D(\zeta, r)$ with center ζ and radius $r = \frac{d}{2}(1-|\zeta|^2)$ is contained in the pseudohyperbolic disk

$$\Delta(\zeta, d) = \{z \,:\, \rho(z,\zeta) < d\}.$$

Thus because the disks $\Delta(z_k, \frac{\delta}{2})$ are disjoint by the uniform discreteness of $\{z_k\}$, it follows that the disks $D_k = D(z_k, r_k)$ are disjoint, where $r_k = \frac{\delta}{2}(1-|z_k|^2)$. Now let $g \in A_1^2$ and apply the submean-value property of $|g|^2$ to write

$$|g(z_k)|^2 \le \frac{1}{r_k^2} \int_{D_k} |g(z)|^2 \, d\sigma \,,$$

so that

$$\begin{aligned}(1-|z_k|^2)^3 |g(z_k)|^2 &\le \frac{4}{\delta^2} \int_{D_k} (1-|z_k|^2)|g(z)|^2 \, d\sigma \\ &\le \frac{8}{\delta^2} \int_{D_k} (1-|z|^2)|g(z)|^2 \, d\sigma \,,\end{aligned}$$

if we assume (as we may) that $\delta < \frac{1}{2}$. Since the disks D_k are disjoint, we conclude that

$$\sum_{k=1}^{\infty} (1-|z_k|^2)^3 |g(z_k)|^2 \le \frac{8}{\delta^2} \|g\|_{2,1}^2 \,.$$

Finally, for any function $f \in H^2$, observe that $g = (Bf)' \in A_1^2$ and

$$|g(z_k)| = |B'(z_k) f(z_k)| = \frac{|B_k(z_k) f(z_k)|}{1-|z_k|^2} \,,$$

where $B_k = B/b_{z_k}$. But $|B_k(z_k)| \ge \delta$ since $\{z_k\}$ is uniformly separated, and so

$$\begin{aligned}\delta^2 \sum_{k=1}^{\infty} (1-|z_k|^2)|f(z_k)|^2 &\le \sum_{k=1}^{\infty} (1-|z_k|^2)^3 |g(z_k)|^2 \le \frac{8}{\delta^2} \|g\|_{2,1}^2 \\ &= \frac{8}{\delta^2} \|(Bf)'\|_{2,1}^2 \le \frac{8}{\delta^2} \|Bf\|_{H^2}^2 = \frac{8}{\delta^2} \|f\|_{H^2}^2 \,,\end{aligned}$$

which completes the first step of the proof.

(ii) $\implies$ *(iii)*. Now we assume that

$$\sum_{k=1}^{\infty} (1-|z_k|^2)|f(z_k)|^2 \le C \, \|f\|_{H^2}^2$$

for all functions $f \in H^2$. Choosing

$$f(z) = f_\zeta(z) = \frac{1-|\zeta|^2}{1-\overline{\zeta} z} \,, \qquad \zeta \in \mathbb{D} \,,$$

with $\|f_\zeta\|_{H^2}^2 = 1 - |\zeta|^2$, we infer that

$$\begin{aligned}\sum_{k=1}^{\infty} (1-|\varphi_\zeta(z_k)|^2) &= \sum_{k=1}^{\infty} \frac{(1-|\zeta|^2)(1-|z_k|^2)}{|1-\overline{\zeta} z_k|^2} \\ &= \frac{1}{1-|\zeta|^2} \sum_{k=1}^{\infty} (1-|z_k|^2)|f_\zeta(z_k)|^2 \le \frac{C}{1-|\zeta|^2} \|f_\zeta\|_{H^2}^2 = C\end{aligned}$$

for all $\zeta \in \mathbb{D}$, which gives (iii).

$(iii) \implies (iv)$. Suppose now that (iii) holds, let B be the Blaschke product of $\{z_k\}$, and let f be a function in A^p which vanishes on $\{z_k\}$. By the corollary to Theorem 5, division by the product H^* produces a function $f/H^* \in L^p(\mathbb{D})$. But now

$$\frac{|f(z)|}{|B(z)|} = \frac{|f(z)|}{H^*(z)} \prod_{k=1}^{\infty} (2 - |\varphi_z(z_k)|) ,$$

so it will be sufficient to show that the infinite product is uniformly bounded for $z \in \mathbb{D}$. But this is a consequence of property (iii), since

$$\log\left(\prod_{k=1}^{\infty} (2 - |\varphi_z(z_k)|) \right) \le \sum_{k=1}^{\infty} (1 - |\varphi_z(z_k)|) .$$

Thus (iv) holds if (iii) does.

$(iv) \implies (v)$. In Section 2.9 (Theorem 13) we showed, using the closed graph theorem, that multiplication by a nonzero function $\psi \in H^\infty$ is bounded below on A^p if and only if the set ψA^p is closed. On the other hand, it is true for any Blaschke product B that BA^p is contained in the subspace N^p of all functions in A^p that vanish on the zero-set of B. Property (iv) says that $N^p \subset BA^p$, so that $BA^p = N^p$ and BA^p is therefore closed. Thus a sequence $\{z_k\}$ has property (v) whenever it has property (iv). (Note that one could also prove directly that $(v) \implies (iv)$. Theorem 6 in Section 4.5 says that $[B] = N^p$ for any Blaschke product, whereas $[B] \subset BA^p \subset N^p$ if BA^p is closed. Thus $BA^p = N^p$ if BA^p is closed, which shows that $(v) \implies (iv)$.)

$(v) \implies (iii)$. This part of the argument follows Horowitz [3]. Under the assumption that (iii) does *not* hold, or equivalently that

$$\sup_{z \in \mathbb{D}} \sum_{k=1}^{\infty} (1 - |\varphi_{z_k}(z)|^2) = \infty ,$$

we shall produce a family of functions $f_\zeta \in A^p$ with $\|f_\zeta\|_p = 1$ for all $\zeta \in \mathbb{D}$, such that

$$\inf_{\zeta \in \mathbb{D}} \|Bf_\zeta\|_p = 0 .$$

Define $f_\zeta = (\varphi_\zeta')^{2/p}$ and observe that

$$\|f_\zeta\|_p^p = \int_{\mathbb{D}} |\varphi_\zeta'(z)|^2 \, d\sigma(z) = \int_{\mathbb{D}} d\sigma(w) = 1 ,$$

where $w = \varphi_\zeta(z)$. Also note that

$$\|Bf_\zeta\|_p = \|B \circ \varphi_\zeta\|_p ,$$

since $\varphi_\zeta^{-1} = \varphi_\zeta$. But the composition $B \circ \varphi_\zeta$ is, up to rotation, a Blaschke product with zeros at the points $\zeta_k = \varphi_\zeta(z_k)$. Thus

$$\begin{aligned}\log |B(\varphi_\zeta(z))| &= \tfrac{1}{2}\sum_{k=1}^{\infty} \log |\varphi_{\zeta_k}(z)|^2 \le \tfrac{1}{2}\sum_{k=1}^{\infty}\{|\varphi_{\zeta_k}(z)|^2 - 1\} \\ &= -\tfrac{1}{2}(1-|z|^2)\sum_{k=1}^{\infty}\frac{1-|\zeta_k|^2}{|1-\overline{\zeta_k}z|^2} \le -\tfrac{1}{8}(1-r^2)\sum_{k=1}^{\infty}(1-|\zeta_k|^2)\end{aligned}$$

for $|z| \le r < 1$. Consequently, given any $\varepsilon > 0$ we may choose $M > 0$ so large that $\|B \circ \varphi_\zeta\|_p < \varepsilon$ when

$$\sum_{k=1}^{\infty}(1-|\zeta_k|^2) = \sum_{k=1}^{\infty}(1-|\varphi_{z_k}(\zeta)|^2) > M\,.$$

Specifically, we have only to choose $r < 1$ so that

$$\int_{r<|z|<1} |B(\varphi_\zeta(z))|^p \, d\sigma \le \int_{r<|z|<1} d\sigma = 1 - r^2 < \tfrac{1}{2}\varepsilon^p$$

for all $\zeta \in \mathbb{D}$, and then to choose M large enough to ensure that

$$\int_{|z|\le r} |B(\varphi_\zeta(z))|^p \, d\sigma < \tfrac{1}{2}\varepsilon^p$$

for any ζ with $\zeta_k = \varphi_\zeta(z_k)$ satisfying $\sum(1-|\zeta_k|^2) > M$. Adding the two integrals, we see that $\|B \circ \varphi_\zeta\|_p^p < \varepsilon^p$ for some $\zeta \in \mathbb{D}$. Thus $\|Bf_\zeta\|_p < \varepsilon$, while $\|f_\zeta\|_p = 1$. This shows that (v) does not hold if (iii) does not.

$(iii) \implies (i)$. This final stage of the proof will consist of two main steps. The first step is to show that a sequence $\{z_k\}$ with property (iii) is a finite union of uniformly discrete sequences. The second step is to show that any uniformly discrete sequence with property (iii) is in fact uniformly separated. To carry out the first step, we begin by recalling that if for some $\delta > 0$ each pseudohyperbolic disk $\Delta(\zeta, \delta)$ contains at most N points of the sequence $\{z_k\}$, then $\{z_k\}$ is the union of at most N uniformly discrete sequences. (See Section 2.11, Lemma 16.) Thus if $\{z_k\}$ is *not* a finite union of uniformly discrete sequences, there must exist a sequence $\{\zeta_m\}$ of points in $\mathbb{D}$ for which $N_m \to \infty$ as $m \to \infty$, where N_m is the number of points z_k in the disk $\Delta(\zeta_m, \frac{1}{2})$. Consequently,

$$\sum_{k=1}^{\infty}(1-|\varphi_{z_k}(\zeta_m)|^2) \ge \sum_{z_k \in \Delta(\zeta_m, \frac{1}{2})}(1-|\varphi_{z_k}(\zeta_m)|^2) \ge \tfrac{3}{4}N_m \to \infty$$

as $m \to \infty$, contradicting the property (iii). Thus (iii) implies that $\{z_k\}$ is a finite union of uniformly discrete sequences.

Finally, suppose that (iii) holds and $\{z_k\}$ is uniformly discrete, so that

$$\sum_{k=1}^{\infty}(1-|\varphi_{z_k}(z)|^2) \leq M < \infty$$

and $|\varphi_{z_k}(z_j)| \geq \delta > 0$ for $j \neq k$. Then in view of the inequality $-\log x \leq \frac{1}{x}(1-x)$ for $x > 0$, we see that

$$\sum_{k\neq j} -\log\left|\varphi_{z_k}(z_j)\right|^2 \leq \frac{1}{\delta^2}\sum_{k\neq j}\left(1-\left|\varphi_{z_k}(z_j)\right|^2\right) \leq \frac{M}{\delta^2}\,.$$

Exponentiating, we conclude that

$$\prod_{k\neq j}\left|\varphi_{z_k}(z_j)\right| \geq e^{-M/2\delta^2} > 0\,, \qquad j = 1, 2, \dots\,,$$

which says that $\{z_k\}$ is uniformly separated. Combining the two main steps of the proof, we conclude that any sequence $\{z_k\}$ with the property (iii) is a finite union of uniformly separated sequences. This completes the proof of Theorem 10. □

For an alternate proof that properties (ii) and (iii) are equivalent, one can invoke the theory of Carleson measures. By appeal to Carleson's geometric description, it is not difficult to show (*cf.* Garnett [1], p. 239) that a positive measure μ is a Carleson measure if and only if

$$\sup_{z\in\mathbb{D}}\int_{\mathbb{D}}\frac{1-|z|^2}{|1-\overline{\zeta}z|^2}\,d\mu(\zeta) < \infty\,.$$

Applying this criterion to the discrete measure

$$\mu = \sum_{k=1}^{\infty}(1-|z_k|^2)\delta_{z_k}$$

and using the standard identity

$$1-|\varphi_{z_k}(z)|^2 = \frac{(1-|z|^2)(1-|z_k|^2)}{|1-\overline{z_k}z|^2}\,,$$

one sees immediately that (ii) and (iii) are equivalent.

McDonald and Sundberg [1] gave a stronger form of the implication $(v) \implies (i)$, showing that multiplication by an arbitrary inner function ψ is bounded below on A^p if and only if ψ is a Blaschke product whose zero-set is a finite union of uniformly separated sequences. Luecking [1] carried the problem further, showing that multiplication by a function $\psi \in H^\infty$ is bounded below on A^p if and only if

$$\sigma\big(z \in \mathbb{D} \cap \Delta \;:\; |\psi(z)| > \varepsilon\big) > \delta\, \sigma(\mathbb{D} \cap \Delta)$$

for some positive constants δ, ε and for all disks Δ centered on the unit circle.

A similar discussion of these topics may be found in Duren and Schuster [1]. Theorem 10 should be compared with Theorem 15 in Section 2.11, which gives corresponding characterizations of finite unions of uniformly *discrete* sequences.

§4.7. Perturbations of zero-sets.

Our objective is now to show that A^p zero-sets are preserved under certain perturbations. The first result of this type is due to Horowitz [3].

THEOREM 11. *If $\{z_k\}$ is an A^p zero-set for some p, $0 < p < \infty$, then a sequence $\{w_k\}$ of points in $\mathbb{D}$ is also an A^p zero-set if*

$$\sum_{k=1}^{\infty} \left| \frac{w_k - z_k}{1 - \overline{w_k} z_k} \right| < \infty .$$

PROOF. Given a function $f \in A^p$ with zero-set $\{z_k\}$, the idea is to show that

$$g(z) = f(z) \prod_{k=1}^{\infty} \frac{b_{w_k}(z)}{b_{z_k}(z)}$$

belongs to A^p and has zero-set $\{w_k\}$. It can be verified that the infinite product converges uniformly on compact subsets of the disk with punctures at the points z_k, under the weaker assumption that $\sum |z_k - w_k| < \infty$. Indeed, if $|z| \le R < 1$ and

$$\left| \frac{z_k - z}{1 - \overline{z_k} z} \right| \ge \delta > 0 \qquad k = 1, 2, \dots ,$$

a straightforward but laborious calculation shows that

$$1 - \left| \frac{b_{w_k}(z)}{b_{z_k}(z)} \right|^2 \le \frac{6\,|z_k - w_k|}{|z_k - z|^2 |1 - \overline{w_k} z|^2} \le \frac{6\,|z_k - w_k|}{\delta^2 (1 - R)^4} .$$

Since the hypothesis of the theorem implies that $\sum |z_k - w_k| < \infty$, this shows that the infinite product converges absolutely and uniformly on each closed subset of $\mathbb{D} \setminus \{z_1, z_2, \dots\}$, and so represents a meromorphic function with zeros w_k and poles z_k, multiplicities respected. But the function f has zero-set $\{z_k\}$, so it follows that g is analytic in $\mathbb{D}$ and has zero-set $\{w_k\}$.

In order to show that $g \in A^p$, we appeal to the corollary to Theorem 5, which asserts that the function f/H^* belongs to $L^p(\mathbb{D})$. Now

$$|g(z)| = \frac{|f(z)|}{H^*(z)} \prod_{k=1}^{\infty} |b_{w_k}(z)|(2 - |b_{z_k}(z)|)\,,$$

so it will suffice to show that the function

$$\prod_{k=1}^{\infty} |b_{w_k}(z)|(2 - |b_{z_k}(z)|)$$

is bounded on the unit disk. But the inequality $\log x \le x - 1$ shows that

$$\log\left\{\prod_{k=1}^{\infty} |b_{w_k}(z)|(2 - |b_{z_k}(z)|)\right\} = \sum_{k=1}^{\infty}\{\log|b_{w_k}(z)| + \log(2 - |b_{z_k}(z)|)\}$$

$$\le \sum_{k=1}^{\infty}\{|b_{w_k}(z)| - |b_{z_k}(z)|\} = \sum_{k=1}^{\infty}\{|b_z(w_k)| - |b_z(z_k)|\}$$

$$\le 2\sum_{k=1}^{\infty}\left|\frac{b_z(w_k) - b_z(z_k)}{1 - \overline{b_z(w_k)}b_z(z_k)}\right| = 2\sum_{k=1}^{\infty}\left|\frac{w_k - z_k}{1 - \overline{w_k}z_k}\right| < \infty\,,$$

by the conformal invariance of the pseudohyperbolic metric

$$\rho(z, w) = \left|\frac{w - z}{1 - \overline{w}z}\right|.$$

This completes the proof. □

Theorem 11 can be improved. Luecking [4] has developed new descriptions of A^p zero-sets that allow Horowitz' result to be sharpened as follows.

THEOREM 12. *If $\{z_k\}$ is an A^p zero-set for some p, $0 < p < \infty$, then a sequence $\{w_k\}$ of points in $\mathbb{D}$ is also an A^p zero-set if*

$$\sum_{k=1}^{\infty} |w_k - z_k| < \infty \qquad \text{and} \qquad \sup_k \rho(z_k, w_k) < 1\,.$$

One of Luecking's characterizations of A^p zero-sets is stated in terms of the subharmonic function

$$k(z) = \frac{|z|^2}{2}\sum_{k=1}^{\infty}\frac{(1 - |z_k|^2)^2}{|1 - \overline{z_k}z|^2}\,, \qquad z \in \mathbb{D}\,.$$

He shows that $\{z_k\}$ is an A^p zero-set ($0 < p < \infty$) if and only if there is a harmonic function $h(z)$ such that $e^{pk+h} \in L^1(\mathbb{D})$. The deduction of Theorem

12 then proceeds by showing that an admissible perturbation of a zero-set $\{z_k\}$ alters the function $k(z)$ by an additive harmonic function which does not destroy the integrability of e^{pk+h}. Details may be found in Luecking's paper.

The condition in Theorem 11 clearly implies both of the conditions in Theorem 12. Examples show, however, that neither of the conditions in Theorem 12 is alone sufficient to guarantee that the perturbed set $\{w_k\}$ is again an A^p zero-set.

§4.8. Zeros on a radial line.

Although every Blaschke sequence is a zero-set for every Bergman space A^p, we know that the converse is false. We shall now show, however, that if the zeros of a function in a Bergman space lie on a single radial line, they must form a Blaschke sequence. Since zero-sets need not be Blaschke sequences, this shows that a complete description cannot be given in terms of modulus alone, but must take the angular distribution into account. The result is actually a very special consequence of the following more general theorem, due to Shapiro and Shields [2].

THEOREM 13. *Let f be analytic in $\mathbb{D}$, with $f(z) \not\equiv 0$, satisfying a growth condition of the form*

$$|f(z)| \le C \exp\left\{ \frac{1}{(1-|z|)^\alpha} \right\},$$

for some constants $\alpha < \frac{1}{2}$ and $C > 0$. Suppose the zero-set of f contains a sequence $\{z_k\}$ on the positive real axis: $0 < z_1 \le z_2 \le \ldots < 1$. Then $\sum_{k=1}^\infty (1 - z_k) < \infty$.

COROLLARY. *If $f \in A^p$ for some $p > 0$ and its zeros z_k all lie on one ray emanating from the origin, then they form a Blaschke sequence: $\sum_{k=1}^\infty (1-|z_k|) < \infty$.*

DEDUCTION OF COROLLARY. After a rotation we may assume that all zeros are real and positive. If $f \in A^p$, then

$$|f(z)| \le \frac{c}{(1-|z|)^{\frac{2}{p}}}, \qquad z \in \mathbb{D},$$

for some constant $c > 0$, as shown in Section 3.2, so the growth condition of the theorem is satisfied with plenty to spare.

PROOF OF THEOREM. Let D be the disk bounded by the circle $|z - \frac{1}{2}| = \frac{1}{2}$, which is internally tangent to the unit circle at the point $z = 1$. If $z = \frac{1}{2} + re^{i\theta}$ is a point on the concentric circle C_r defined by $|z - \frac{1}{2}| = r < \frac{1}{2}$, then

$$1 - |z|^2 = \tfrac{3}{4} - r^2 - r\cos\theta \ge \tfrac{1}{2}(1-\cos\theta) \ge k\theta^2, \qquad -\pi < \theta < \pi,$$

for some constant $k > 0$. Therefore,

$$\int_{C_r} \log^+ |f(z)|\,|dz| \le b \int_{C_r} \frac{|dz|}{(1-|z|^2)^\alpha} \le b \int_{-\pi}^{\pi} \frac{d\theta}{(k\theta^2)^\alpha}$$

for some constant $b > 0$, and the last integral converges since $\alpha < \frac{1}{2}$. Thus f belongs to the Nevanlinna class for the disk D, and so (*cf.* Section 3.1) its zeros form a Blaschke sequence with respect to the boundary circle $C_{\frac{1}{2}}$. In particular, its positive zeros satisfy $\sum (1 - z_k) < \infty$. □

The hypothesis that the zeros lie on a ray can be replaced by the weaker assumption that they lie in a Stolz angle with vertex at a point of the unit circle. Moreover, the theorem remains true for $\alpha < 1$. The proof is essentially the same, as Shapiro and Shields observe, except that the disk D must now be replaced by a region making a lower order of contact with the unit circle. Shvedenko [2] has supplied a detailed construction. In fact, the result for $\alpha < 1$ is very nearly best possible. Hayman and Korenblum [1,2] considered the more general family A^ψ of analytic functions f with growth restriction $\log^+ |f(z)| \le \psi(|z|)$, where $\psi(r)$ is a nondecreasing function with $\psi(r) \to \infty$ as $r \to 1$. They showed that if $f \in A^\psi$ and

$$\int_0^1 \sqrt{\frac{\psi(r)}{1-r}}\, dr < \infty\,,$$

then any subset of zeros of f which lie in a Stolz angle must be a Blaschke sequence. Conversely, they show that the condition is best possible. This means that whenever the integral diverges, there is a function of class A^ψ whose positive zeros fail to satisfy the Blaschke condition.

Korenblum [1] made a penetrating study of A^p zero-sets and found further conditions that take account of angular distribution. Seip [4,5] used more refined methods to obtain sharper results. An exposition of these developments can be found in the book of Hedenmalm, Korenblum, and Zhu [2]. Along similar lines, Seip [3] characterized the A^p interpolation sets, which are in particular A^p zero-sets, in terms of certain densities. An account of this theory will be given in Chapters 6 and 7.

CHAPTER 5

Contractive Zero-Divisors

For $0 < p \leq \infty$, the zeros of a function f in the Hardy space H^p satisfy the Blaschke condition $\sum(1 - |z_k|) < \infty$. The corresponding Blaschke product $B(z)$ has a radial limit of unit modulus in almost every direction, and so B has unit H^p norm. Furthermore, it has precisely the same zeros as the function f, multiplicities counted. A basic theorem of F. Riesz (1923) asserts that the nonvanishing function f/B again belongs to H^p and its norm is the same as that of f. In other words, the Blaschke product is an *isometric zero-divisor* of unit norm. This theorem of Riesz is not difficult to prove, but it plays an essential role in the theory of H^p spaces. It would be very useful to have a similar result for Bergman spaces, but it turns out that no Bergman space A^p can have isometric zero-divisors. Nevertheless, there is a canonical contractive zero-divisor G associated with each A^p zero-set. It vanishes precisely on the given zero-set, has norm $\|G\|_p = 1$, and has the property $\|f/G\|_p \leq \|f\|_p$ for every function $f \in A^p$ that vanishes on the zero-set of G. It is unique up to rotation and has analytic continuation properties similar to those of a Blaschke product. The present chapter will be devoted to the construction of these canonical divisors, which can be viewed as analogues of Blaschke products for the Bergman spaces.

§5.1. An extremal problem.

In searching for an appropriate analogue of Blaschke products in the Bergman space A^2, Håkan Hedenmalm [1] had the idea of posing an extremal problem whose counterpart in the Hardy space produces Blaschke products. This led ultimately to the construction of a contractive zero-divisor in A^2. Following the main lines of Hedenmalm's argument, Duren, Khavinson, Shapiro, and Sundberg [1,2] then extended the construction to A^p for $0 < p < \infty$. The more general version required some new devices, since Hedenmalm's construction for $p = 2$ had relied heavily on Hilbert space techniques and explicit formulas involving the Bergman kernel function. This chapter will focus on the theory for A^p.

Let $\{z_k\}$ be the sequence of zeros (repeated according to multiplicity) of some function in A^p, for $0 < p < \infty$. We make the standing assumption that $z_k \neq 0$ for all k. Let N^p denote the set of all functions in A^p that vanish to at least the prescribed order at each point z_k. Then the basic extremal problem is

$$\max_{f\in N^p,\,\|f\|_p=1} |f(0)|\,, \tag{1}$$

to find the maximum value of $|f(0)|$ for all functions $f \in A^p$ of unit norm that vanish on the prescribed zero-set. In order to see that an extremal function exists and is unique up to rotation, we can pass to the equivalent problem

$$\min_{f\in N^p,\,f(0)=1} \|f\|_p \tag{2}$$

of minimizing $\|f\|_p$ among all $f \in N^p$ with $f(0) = 1$. The two problems are equivalent in the sense that a function g of unit norm is an extremal function for problem (1) if and only if $g/g(0)$ is an extremal function for problem (2). For $1 < p < \infty$, the alternate form (2) of the problem has the advantage of being posed over a closed convex subset of a uniformly convex Banach space, so a standard theorem of functional analysis (*cf.* Section 2.2) guarantees the existence and uniqueness of a function of minimum norm.

An alternate proof of existence, valid for $0 < p < \infty$, proceeds by a normal family argument. Specifically, let M denote the supremum of $|f(0)|$ over all $f \in N^p$ with $\|f\|_p = 1$. Note that $M \le 1$, since $|f(0)| \le \|f\|_p$ for all $f \in A^p$. Let $\{f_n\}$ be a sequence of functions $f_n \in N^p$ with $\|f_n\|_p = 1$ and $|f_n(0)| \to M$ as $n \to \infty$. Since the functions f_n are locally uniformly bounded (*cf.* Theorem 1 of Chapter 1), they form a normal family by Montel's theorem, and so a subsequence $\{f_{n_k}\}$ converges locally uniformly to a function f analytic in $\mathbb{D}$. By Fatou's lemma, $f \in A^p$ and $\|f\|_p \le 1$. It is clear also that $f \in N^p$, and that $|f(0)| = M$. Thus $\|f\|_p = 1$, and f is an extremal function for the problem (1).

Uniqueness of an extremal function can be established for $p = 1$ by passing to the equivalent problem (2) and appealing to the fact that A^1 is a strictly convex space (see Section 2.2). For $0 < p < 1$ the proof of uniqueness is more delicate, but it will follow ultimately from the "contractive divisor" property. The unique solution G to the original extremal problem, normalized so that $G(0) > 0$, is called the *canonical divisor* of the zero-set $\{z_k\}$. It has the properties $G \in N^p$ and $\|G\|_p = 1$.

A few more preliminary remarks are in order. First observe that if the original extremal problem (1) is posed with respect to the H^p norm, where $0 < p < \infty$, the extremal function is a Blaschke product. This is best seen by referring to the canonical factorization (see Section 3.1 and Duren [5], Chapter 2). Given a Blaschke sequence $\{z_k\}$ with $z_k \neq 0$, let $\mathcal{N}^p$ denote the subspace of functions in H^p that vanish on $\{z_k\}$ to at least the prescribed multiplicity. The functions $f \in \mathcal{N}^p$ are characterized as products $f = e^{i\gamma}BSF$, where $e^{i\gamma}$ is a unimodular constant, B is a Blaschke product that vanishes on $\{z_k\}$ to at least the prescribed multiplicity, S is an arbitrary singular inner function, and F is an arbitrary outer function for H^p. Thus $|f(0)|$ is maximized by making each of the factors $B(0)$, $S(0)$, and $F(0)$ as large as possible, subject only to the constraints that $B \in \mathcal{N}^p$

and $\|f\|_{H^p} = 1$. But $S(0)$ is clearly maximized by choosing the singular measure $\mu = 0$; while the arithmetic–geometric mean inequality shows that $F(0) \le \|f\|_{H^p} = 1$, with equality if and only if the boundary function $f(e^{it})$ is constant almost everywhere. Thus the maximum occurs for a Blaschke product, and since $B(0)$ is the product of the moduli of zeros, it will be largest when all but the required zeros are discarded.

The stipulation that $z_k \neq 0$ is not essential and can be removed by maximizing a suitable derivative at the origin. For $p = 2$ and a *finite* zero-set, the extremal problem was discussed in more general form in Chapter 1 (Section 1.4). Appealing to the more general result established there, we see that the extremal function for a finite set of simple zeros $z_1, z_2, \dots, z_n$ is a linear combination of the Bergman kernel functions $K(z, 0) \equiv 1$ and $K(z, z_1), \dots, K(z, z_n)$. If a zero of order m is prescribed at some point z_k, the spanning set will include the derivatives $K^{(j)}(z, z_k)$ of the corresponding kernel function for $j = 0, 1, \dots, m-1$. In particular, the canonical divisor in A^2 of a *finite* zero-set has an analytic continuation to a larger disk.

This property of analytic continuation generalizes to $0 < p < \infty$, as was shown by DKSS [1,2] (shorthand for Duren–Khavinson–Shapiro–Sundberg). However, their proof depended heavily on a deep result in partial differential equations, the phenomenon of elliptic regularity. In order to make the present exposition self-contained, we shall take a more circuitous route, arriving eventually at the analytic continuation result by exploiting a representation of canonical divisors in terms of certain reproducing kernels. Meanwhile, our preliminary results will utilize a basic integral formula, proved in Section 5.3, that makes no assumption of boundary regularity and actually applies to arbitrary invariant subspaces of A^p. A consequence of the integral formula is that every canonical divisor is an expansive multiplier.

Still in force is the convention that σ denotes *normalized* area measure, with $\int_{\mathbb{D}} d\sigma = 1$. The A^p norm of f is defined by $\|f\|_p^p = \int_{\mathbb{D}} |f(z)|^p \, d\sigma$. In the following lemma, h^∞ denotes the space of real-valued bounded harmonic functions in the disk. We shall speak of *the* canonical divisor although its uniqueness for $0 < p < 1$ will not be established until later in the chapter. Meanwhile, assertions about the canonical divisor in A^p for $0 < p < 1$ should be interpreted to mean that *every* canonical divisor has the stated property.

LEMMA 1. *For $0 < p < \infty$, let $\{z_k\}$ be an arbitrary A^p zero-set with $z_k \neq 0$, and let G be its canonical divisor in A^p. Then*

$$\int_{\mathbb{D}} \big(|G(z)|^p - 1\big) u(z) \, d\sigma = 0 \tag{3}$$

for every $u \in h^\infty$.

PROOF. Consider the equivalent extremal problem (2) of minimizing f among all $f \in A^p$ with the prescribed zeros z_k and $f(0) = 1$. The unique extremal function is $H = G/G(0)$. For any polynomial f with $f(0) = 0$, the function $H^* = H(1 + \lambda f)$ is in contention for each $\lambda \in \mathbb{C}$, since it vanishes on $\{z_k\}$, $H^*(0) = 1$, and $H^* \in A^p$. Thus by the extremal property of H,

$$\begin{aligned}\|H\|_p^p \le \|H^*\|_p^p &= \int_{\mathbb{D}} |H(z)|^p |1 + \lambda f(z)|^p \, d\sigma \\ &= \int_{\mathbb{D}} |H(z)|^p \big[1 + 2\,\mathrm{Re}\{\lambda f(z)\} + |\lambda f(z)|^2\big]^{p/2} \, d\sigma \\ &= \|H\|_p^p + p\,\mathrm{Re}\Big\{\lambda \int_{\mathbb{D}} |H(z)|^p f(z)\, d\sigma\Big\} + O(\lambda^2)\,, \qquad \lambda \to 0\,.\end{aligned}$$

It follows that

$$\mathrm{Re}\Big\{\lambda \int_{\mathbb{D}} |H(z)|^p f(z)\, d\sigma\Big\} + O(\lambda^2) \ge 0\,.$$

Now divide by $|\lambda|$ and let $\lambda \to 0$ along an arbitrary ray to conclude that

$$\int_{\mathbb{D}} |G(z)|^p f(z)\, d\sigma = 0$$

for all polynomials f with $f(0) = 0$. For arbitrary polynomials f, use the formula

$$f(0) = \int_{\mathbb{D}} f(z)\, d\sigma$$

and the fact that $\|G\|_p = 1$ to deduce that

$$\int_{\mathbb{D}} \big(|G(z)|^p - 1\big) f(z)\, d\sigma = 0\,.$$

Now take real parts to conclude that (3) holds for all harmonic polynomials u. (Apply the Cauchy–Riemann equations to see that every harmonic polynomial is the real part of an analytic polynomial.)

The extension of (3) to general $u \in h^\infty$ now proceeds by observing that h^∞ is the weak-star closure of the harmonic polynomials. This is the content of the following argument. Let $u \in h^\infty$, and apply the Lebesgue dominated convergence theorem to see that the integral (3) can be approximated by the same integral with u replaced by a dilation $u_r(z) = u(rz)$ for some $r < 1$ sufficiently close to 1. Now observe that u_r is the real part of a function f analytic in the larger disk $|z| < 1/r$, so the partial sums of its power-series expansion (real parts of the power-series expansion of f) are harmonic polynomials that approximate u_r uniformly in the closed disk. Thus the validity of (3) for harmonic polynomials implies its validity for bounded harmonic functions. Here the only relevant property of G is that $|G|^p - 1 \in L^1$. □

With suitable interpretation, the above variational argument applies also to the excluded case where the origin is prescribed as a zero of order

$m \geq 1$. The canonical divisor G is then defined by maximizing $|f^{(m)}(0)|$ over all functions $f \in N^p$ with norm $\|f\|_p = 1$, taking the derivative $G^{(m)}(0) > 0$. An equivalent problem is to minimize $\|f\|_p = 1$ for $f \in N^p$ with $f^{(m)}(0) = 1$. If H solves the latter extremal problem, then $H^* = H(1+\lambda f)$ is again in contention for each $\lambda \in \mathbb{C}$ and every polynomial f with $f(0) = 0$. Consequently, the inequality $\|H\|_p \leq \|H^*\|_p$ again holds, and the above calculations lead in the same way to the conclusion (3).

For $0 < p < \infty$, a function $\varphi \in A^p$ will be called an A^p *inner function* if $\|\varphi\|_p = 1$ and

$$\int_{\mathbb{D}} |\varphi(z)|^p z^n \, d\sigma = 0\,, \qquad n = 1, 2, \dots .$$

An equivalent requirement is that $\varphi \in A^p$ and $\int_{\mathbb{D}} (|\varphi|^p - 1) f \, d\sigma = 0$ for every polynomial f. It is the content of Lemma 1 that every canonical divisor of an A^p zero-set is an inner function. The term "A^p inner function" will be justified in Chapter 9, where it is observed that the classical inner functions for H^p spaces can be characterized in an analogous way.

A similar variational argument will now be applied to obtain further information about the canonical divisor, but only for $1 \leq p < \infty$.

LEMMA 2. *For $1 \leq p < \infty$, let $\{z_k\}$ be an A^p zero-set and let G be its canonical divisor in A^p. Let N^p denote the subspace of functions in A^p that vanish on $\{z_k\}$. Then*

$$\int_{\mathbb{D}} |G(z)|^{p-2} \overline{G(z)} h(z) \, d\sigma = 0 \tag{4}$$

for every $h \in N^p$ with $h(0) = 0$.

PROOF. Passing as before to the alternate form (2) of the extremal problem, let $H = G/G(0)$ be the unique solution. Thus H minimizes $\|f\|_p$ among all $f \in N^p$ with $f(0) = 1$. Now let $h \in N^p$ with $h(0) = 0$, let λ be a (small) complex parameter, and consider the variation $H^* = H + \lambda h$. Note that $H^* \in N^p$ and $H^*(0) = 1$, so that $\|H^*\|_p \geq \|H\|_p$ by the extremal property of H. Now write

$$\begin{aligned} |H + \lambda h|^p &= \big[\, |H|^2 + 2\,\mathrm{Re}\{\lambda \overline{H} h\} + |\lambda|^2 |h|^2 \,\big]^{p/2} \\ &= |H|^p + p\,|H|^{p-2}\,\mathrm{Re}\{\lambda \overline{H} h\} + O(\lambda^2)\,, \end{aligned}$$

and integrate to obtain

$$\|H\|_p^p \leq \|H\|_p^p + p\,\mathrm{Re}\Big\{\lambda \int_{\mathbb{D}} |H|^{p-2}\, \overline{H} h \, d\sigma\Big\} + O(\lambda^2)\,,$$

so that

$$\int_{\mathbb{D}} |H|^{p-2}\, \overline{H} h \, d\sigma = 0\,,$$

which is equivalent to the property (4) claimed for the canonical divisor. □

Note that the integral in (4) makes sense because $|G|^{p-2}\overline{G} \in L^q$, where $q = p/(p-1)$ is the conjugate index. For $1 < p < 2$, the expression $|g|^{p-2}\overline{g}$ should be taken to vanish wherever g does. It may also be written as $|g|^{p-1}\overline{\operatorname{sgn} g}$.

On the other hand, Lemma 2 runs into trouble if $0 < p < 1$, since it is then no longer clear that $|G|^{p-2}\overline{G}h \in L^1$. Problems may conceivably arise from extraneous zeros of G, or from bad boundary behavior. We will see eventually that G has no extraneous zeros, and is well behaved at the boundary when the prescribed zero-set is *finite*, so that Lemma 2 is actually valid for finite zero-sets when $0 < p < 1$.

The property (4) is a necessary condition for G to be an extremal function of the problem (1). Surprisingly, it turns out to be sufficient as well. The following lemma is essentially a converse of Lemma 2.

LEMMA 3. *For $1 \le p < \infty$, let $\{z_k\}$ be an A^p zero-set. Suppose that a function $g \in N^p$ satisfies*

$$\int_{\mathbb{D}} |g(z)|^{p-2}\,\overline{g(z)}h(z)\,d\sigma = 0 \tag{5}$$

for all $h \in N^p$ with $h(0) = 0$. Then g is a constant multiple of the canonical divisor of $\{z_k\}$.

The proof uses the following known theorem of functional analysis, or approximation theory.

THEOREM A. *For $1 \le p < \infty$, let M be a closed subspace of L^p over $\mathbb{D}$. Suppose that a function $g \in L^p$ has the property (5) for all $h \in M$. Then $\|g\|_p = d(g, M)$, the distance from g to M.*

PROOF OF THEOREM A. There is no loss of generality in assuming $\|g\|_p = 1$. Let $F = |g|^{p-2}\overline{g}$. Then $F \in L^q$ and $\|F\|_q = 1$, where $q = p/(p-1)$ is the conjugate index. Thus the condition (5) implies

$$\|g\|_p = \int_{\mathbb{D}} Fg\,d\sigma = \int_{\mathbb{D}} F(g+h)\,d\sigma \le \|g+h\|_p\|F\|_q = \|g+h\|_p$$

for all $h \in M$. This shows that $\|g\|_p = d(g, M)$. □

DEDUCTION OF LEMMA 3. Let M be the subspace of functions $h \in N^p$ with $h(0) = 0$. Let $g \in N^p$ satisfy (5) for all $h \in M$, and suppose that $g(0) = 1$. Let f be any other function in N^p with $f(0) = 1$. Then $g - f \in M$, so by Theorem A,

$$\|g\|_p = d(g, M) \le d(g, g-f) = \|f\|_p\,.$$

This says that g is the (unique) function $H \in N^p$ of minimal norm with $H(0) = 1$. Thus g is a constant multiple of G. □

Lemma 3 can be applied to obtain an explicit formula for the canonical divisor G of a zero-set consisting of a single point $\alpha \in \mathbb{D}$, $\alpha \neq 0$. This was found in a different context by Osipenko and Stessin [1] and was rediscovered by DKSS [1]. The formula, actually valid for $0 < p < \infty$, is

$$G_\alpha(z) = C\frac{\alpha - z}{1 - \overline{\alpha}z}\left\{1 + \frac{p}{2}\left(1 - \overline{\alpha}\frac{\alpha - z}{1 - \overline{\alpha}z}\right)\right\}^{2/p}, \tag{6}$$

where

$$C = \frac{\overline{\alpha}}{|\alpha|}\left\{1 + \frac{p}{2}(1 - |\alpha|^2)\right\}^{-1/p}.$$

Because the formula (6) plays a key role in further developments, we now outline its derivation for $1 \leq p < \infty$. In view of Lemma 3, the canonical divisor G_α of the singleton zero-set $\{\alpha\}$ can be constructed up to normalization by producing a function $g \in A^p$ that vanishes at α and satisfies the condition (5) for all $h \in A^p$ with $h(\alpha) = h(0) = 0$. It would be sufficient to *verify* that the function exhibited in (6) has the required property, but it is more instructive to show how the formula can be found. First rewrite (5) in the equivalent form

$$\int_{\mathbb{D}} |g(z)|^{p-2}\,\overline{g(z)}z\varphi_\alpha(z)h(z)\,d\sigma(z) = 0 \tag{7}$$

for all $h \in A^p$, where $\varphi_\alpha(z) = \frac{\alpha - z}{1 - \overline{\alpha}z}$. *Suppose* now that g has the structure $g(z) = \varphi_\alpha(z)f(z)$, where f is analytic in $\overline{\mathbb{D}}$ and $f(z) \neq 0$ there. Insert this formula into (7) and make the substitution $w = \varphi_\alpha(z)$, with inverse $z = \varphi_\alpha(w)$ and

$$d\sigma(z) = |\varphi_\alpha{}'(w)|^2\,d\sigma(w) = \frac{(1 - |\alpha|^2)^2}{|1 - \overline{\alpha}w|^4}\,d\sigma(w),$$

to transform the resulting condition to

$$\int_{\mathbb{D}} \overline{\Phi(w)}\,|w|^p\varphi_\alpha(w)\Psi(w)\,d\sigma(w) = 0, \tag{8}$$

where

$$\Phi(w) = F(w)^{p/2}\varphi_\alpha{}'(w), \qquad F(w) = f(\varphi_\alpha(w)),$$
$$\Psi(w) = \frac{\Phi(w)H(w)}{F(w)}, \qquad H(w) = h(\varphi_\alpha(w)).$$

Because we have supposed f to be analytic and nonvanishing in the closed disk, it is clear that F and Φ have the same properties, so we may take Ψ to be an arbitrary function in A^p. Let

$$\Phi(w) = \sum_{n=0}^{\infty} c_n w^n$$

and choose $\Psi(w) = (1 - \overline{\alpha} w)w^k$ to calculate from (8)

$$c_{k+1} = \overline{\alpha}\, \frac{2(k+2)+p}{2(k+1)+p}\, c_k\,, \qquad k = 0, 1, 2, \dots,$$

with solution $c_n = (2n + p + 2)\overline{\alpha}^n$ for $n = 0, 1, 2, \dots$, up to a constant multiple. Summation of the power series then gives the expression

$$\Phi(w) = \frac{p(1 - \overline{\alpha} w) + 2}{(1 - \overline{\alpha} w)^2}\,.$$

But the definition of Φ says that $F(w) = \{\Phi(w)/\varphi_\alpha{}'(w)\}^{\frac{2}{p}}$, whereupon a calculation gives

$$f(z) = \left\{1 + \tfrac{p}{2}\left(1 - \overline{\alpha}\, \frac{\alpha - z}{1 - \overline{\alpha} z}\right)\right\}^{2/p}$$

up to a constant multiple. Observe that this function f is analytic and nonvanishing in the closed disk. Thus our calculations have produced a function g that vanishes at α and has the property (5) for all functions $h \in A^p$ with $h(\alpha) = h(0) = 0$. Lemma 3 then tells us that g is a constant multiple of G_α. In other words, the canonical divisor G_α has the form (6) for some constant C. The constant C is determined by the requirements $\|G_\alpha\|_p = 1$ and $G_\alpha(0) > 0$, and is found to have the given form.

If $\alpha = 0$, so that the origin is prescribed as a single zero, the canonical divisor in A^p is found to be

$$G_0(z) = \left(1 + \tfrac{p}{2}\right)^{1/p} z\,.$$

It will be useful to note that the function G_α defined by (6) is an A^p inner function even for the case $0 < p < 1$, when the foregoing calculations (based on Lemma 3) do not apply to show that the canonical divisor has this form. Since $\|G_\alpha\|_p = 1$, it is enough to verify that

$$\int_{\mathbb{D}} |G_\alpha(z)|^p z^n \, d\sigma(z) = 0\,, \qquad n = 1, 2, \dots.$$

But in view of (6), the substitution $w = \varphi_\alpha(z)$ transforms this condition to

$$\int_{\mathbb{D}} |w|^p \, |1 + \tfrac{p}{2}(1 - \overline{\alpha} w)|^2 \, [\varphi_\alpha(w)]^n \, |\varphi_\alpha'(w)|^2 \, d\sigma(w) = 0$$

for $n = 1, 2, \dots$. Equivalently, it is to be shown that

$$\int_{\mathbb{D}} |w|^p \, |k(w)|^2 \, [\varphi_\alpha(w)]^n \, d\sigma(w) = 0\,, \qquad n = 1, 2, \dots,$$

where $k(w) = [1 + \frac{p}{2}(1 - \overline{\alpha}w)](1 - \overline{\alpha}w)^{-2}$. Now let $k(w) = \sum b_j w^j$ and

$$f(w) = k(w)[\varphi_\alpha(w)]^n = \sum_{j=0}^{\infty} c_j w^j \,,$$

and integrate by polar coordinates to calculate the integral as

$$\begin{aligned}\int_{\mathbb{D}} |w|^p \, \overline{k(w)} f(w) \, d\sigma(w) &= \sum_{j=0}^{\infty} \frac{2}{2j + p + 2} \overline{b_j} \, c_j \\ &= \sum_{j=0}^{\infty} c_j \alpha^j = f(\alpha) = 0 \,,\end{aligned}$$

since $b_j = (j + \frac{p}{2} + 1)\, \overline{\alpha}^j$ and $\varphi_\alpha(\alpha) = 0$.

The explicit formula (6) shows that G_α is analytic in $\overline{\mathbb{D}}$, and it has only a simple zero at α and no other zeros in $\mathbb{D}$. It also shows that $|G_\alpha(0)| < 1$ and $|G_\alpha(z)| \geq 1$ on the unit circle $\mathbb{T} = \partial\mathbb{D}$. The last property comes down to the inequalities

$$\left|1 + \frac{p}{2}(1 - \overline{\alpha}w)\right|^2 \geq \left(1 + \frac{p}{2}(1 - |\alpha|)\right)^2 \geq 1 + \frac{p}{2}(1 - |\alpha|^2)$$

for $|w| = 1$. The inequality $|G_\alpha(0)| < 1$ is a general property of nonconstant A^p functions of unit norm. In the case $p \geq 1$ we can apply Hölder's inequality to the mean-value identity $f(0) = \int_{\mathbb{D}} f(z) d\sigma$ to see that $|f(0)| \leq \|f\|_p = 1$, with equality only if f is constant. For any $p > 0$, the same conclusion is reached by observing that $|f|^p$ is subharmonic. We shall see later that all of these properties of G_α generalize to canonical divisors of arbitrary finite zero-sets.

§5.2. Expansive multipliers.

Our main objective is now to prove that the canonical divisor G of an arbitrary A^p zero-set is an *expansive multiplier* in the sense that $\|Gf\|_p \geq \|f\|_p$ for all polynomials f. If G is bounded in $\mathbb{D}$, this inequality will extend to all $f \in A^p$, since the polynomials are dense.

The expansive multiplier property follows from a basic identity due to DKSS [2], known as the *integral formula.* It is

$$\int_{\mathbb{D}} (|G(z)|^p - 1)|f(z)|^p \, d\sigma = \pi \int_{\mathbb{D}} \int_{\mathbb{D}} \Gamma(z,\zeta) \Delta(|G(z)|^p) \Delta(|f(\zeta)|^p) \, d\sigma(z) d\sigma(\zeta), \tag{9}$$

where

$$\Gamma(z,\zeta) = \frac{1}{16}\left\{ |z - \zeta|^2 \log\left|\frac{z - \zeta}{1 - \overline{\zeta}z}\right|^2 + (1 - |z|^2)(1 - |\zeta|^2) \right\} \tag{10}$$

is the biharmonic Green function of the unit disk and f is an arbitrary polynomial. Since

$$\Delta(|g(z)|^p) = p^2\,|g(z)|^{p-2}\,|g'(z)|^2 \ge 0\,,$$

for every analytic function g, and we saw in Chapter 1 that $\Gamma(z,\zeta) > 0$ for all $z,\zeta \in \mathbb{D}$, it will follow from (9) that $\|Gf\|_p \ge \|f\|_p$ for all polynomials f. It should be remarked that the Laplacian $\Delta(|g|^p)$ is integrable even for $0 < p < 2$, when it has singularities at the (isolated) zeros of g.

The proof of the integral formula (9) is deferred to the next section. Meanwhile, we propose to give a heuristic derivation through a boundary value problem considered by Hedenmalm [1] and DKSS [1]. We will also explore some consequences of the expansive multiplier property. The only property of the canonical divisor actually used in the proof of (9) will be that it is an A^p inner function. Equivalently, $\int_{\mathbb{D}}(|G|^p - 1)u\,d\sigma = 0$ for all functions $u \in h^\infty$, as Lemma 1 asserts.

Given an A^p inner function G, where $0 < p < \infty$, consider the boundary value problem

$$\begin{cases} \Delta\varphi(z) = |G(z)|^p - 1 & \text{in } \mathbb{D}\,, \\ \varphi(z) = 0 & \text{on } \mathbb{T}\,, \end{cases}$$

where $\mathbb{T} = \partial\mathbb{D}$ denotes the unit circle. Supposing that φ has a smooth extension to $\overline{\mathbb{D}}$, we can infer that the normal derivative $\partial\varphi/\partial n = 0$ on $\mathbb{T}$. For this it suffices to show that

$$\int_{\mathbb{T}} u\frac{\partial\varphi}{\partial n}\,ds = 0$$

for every function $u \in C^2(\overline{\mathbb{D}})$ that is harmonic in $\mathbb{D}$. But by Green's formula,

$$\begin{aligned}\int_{\mathbb{T}} u\frac{\partial\varphi}{\partial n}\,ds &= \int_{\mathbb{T}}\left(u\frac{\partial\varphi}{\partial n} - \varphi\frac{\partial u}{\partial n}\right)ds = \pi\int_{\mathbb{D}}(u\Delta\varphi - \varphi\Delta u)\,d\sigma \\ &= \pi\int_{\mathbb{D}}(|G|^p - 1)u\,d\sigma = 0\,,\end{aligned}$$

since G is an A^p inner function. Now observe that because φ and its normal derivative both vanish on the boundary, the function can be recovered from its bilaplacian by the formula

$$\varphi(\zeta) = \pi\int_{\mathbb{D}}\Gamma(z,\zeta)\Delta^2\varphi(z)\,d\sigma(z) = \pi\int_{\mathbb{D}}\Gamma(z,\zeta)\Delta(|G(z)|^p)\,d\sigma(z)\,, \qquad \zeta \in \mathbb{D}\,.$$

(See Section 1.6.) Finally, another application of Green's formula gives

$$\begin{aligned}&\int_{\mathbb{D}}(|G(\zeta)|^p - 1)\,|f(\zeta)|^p\,d\sigma(\zeta) = \int_{\mathbb{D}}|f(\zeta)|^p\Delta\varphi(\zeta)\,d\sigma(\zeta) \\ &= \int_{\mathbb{D}}\varphi(\zeta)\Delta(|f(\zeta)|^p)d\sigma(\zeta) = \pi\int_{\mathbb{D}}\int_{\mathbb{D}}\Gamma(z,\zeta)\Delta(|G(z)|^p)\Delta(|f(\zeta)|^p)d\sigma(z)d\sigma(\zeta)\end{aligned}$$

for any polynomial f, which "proves" the integral formula (9).

This heuristic proof is not valid because G may be badly behaved at the boundary, and so φ need not have a smooth extension to $\overline{\mathbb{D}}$. However, the argument can be made rigorous for canonical divisors of finite zero-sets in the special case where $p = 2$, as was shown by Hedenmalm [5]. Then the canonical divisor is a finite linear combination of kernel functions, and $|G|^p$ has a smooth extension to a larger disk. Khavinson and Shapiro [1] extended this approach to general A^p spaces, for $0 < p < \infty$, using the theory of distributions. The above manipulations can also be justified directly for the function G_α given by the explicit formula (6), which is known to be an A^p inner function, since it is clear by inspection that $|G_\alpha|^p$ has a smooth extension to $\overline{\mathbb{D}}$. The formula (9) with $G = G_\alpha$ can be viewed as a very special case of the general result, to be established in the next section. Thus we can state the following lemma.

LEMMA 4. *For $0 < p < \infty$, the function G_α defined by (6) is an expansive multiplier. In fact, $\|G_\alpha f\|_p \geq \|f\|_p$ for all $f \in A^p$. Furthermore, it is a contractive divisor in the sense that $f/G_\alpha \in A^p$ and $\|f/G_\alpha\|_p \leq \|f\|_p$ for every function $f \in A^p$ with $f(\alpha) = 0$.*

PROOF. The expansive multiplier property follows from the integral formula (9). It extends to all functions $f \in A^p$ because $G_\alpha \in H^\infty$. To deduce the contractive divisor property, recall that G_α has only a simple zero at α and $|G_\alpha(z)| \geq 1$ for all $z \in \mathbb{T}$. Thus $g = f/G_\alpha \in A^p$ if $f \in A^p$ and $f(\alpha) = 0$, so the expansive multiplier property implies

$$\|f/G_\alpha\|_p = \|g\|_p \leq \|G_\alpha g\|_p = \|f\|_p . \qquad \square$$

We can now apply our information about G_α to establish an important general property of canonical divisors.

THEOREM 1. *For $0 < p < \infty$, let G be the canonical divisor of an A^p zero-set $\{z_k\}$. Then G has no extraneous zeros; it has a zero at each z_k of precisely the order prescribed, and $G(z) \neq 0$ elsewhere in $\mathbb{D}$.*

PROOF. Suppose on the contrary that $G(\alpha) = 0$ for some point $\alpha \neq z_k$, or that G vanishes to higher order than required at some point $\alpha = z_k$. Let G_α be the canonical divisor of the singleton set $\{\alpha\}$, as displayed in formula (6). Lemma 4 tells us that $\|G/G_\alpha\|_p \leq \|G\|_p = 1$. But $G/G_\alpha \in N^p$, so the defining extremal property (1) of G ensures that $|G(0)/G_\alpha(0)| \leq |G(0)|$. However, $G(0) > 0$ and $G_\alpha(0) < 1$, so this is a contradiction. $\square$

Since we have not yet established the uniqueness of a canonical divisor in A^p when $p < 1$, the theorem requires interpretation in this case. However, the proof shows that no canonical divisor can have extraneous zeros.

One further consequence of the expansive multiplier property can now be derived.

LEMMA 5. *Let $G \in A^p$ with $\|G\|_p = 1$, and suppose that G is continuous in $\overline{\mathbb{D}}$. If G is an expansive multiplier of A^p, then $|G(z)| \geq 1$ for all $z \in \mathbb{T}$.*

PROOF. By hypothesis, $\|Gf\|_p \geq \|f\|_p$ for all polynomials f. The idea of the proof is to exploit that property by introducing a special family of "peaking polynomials". Fix any point $z_0 \in \mathbb{T}$, define $q(z) = \frac{1}{2}(z + z_0)$, and note that

$$\|q^n\|_p^p = \int_{\mathbb{D}} \left|\tfrac{1}{2}(z+z_0)\right|^{np} d\sigma(z) = \frac{1}{\pi\, 2^{np}} \int_{-\frac{\pi}{2}}^{\frac{\pi}{2}} \int_0^{2\cos\theta} r^{np+1}\, dr\, d\theta$$

$$= \frac{4}{\pi(np+2)} \int_{-\frac{\pi}{2}}^{\frac{\pi}{2}} (\cos\theta)^{np+2} d\theta = \frac{4}{\pi(np+2)} B\left(\tfrac{1}{2}, \tfrac{np+3}{2}\right) \sim C\, n^{-3/2}$$

as $n \to \infty$, by Stirling's formula, where $B(x,y)$ is the beta function. Now let $Q_n(z) = q(z)^n / \|q^n\|_p$. Then $\|Q_n\|_p = 1$ and $Q_n(z) \to 0$ uniformly in each closed subset of $\overline{\mathbb{D}} \backslash \{z_0\}$ as $n \to \infty$, because $|q(z)| \leq M < 1$ in each such region, while $\|q^n\|_p \geq c\, n^{-3/2}$ for some constant $c > 0$. Since G is continuous at z_0, it follows that the integrals $\int |GQ_n|^p d\sigma$ tend to $|G(z_0)|^p$ as $n \to \infty$. To see this, let $\varepsilon > 0$ be given and choose $\delta > 0$ so that

$$\left||G(z)|^p - |G(z_0)|^p\right| < \varepsilon \quad \text{for all } z \in \Delta = \{z \in \overline{\mathbb{D}} \;:\; |z - z_0| < \delta\}\,.$$

Then

$$\left|\int_{\mathbb{D}} |GQ_n|^p d\sigma - |G(z_0)|^p\right| \leq \int_{\mathbb{D}} \left||G(z)|^p - |G(z_0)|^p\right| |Q_n(z)|^p\, d\sigma(z)$$

$$= \int_{\Delta} + \int_{\mathbb{D}\backslash\Delta} \;<\; \varepsilon + C \int_{\mathbb{D}\backslash\Delta} |Q_n|^p\, d\sigma\,,$$

and the last term tends to zero as $n \to \infty$, since $Q_n(z) \to 0$ uniformly in $\mathbb{D} \setminus \Delta$. This proves that

$$\int_{\mathbb{D}} |G\, Q_n|^p\, d\sigma \to |G(z_0)|^p\,.$$

But by the expansive multiplier property,

$$\int_{\mathbb{D}} |G\, Q_n|^p\, d\sigma \geq \int_{\mathbb{D}} |Q_n|^p\, d\sigma = 1\,.$$

Hence $|G(z_0)| \geq 1$. □

Finally, by appeal to Theorem 1 we can establish a general relation between certain canonical divisors.

LEMMA 6. *Let $0 < p = mq$ for some integer $m \geq 2$. Let G be the canonical divisor in A^p of a zero-set $\{z_k\}$, and let H be the canonical divisor in A^q of the zero-set $\{\zeta_k\}$ obtained from $\{z_k\}$ by including each point z_k exactly m times. Then $H = G^m$.*

PROOF. Let $F = G^m$. Observe that $F \in A^q$, $\|F\|_q = 1$, $F(0) > 0$, and F has the required zeros ζ_k. Thus $F(0) \le H(0)$, by the extremal property of H, and so $G(0) \le H(0)^{1/m}$. But H has no extraneous zeros, by Theorem 1, so the function $H^{1/m}$ belongs to A^p and vanishes on the prescribed zero-set $\{z_k\}$, while $H^{1/m}(0) > 0$ and $\|H^{1/m}\|_p^p = \|H\|_q^q = 1$, so $H(0)^{1/m} \le G(0)$. Thus $F(0) = H(0)$, and so $F = H$. □

As in the statement of Theorem 1, the last lemma requires interpretation if $p < 1$ or $q < 1$. In such cases it can be taken to say that if G is *any* canonical divisor in A^p of $\{z_k\}$, then G^m is a canonical divisor in A^q of $\{\zeta_k\}$.

Lemma 6 can be applied to generalize the formula (6) for the canonical divisor of a singleton zero-set $\{\alpha\}$. If the zero-set consists of α with repetitions, so that functions are required to have a zero at α of order at least m, then the canonical divisor in A^p takes the form

$$G(z) = C\,[\varphi_\alpha(z)]^m \big\{1 + \tfrac{mp}{2}\,(1 - \overline{\alpha}\,\varphi_\alpha(z))\big\}^{2/p},$$

where $\varphi_\alpha(z) = \frac{\alpha - z}{1 - \overline{\alpha} z}$ and

$$C = \big(\tfrac{\overline{\alpha}}{|\alpha|}\big)^m \big\{1 + \tfrac{mp}{2}(1 - |\alpha|^2)\big\}^{-1/p}.$$

§5.3. Proof of the integral formula.

Our aim is now to prove the integral formula (9) under the general hypothesis that G is an inner function for the space A^p. This means that $\int_{\mathbb{D}}(|G|^p - 1)u\,d\sigma = 0$ for all polynomials u, or equivalently for all $u \in h^\infty$, the space of bounded harmonic functions. Thus by Lemma 1, the canonical divisor of any A^p zero-set is an A^p inner function.

THEOREM 2. *Suppose G is an A^p inner function, where $0 < p < \infty$. Then the integral formula*

$$\int_{\mathbb{D}} (|G(z)|^p - 1)|f(z)|^p\,d\sigma = \pi \int_{\mathbb{D}}\int_{\mathbb{D}} \Gamma(z,\zeta)\Delta(|G(z)|^p)\Delta(|f(\zeta)|^p)\,d\sigma(z)d\sigma(\zeta)$$

holds for every polynomial f, where $\Gamma(z,\zeta)$ is the biharmonic Green function of $\mathbb{D}$.

In the previous section, we "derived" the integral formula by straightforward applications of Green's identity, together with a basic property of the biharmonic Green function. The argument was not rigorous, however, because it assumed that all functions were smooth up to the boundary. To deal with the problem of boundary regularity, our formal proof will employ certain mollifiers. We begin with a standard construction. Here $C_0^\infty(\mathbb{D})$ denotes the class of real-valued infinitely differentiable functions with compact support in $\mathbb{D}$, and $\nabla\psi$ denotes the gradient of a function ψ.

PROPOSITION 1. *For $0 < \varepsilon < 1$, there is a function $\psi_\varepsilon \in C_0^\infty(\mathbb{D})$ with the properties $0 \le \psi_\varepsilon(z) \le 1$, $\psi_\varepsilon(z) = 1$ for $|z| \le 1-\varepsilon$, $|\nabla\psi_\varepsilon(z)| \le c/\varepsilon$, and $|\Delta\psi_\varepsilon(z)| \le c/\varepsilon^2$, where c is an absolute constant. Also, for each fixed $z \in \mathbb{D}$, the functions $\psi_\varepsilon(z)$ increase to 1 as ε decreases to 0.*

These functions ψ_ε can be constructed with radial symmetry, appealing to familiar properties of the function e^{-1/x^2} and convolving with a smooth peaking kernel. Details may be found for instance in Hörmander [1], p. 25.

We will also use a result known as Weyl's lemma. A locally integrable function u is said to be *weakly harmonic* in $\mathbb{D}$ if $\int_{\mathbb{D}} u\Delta\varphi\, d\sigma = 0$ for all functions $\varphi \in C_0^\infty(\mathbb{D})$. Green's identity shows that harmonic functions are weakly harmonic. It is a remarkable fact that the converse is also true.

PROPOSITION 2 (WEYL'S LEMMA). *If a function is locally integrable in $\mathbb{D}$ and is weakly harmonic, then it is harmonic.*

This is a special case of a more general result on weak solutions to elliptic partial differential equations. For completeness, we sketch a proof here. Let $\varphi \in C_0^\infty(\mathbb{D})$ be nonnegative and radially symmetric, with $\int_{\mathbb{D}} \varphi\, d\sigma = 1$. Define $\varphi_\varepsilon(z) = \varepsilon^{-2}\varphi(z/\varepsilon)$ for $0 < \varepsilon < 1$, and let $\mathbb{D}_\varepsilon$ be the disk $|z| < 1-\varepsilon$. Then for each $\psi \in C_0^\infty(\mathbb{D}_\varepsilon)$, the convolution $\psi * \varphi_\varepsilon \in C_0^\infty(\mathbb{D})$, and also $u * \varphi_\varepsilon \in C^\infty(\mathbb{D}_\varepsilon)$. Thus

$$\langle \Delta(u * \varphi_\varepsilon), \psi\rangle = \langle u * \varphi_\varepsilon, \Delta\psi\rangle = \langle u, \varphi_\varepsilon * \Delta\psi\rangle = \langle u, \Delta(\varphi_\varepsilon * \psi)\rangle = 0\,,$$

since u is weakly harmonic. This shows that $\Delta(u * \varphi_\varepsilon) = 0$ on $\mathbb{D}_\varepsilon$, so $u * \varphi_\varepsilon$ is harmonic and therefore has the local mean-value property. But $\|u * \varphi_\varepsilon - u\|_1 \to 0$ as $\varepsilon \to 0$, where the L^1 norm is computed locally, so the mean-value property implies that $u * \varphi_\varepsilon \to u$ locally uniformly. It follows that u is continuous and has the local mean-value property, so it is harmonic. For further details and generalizations, see Folland [1], p. 76.

The proof of Theorem 2 is based on two technical lemmas. A function is said to be *locally bounded* in a given domain if it is bounded in each compact subset. We write $F \perp h^\infty$ to indicate that a function $F \in L^1$ is orthogonal to every function $u \in h^\infty$, meaning that $\int_{\mathbb{D}} Fu\, d\sigma = 0$.

LEMMA 7. *If $F \in L^1$, $F \perp h^\infty$, and F is locally bounded in $\mathbb{D}$, then there is a sequence of functions $\varphi_j \in C_0^\infty(\mathbb{D})$ such that*

(*a*) $\|\Delta\varphi_j - F\|_1 \to 0$ *and*

(*b*) $\|\varphi_j - \Phi\|_1 \to 0$ *as* $j \to \infty$,

where

$$\Phi(z) = \pi \int_{\mathbb{D}} F(\zeta)\, \Delta_\zeta \Gamma(z,\zeta)\, d\sigma(\zeta)\,.$$

LEMMA 8. *If $F \in L^1$ and F is locally bounded in $\mathbb{D}$, then*

$$\lim_{\varepsilon\to 0} \pi \int_{\mathbb{D}} F(\zeta)\, \Delta_\zeta\big(\psi_\varepsilon(\zeta)\Gamma(z,\zeta)\big)\, d\sigma(\zeta) = \Phi(z)\,, \qquad z \in \mathbb{D}\,,$$

where Φ is defined in Lemma 7.

Recall that the Laplacian of a biharmonic Green function has a logarithmic singularity. For the unit disk the explicit formula (10) leads to the expression

$$\Delta_\zeta \Gamma(z,\zeta) = \frac{1}{4}\left\{\log\left|\frac{z-\zeta}{1-\overline{\zeta}z}\right|^2 + \frac{(1-|z|^2)(1-|\zeta z|^2)}{|1-\overline{\zeta}z|^2}\right\}. \tag{11}$$

Thus the extra assumption of local boundedness ensures that $F(\zeta)\Delta_\zeta\Gamma(z,\zeta)$ is integrable when $F \in L^1$, so that the function Φ is well defined. If $G \in A^p$, it is clear that $F = |G|^p - 1 \in L^1$ and F is locally bounded.

It will also be useful to observe that

$$\sup_{\zeta\in\mathbb{D}} \int_{\mathbb{D}} |\Delta_\zeta \Gamma(z,\zeta)\, h(z)|\, d\sigma(z) < \infty \tag{12}$$

if $h \in L^q$ for some $q > 1$. For this purpose, we note first that

$$\frac{(1-|z|^2)(1-|\zeta z|^2)}{|1-\overline{\zeta}z|^2} \le 4, \qquad z,\zeta \in \mathbb{D},$$

since

$$|1-\overline{\zeta}z| \ge \tfrac{1}{2}(1-|\zeta z|^2) \ge \tfrac{1}{2}(1-|z|^2).$$

Hence in view of (11), it suffices to prove (12) with $\Delta_\zeta\Gamma(z,\zeta)$ replaced by $\log\left|\frac{z-\zeta}{1-\overline{\zeta}z}\right|$. Then an appeal to Hölder's inequality reduces the proof to showing that the integrals

$$\int_{\mathbb{D}} \left|\log|z-\zeta|\right|^{q'} d\sigma(z) \qquad \text{and} \qquad \int_{\mathbb{D}} \left|\log|1-\overline{\zeta}z|\right|^{q'} d\sigma(z)$$

are bounded for $\zeta \in \mathbb{D}$, where $q' = q/(q-1)$ is the conjugate index. But clearly,

$$\int_{\mathbb{D}} \left|\log|z-\zeta|\right|^{q'} d\sigma(z) \le \int_{2\mathbb{D}} \left|\log|z|\right|^{q'} d\sigma(z) < \infty,$$

where $2\mathbb{D} = \{z : |z| < 2\}$. Also,

$$\int_{\mathbb{D}} \left|\log|1-\overline{\zeta}z|\right|^{q'} d\sigma(z) \le 4\int_{\mathbb{D}} \left|\log|1-w|\right|^{q'} d\sigma(w) < \infty \qquad \text{for } |\zeta| \ge \tfrac{1}{2},$$

while $|\log|1-\overline{\zeta}z|| \le \log 2$ for $|\zeta| \le \frac{1}{2}$ and all $z \in \mathbb{D}$. This proves (12), which shows in particular that the integral

$$\int_{\mathbb{D}} |\Delta_\zeta\Gamma(z,\zeta)|\, d\sigma(z)$$

is bounded for $\zeta \in \mathbb{D}$.

After these preliminaries, we are prepared to prove the two lemmas.

Proof of Lemma 7. Proposition 2 says that h^∞ is precisely the set of functions $u \in L^\infty$ such that $u \perp \Delta\varphi$ for all $\varphi \in C_0^\infty$. Consequently, the hypothesis of the lemma asserts that $F \perp u$ whenever $u \in L^\infty$ and $u \perp \Delta\varphi$ for all $\varphi \in C_0^\infty$. In other words, if Λ is a bounded linear functional on L^1 and $\Lambda(\Delta\varphi) = 0$ for all $\varphi \in C_0^\infty$, then $\Lambda(F) = 0$. By the Hahn–Banach theorem, this says that F is in the L^1-closure of $\{\Delta\varphi : \varphi \in C_0^\infty\}$, as asserted in (a).

To prove (b), observe that each $\varphi \in C_0^\infty$ has the representation

$$\varphi(z) = \pi \int_{\mathbb{D}} \Delta^2\varphi(\zeta)\,\Gamma(z,\zeta)\,d\sigma(\zeta) = \pi \int_{\mathbb{D}} \Delta\varphi(\zeta)\,\Delta_\zeta\Gamma(z,\zeta)\,d\sigma(\zeta)\,,$$

where Green's formula has been applied. Thus by the definition of Φ,

$$\varphi(z) - \Phi(z) = \pi \int_{\mathbb{D}} \big(\Delta\varphi(\zeta) - F(\zeta)\big)\Delta_\zeta\Gamma(z,\zeta)\,d\sigma(\zeta)\,,$$

so that

$$\|\varphi - \Phi\|_1 \le \pi \int_{\mathbb{D}}\int_{\mathbb{D}} |\Delta\varphi(\zeta) - F(\zeta)|\,|\Delta_\zeta\Gamma(z,\zeta)|\,d\sigma(\zeta)d\sigma(z) \le C\|\Delta\varphi - F\|_1\,,$$

by Fubini's theorem and the boundedness of $\int_{\mathbb{D}} |\Delta_\zeta\Gamma(z,\zeta)|d\sigma(z)$. This shows that $\|\varphi_j - \Phi\|_1 \to 0$ whenever $\|\Delta\varphi_j - F\|_1 \to 0$, completing the proof of Lemma 7. $\square$

Proof of Lemma 8. Applying the standard formula

$$\Delta(uv) = u\Delta v + v\Delta u + 2\,\nabla u \cdot \nabla v\,,$$

we have

$$\Delta_\zeta\big(\psi_\varepsilon(\zeta)\Gamma(z,\zeta)\big) = \psi_\varepsilon(\zeta)\Delta_\zeta\Gamma(z,\zeta) + \Gamma(z,\zeta)\Delta\psi_\varepsilon(\zeta) + 2\,\nabla\psi_\varepsilon(\zeta)\cdot\nabla_\zeta\Gamma(z,\zeta)\,.$$

Observe now that

$$\Phi(z) = \lim_{\varepsilon\to 0} \pi \int_{\mathbb{D}} F(\zeta)\,\psi_\varepsilon(\zeta)\,\Delta_\zeta\Gamma(z,\zeta)\,d\sigma(\zeta)\,,$$

by the Lebesgue dominated convergence theorem. Thus it will suffice to show that the other two integrals tend to zero. But by construction (Proposition 1), the mollifiers ψ_ε have the properties

$$|\nabla\psi_\varepsilon(\zeta)| \le c/\varepsilon \qquad \text{and} \qquad |\Delta\psi_\varepsilon(\zeta)| \le c/\varepsilon^2\,, \qquad \zeta \in \mathbb{D}\,,$$

and $|\nabla\psi_\varepsilon(\zeta)| = \Delta\psi_\varepsilon(\zeta) = 0$ for $0 \le |\zeta| \le 1-\varepsilon$; while direct calculations based on the explicit formula (10) show that

$$0 < \Gamma(z,\zeta) \le C(1-|\zeta|)^2 \qquad \text{and} \qquad |\nabla_\zeta\Gamma(z,\zeta)| \le C(1-|\zeta|)\,, \qquad z,\zeta \in \mathbb{D}\,,$$

for some absolute constant $C > 0$. Consequently, both of the error terms

$$\Gamma(z,\zeta)\Delta\psi_\varepsilon(\zeta) \qquad \text{and} \qquad 2\,\nabla\psi_\varepsilon(\zeta)\cdot\nabla_\zeta\Gamma(z,\zeta)$$

vanish in the disks $|\zeta| \le 1-\varepsilon$ and are uniformly bounded by absolute constants in the thin annuli $\Omega_\varepsilon = \{z : 1-\varepsilon \le |z| < 1\}$. Thus the two corresponding integrals are bounded by constant multiples of the integral

$$\int_{\Omega_\varepsilon} |F(\zeta)|\,d\sigma(\zeta)\,.$$

But since $F \in L^1$, this last integral tends to 0 as $\varepsilon \to 0$. This shows that the two other integrals tend to zero, which completes the proof of Lemma 8. □

PROOF OF THEOREM 2. Let $F = |G|^p - 1$ and apply Green's formula to write

$$\int_{\mathbb{D}} F(\zeta)\,\Delta_\zeta\big(\psi_\varepsilon(\zeta)\Gamma(z,\zeta)\big)\,d\sigma(\zeta) = \int_{\mathbb{D}} \psi_\varepsilon(\zeta)\,\Gamma(z,\zeta)\,\Delta F(\zeta)\,d\sigma(\zeta)\,.$$

In using Green's formula, one must place small "safety circles" around the finitely many zeros of G that lie in the support of ψ_ε, then observe that because $G(z) = (z-z_0)^n\widetilde{G}(z)$ near a zero of order n, where $\widetilde{G}$ is a nonvanishing analytic function, the normal derivative of $|G|^p$ on the safety circle surrounding z_0 behaves like r^{np-1} as the radius $r \to 0$, whereas the length of the circle is $2\pi r$, so the product tends to zero. Applying Lemma 8 and the Lebesgue monotone convergence theorem, we may therefore conclude that

$$\Phi(z) = \lim_{\varepsilon\to 0}\pi\int_{\mathbb{D}} \psi_\varepsilon(\zeta)\,\Gamma(z,\zeta)\,\Delta F(\zeta)\,d\sigma(\zeta) = \pi\int_{\mathbb{D}} \Gamma(z,\zeta)\,\Delta F(\zeta)\,d\sigma(\zeta)$$

for $z \in \mathbb{D}$, where Φ is the function defined in Lemma 7. This justifies a formal application of Green's formula to the function Φ.

Next let $s = |f|^p$, where f is a polynomial. Since G is an A^p inner function, we have $F \perp h^\infty$, so by Lemma 7 there is a sequence of functions $\varphi_j \in C_0^\infty(\mathbb{D})$ with $\Delta\varphi_j \to F$ and $\varphi_j \to \Phi$ in L^1 norm. But another application of Green's formula, with similar precautions near the zeros of f, gives the relation

$$\int_{\mathbb{D}} s\Delta\varphi_j\,d\sigma = \int_{\mathbb{D}} \varphi_j\Delta s\,d\sigma\,.$$

Consequently,

$$\begin{aligned}\int_{\mathbb{D}} Fs\,d\sigma &= \lim_{j\to\infty}\int_{\mathbb{D}} s\Delta\varphi_j\,d\sigma = \lim_{j\to\infty}\int_{\mathbb{D}} \varphi_j\Delta s\,d\sigma \\ &= \int_{\mathbb{D}} \Phi\Delta s\,d\sigma = \pi\int_{\mathbb{D}} \Delta s(z)\int_{\mathbb{D}} \Gamma(z,\zeta)\,\Delta F(\zeta)\,d\sigma(\zeta)d\sigma(z)\,,\end{aligned}$$

which proves the theorem, at least for $p \geq 2$.

For $0 < p < 2$, the assertion that

$$\lim_{j\to\infty} \int_{\mathbb{D}} \varphi_j \Delta s \, d\sigma = \int_{\mathbb{D}} \Phi \Delta s \, d\sigma$$

requires further justification because $\Delta s = p^2 |f|^{p-2} |f'|^2$ is unbounded near the zeros of the polynomial f. Observe, however, that $\Delta s \in L^q$ for q in the interval $1 < q < \frac{2}{2-p}$, because then $q(p-2) > -2$. Thus it follows from (12) that

$$\sup_{\zeta \in \mathbb{D}} \int_{\mathbb{D}} |\Delta_\zeta \Gamma(z,\zeta) \, \Delta s(z)| \, d\sigma(z) < \infty \, ,$$

and the proof of Lemma 7 shows that

$$\left| \int_{\mathbb{D}} (\varphi_j - \Phi) \Delta s \, d\sigma \right| \leq C \, \|\Delta \varphi_j - F\|_1 \to 0$$

as $j \to \infty$, which completes the proof for $0 < p < 2$. □

For later reference, we remark that the integral formula remains valid if f is any function analytic in $\overline{\mathbb{D}}$, not necessarily a polynomial. In fact, $|f|^p$ can be replaced by $|f|^q$ for $0 < q < \infty$ and any function $f \in H^\infty$. More generally, it can be replaced by $|f|^q |g|^t$ for $0 < q < \infty$, $0 < t < \infty$, and $f, g \in H^\infty$. Moreover, the same proof works if $|f|^p$ is replaced by any function $v \in C^2(\overline{\mathbb{D}})$.

Having established the integral formula, we can now conclude that every canonical divisor is an expansive multiplier. In fact, the same is true for every A^p inner function.

THEOREM 3. *For $0 < p < \infty$, every A^p inner function G is an expansive multiplier: $\|Gf\|_p \geq \|f\|_p$ for all polynomials f. In particular, the canonical divisor of an arbitrary A^p zero-set has this property.*

Once again the statement requires interpretation for $0 < p < 1$, where we have not yet established the uniqueness of the canonical divisor. The theorem is then taken to say that *every* canonical divisor of the given zero-set is an expansive multiplier.

PROOF OF THEOREM. The integral formula of Theorem 2 applies to any A^p inner function. However, $\Gamma(z,\zeta) > 0$ for all points $z, \zeta \in \mathbb{D}$, while $\Delta(|G(z)|^p) \geq 0$ and $\Delta(|f(\zeta)|^p) \geq 0$. Thus the right-hand side of the integral formula is nonnegative, showing that $\|Gf\|_p \geq \|f\|_p$ for every polynomial f. By Lemma 1, the canonical divisor of any A^p zero-set is an A^p inner function, so it is an expansive multiplier. □

REMARK. The expansive multiplier property $\|Gf\|_p \geq \|f\|_p$ is stated for *polynomials* f, but it applies more generally to functions f that are analytic in $\overline{\mathbb{D}}$, since any such function can be approximated uniformly in $\mathbb{D}$ by polynomials. More generally, the inequality extends to every function $f \in H^\infty$. This follows from a corresponding extension of the integral formula (see remarks after the proof of Theorem 2), but it can also be seen directly. One need only approximate f by a dilation $f_r(z) = f(rz)$ for $r < 1$, which is analytic in $\overline{\mathbb{D}}$. Then $\|Gf_r\|_p \geq \|f_r\|_p$, and one concludes that $\|Gf\|_p \geq \|f\|_p$ by letting $r \to 1$ and invoking the Lebesgue dominated convergence theorem.

The expansive multiplier property of a canonical divisor was generalized by Hedenmalm, Jakobsson, and Shimorin [1,2]. They showed that if A and B are A^p zero-sets with respective canonical divisors G_A and G_B, and if $A \subset B$, then $\|G_A f\|_p \leq \|G_B f\|_p$ for all polynomials f. This result had been conjectured by Hedenmalm [7]. Note that it reduces to the standard expansive multiplier property of a canonical divisor when A is the empty set, whose canonical divisor is the function $G_A(z) \equiv 1$. Aleman and Richter [2] gave a simpler proof for the case $p = 2$. Their argument relies on facts about reproducing kernels from work of McCullough and Richter [1].

§5.4. Representation by kernel functions.

For a weight function $w(z) \geq 0$ that is integrable over $\mathbb{D}$, the *weighted Bergman space* A^2_w consists of all analytic functions f with finite integral

$$\|f\|^2_{2,w} = \int_{\mathbb{D}} |f(z)|^2 w(z)\, d\sigma < \infty\,.$$

Under the hypothesis that point-evaluation functionals are uniformly bounded on compact subsets of $\mathbb{D}$, the space A^2_w is complete and is therefore a Hilbert space with inner product

$$\langle f, g\rangle_w = \int_{\mathbb{D}} f(z)\overline{g(z)} w(z)\, d\sigma\,,$$

as observed in Chapter 1. In particular, the space A^2_w then has a reproducing kernel $J(z,\zeta)$ with the properties $J(\cdot\,,\zeta) \in A^2_w$ for each $\zeta \in \mathbb{D}$, $J(\zeta, z) = \overline{J(z,\zeta)}$, and

$$f(\zeta) = \int_{\mathbb{D}} J(\zeta, z) f(z) w(z)\, d\sigma(z)\,, \qquad \zeta \in \mathbb{D}\,,$$

for every $f \in A^2_w$. Applying the Schwarz inequality, we infer that

$$|f(\zeta)| \leq \|J(\zeta,\cdot)\|_{2,w}\, \|f\|_{2,w} = \sqrt{J(\zeta,\zeta)}\, \|f\|_{2,w}\,,$$

with equality occurring for given $\zeta \in \mathbb{D}$ only when $f(z) = c\,J(z, \zeta)$ for some complex constant c. For fixed $\zeta \in \mathbb{D}$, this shows that among all functions $f \in A^2_w$ of unit norm with $f(\zeta) > 0$, the quantity $f(\zeta)$ attains its maximum value $\sqrt{J(\zeta, \zeta)}$ only for $f(z) = J(z, \zeta)/\sqrt{J(\zeta, \zeta)}$. Note that the reproducing property implies that $J(\zeta, \zeta) = \|J(\cdot\,, \zeta)\|^2_{2,w} > 0$.

In the case of particular interest to us, the weight function will have the form $w(z) = |\psi(z)|^p$ for some function $\psi \in A^p$, where $0 < p < \infty$. It is then clear that the evaluation functional is bounded at any point z_0 where $\psi(z_0) \neq 0$, since $|f|^2|\psi|^p$ is subharmonic. The same argument shows that point evaluations are uniformly bounded on any set where $|\psi(z)| \geq \delta > 0$. Thus by continuity, the point evaluations are uniformly bounded in some neighborhood of any point where $\psi(z_0) \neq 0$. If $\psi(z_0) = 0$, then because the zeros are isolated we can find a small annulus Ω centered at z_0 where $|\psi(z)|^p \geq \delta > 0$. By the sub-mean-value property of $|f|^2$,

$$\begin{aligned} |f(z_0)|^2 &\leq \frac{1}{|\Omega|}\int_\Omega |f(z)|^2\,d\sigma \leq \frac{1}{\delta|\Omega|}\int_\Omega |f(z)|^2|\psi(z)|^p\,d\sigma \\ &\leq \frac{1}{\delta|\Omega|}\int_{\mathbb{D}} |f(z)|^2 w(z)\,d\sigma = \frac{1}{\delta|\Omega|}\|f\|^2_{2,w}\,, \end{aligned}$$

where $|\Omega|$ denotes the area of Ω. By continuity, then, the point evaluations are uniformly bounded in some neighborhood of z_0. Finally, for any compact subset $S \subset \mathbb{D}$, the Heine–Borel theorem shows that point evaluations are uniformly bounded on S. Thus the space A^2_w is complete, and it has a kernel function.

It is important to observe that the kernel function never vanishes. Here we must make a technical distinction between the Hilbert space A^2_w and its subspace $\mathcal{P}^2_w$ defined as the closure of the polynomials in the A^2_w norm. If the weight function w is bounded in $\mathbb{D}$, then $\mathcal{P}^2_w = A^2_w$, since polynomials are dense in the unweighted space A^2. For general weights $w \in L^1$, however, the polynomials need not be dense in A^2_w, and so $\mathcal{P}^2_w$ is a proper subspace.

Here is a simple example. Let

$$S(z) = \exp\left\{-\frac{1+z}{1-z}\right\}$$

be the standard atomic singular inner function (*cf.* Section 3.1), with singular measure μ concentrated at one point, and define $w = |S|^2$. Then $1/S \in A^2_w$, but we claim that $1/S \notin \mathcal{P}^2_w$. Indeed, if

$$\int_{\mathbb{D}} \left|\frac{1}{S} - Q_n\right|^2 w\,d\sigma \to 0$$

for some sequence of polynomials Q_n, then $\|1 - Q_n S\|_2 \to 0$ and it follows that the polynomial multiples of S are dense in A^2. However, we will see in

Chapter 8 (Theorem 1) that this is not the case; in other words, S is not a cyclic element of A^2. Thus $1/S \notin \mathcal{P}^2_w$, and $\mathcal{P}^2_w$ is a proper subspace of A^2_w.

For weight functions of the form $w = |\psi|^p$ with $\psi \in A^p$, the subspace $\mathcal{P}^2_w$ has its own kernel function. The following lemma is crucial for later developments.

LEMMA 9. *Let $J(z,\zeta)$ be the kernel function of the space $\mathcal{P}^2_w$ with weight $w = |\psi|^p$ for some function $\psi \in A^p$, where $0 < p < \infty$. Then $J(z,\zeta) \neq 0$ for all points $z, \zeta \in \mathbb{D}$.*

PROOF. It is enough to show that $J(z,0) \neq 0$ for all $z \in \mathbb{D}$, since the general result then follows from a change of variables $z \mapsto \varphi_\alpha(z) = \frac{\alpha - z}{1-\bar{\alpha}z}$ in the reproducing formula, showing that

$$K(z,\zeta) = J(\varphi_\alpha(z), \varphi_\alpha(\zeta))$$

is the kernel function of the space $\mathcal{P}^2_\omega$ with weight

$$\omega(z) = w(\varphi_\alpha(z))|\varphi'_\alpha(z)|^2 = |\psi(\varphi_\alpha(z))|^p |\varphi'_\alpha(z)|^2,$$

which has the same form $|\Psi(z)|^p$ with $\Psi \in A^p$, since $\varphi'_\alpha(z) \neq 0$. Note also that for any polynomial Q the function $Q(\varphi_\alpha(z))$ can be approximated by polynomials uniformly in $\mathbb{D}$, hence in the A^2_ω norm. Thus $K(z,\zeta)$ is the reproducing kernel for the space $\mathcal{P}^2_\omega$, and a proof that $J(z,0) \neq 0$ will also show that $K(z,0) \neq 0$ in $\mathbb{D}$. In other words, it will follow that $J(\varphi_\alpha(z), \varphi_\alpha(0)) \neq 0$ for all $z, \alpha \in \mathbb{D}$, which is equivalent to saying that $J(z,\zeta) \neq 0$ for all $z, \zeta \in \mathbb{D}$.

To show that $J(z,0) \neq 0$, we recall that the function

$$G(z) = \frac{J(z,0)}{\sqrt{J(0,0)}}$$

uniquely maximizes $\mathrm{Re}\{f(0)\}$ among all functions $f \in \mathcal{P}^2_w$ of unit norm. If $G(\alpha) = 0$ for some $\alpha \in \mathbb{D}$, we consider the function $F = G/G_\alpha$, where G_α is the canonical divisor in A^2 of the singleton zero-set $\{\alpha\}$, as given by (6). According to the remark following the proof of Theorem 2, the integral formula for G_α is valid in the more general form

$$\begin{aligned}&\int_{\mathbb{D}} (|G_\alpha|^2 - 1)|f|^2 |\psi_r|^p \, d\sigma \\ =&\pi \int_{\mathbb{D}} \int_{\mathbb{D}} \Gamma(z,\zeta) \Delta\big(|G_\alpha(z)|^2\big) \Delta\big(|f(\zeta)|^2 |\psi_r(\zeta)|^p\big) \, d\sigma(z) d\sigma(\zeta),\end{aligned}$$

where f is an arbitrary polynomial and $\psi_r(z) = \psi(rz)$ is a dilation of ψ, with $0 < r < 1$. Thus

$$\int_{\mathbb{D}} (|G_\alpha|^2 - 1)|f|^2 |\psi_r|^p \, d\sigma \geq 0, \qquad 0 < r < 1.$$

But the factor $(|G_\alpha|^2 - 1)|f|^2$ is bounded and $\psi_r \to \psi$ in A^p norm as $r \to 1$, so it follows that

$$\int_{\mathbb{D}} (|G_\alpha|^2 - 1)|f|^2 |\psi|^p \, d\sigma \geq 0$$

for every polynomial f, hence for every function $f \in \mathcal{P}^2_w$. This says that G_α is an expansive multiplier of the space $\mathcal{P}^2_w$ with weight $w = |\psi|^p$, meaning that $\|G_\alpha f\|_{2,w} \geq \|f\|_{2,w}$ for all $f \in \mathcal{P}^2_w$. Consequently, G_α is also a contractive divisor, by the argument used in the proof of Lemma 4. In particular,

$$\|F\|_{2,w} = \|G/G_\alpha\|_{2,w} \leq \|G\|_{2,w} \, ,$$

so it follows that $|F(0)| \leq |G(0)|$, by the extremal property of G. However, $G(0) > 0$ and $G_\alpha(0) < 1$, so we have arrived at a contradiction. This proves that $G(z) \neq 0$, or that $J(z,0) \neq 0$ for all $z \in \mathbb{D}$. □

We are now in position to establish a structural formula for canonical divisors of finite zero-sets in terms of kernel functions of weighted Bergman spaces.

THEOREM 4. *In the space A^p for $0 < p < \infty$, let G be the canonical divisor of a finite zero-set and let B be the corresponding Blaschke product. Then*

$$G(z) = J(0,0)^{-1/p} \, B(z) \, J(z,0)^{2/p} \, ,$$

where $J(z,\zeta)$ is the kernel function of the Bergman space A^2_w with weight $w = |B|^p$.

Note that $J(0,0) = \|J(\cdot,0)\|^2_{2,w} > 0$ and $J(z,0) \neq 0$ in $\mathbb{D}$, by Lemma 9. Thus $J(z,0)^{\frac{2}{p}}$ is taken to be the branch that is real and positive at $z = 0$. It is analytic and single-valued in $\mathbb{D}$.

This theorem was proved by DKSS [1] for $1 \leq p < \infty$ by writing the result of Lemma 2 in the equivalent form

$$\int_{\mathbb{D}} |G|^{p-2} \overline{G} B f \, d\sigma = f(0) \int_{\mathbb{D}} |G|^{p-2} \overline{G} B \, d\sigma$$

for all $f \in A^p$, then appealing to Theorem 1 to write $G = Bk^{2/p}$ for some nonvanishing function $k \in A^2_w$ ultimately shown to have a reproducing property that identifies it as a constant multiple of $J(z,0)$. We offer here a more direct proof that works equally well for $0 < p < 1$. This will settle the question of uniqueness of canonical divisors for $0 < p < 1$, at least for finite zero-sets, since the argument will show that *any* canonical divisor has the specified form. The last remark is now recorded for later reference.

COROLLARY. *For $0 < p < \infty$, the canonical divisor in A^p of a finite zero-set is unique.*

PROOF OF THEOREM. As before, let N^p denote the subspace of all functions in A^p that vanish to at least the prescribed order on the given finite zero-set. The canonical divisor G has the properties $\|G\|_p = 1$ and $G(0) > 0$, and it belongs to N^p. According to Theorem 1, it has no further zeros. Thus it has the form $G = Bk^{2/p}$ for some *nonvanishing* function $k \in A^2_w$ of unit norm with $k(0) > 0$. Conversely, every nonvanishing function $\ell \in A^2_w$ of unit norm gives rise to a function $g = B\ell^{2/p} \in N^p$ of unit norm. Recall now that the canonical divisor G was defined as the solution to the extremal problem (1); it maximizes $\operatorname{Re}\{f(0)\}$ among all functions $f \in N^p$ of unit norm. Since the extremal function has no extraneous zeros, it may be characterized as the function that maximizes $\operatorname{Re}\{f(0)\}$ within the class of functions $f \in N^p$ of unit norm that vanish *precisely* on the prescribed zero-set. But these are the functions $f = B\ell^{2/p}$ for some nonvanishing function $\ell \in A^2_w$ of unit norm. Thus k is the function that maximizes $\ell(0)$ in the class of nonvanishing functions $\ell \in A^2_w$ of unit norm with $\ell(0) > 0$.

However, we have already noted that the function $J(z,0)/\sqrt{J(0,0)}$ uniquely maximizes $\ell(0)$ in the full class of functions $\ell \in A^2_w$ of unit norm with $\ell(0) > 0$. Furthermore, Lemma 9 says that $J(z,0) \neq 0$ in $\mathbb{D}$, so the same function is the unique extremal within the subclass of *nonvanishing* functions $\ell \in A^2_w$ of unit norm with $\ell(0) > 0$. Thus since $G(0) > 0$ and $B(0) > 0$, we conclude that

$$k(z) = \frac{J(z,0)}{\sqrt{J(0,0)}},$$

which proves the theorem. □

At the end of this chapter we will see that the representation of Theorem 4 extends to canonical extremal functions of invariant subspaces generated by any function $\psi \in A^p$. A generalization of the theory to arbitrary invariant subspaces will be discussed in Section 5.7.

§5.5. Analytic continuation.

The representation by kernel functions, as described in Theorem 4, is an effective tool for the study of canonical divisors. We now use the representation formula to prove that the canonical divisor of a finite zero-set always has an analytic continuation to a larger disk. The main step in the proof is to show that the kernel function has an analytic continuation, and is in fact a rational function.

LEMMA 10. *Let $J(z,\zeta)$ be the kernel function of the weighted Bergman space A^2_w with weight $w = |B|^p$, where B is a finite Blaschke product with zeros $z_1, z_2, \dots, z_n$. For each fixed $\zeta \in \mathbb{D}$, the function $J(z,\zeta)$ has an analytic continuation to the whole plane except for poles at the reflected points $1/\overline{\zeta}$ and $1/\overline{z_k}$, for $k = 1, 2, \dots, n$.*

PROOF. Suppose first that B has only simple zeros, so that the points z_k are all distinct. Write

$$B(z) = \prod_{k=1}^{n} \frac{z - z_k}{1 - \overline{z_k} z}, \qquad B_k(z) = \frac{1 - \overline{z_k} z}{z - z_k} B(z).$$

Then for any polynomial f, the residue theorem gives

$$\frac{1}{2\pi i} \int_{\mathbb{T}} \frac{B(z)}{B(\zeta)} \frac{f(\zeta)}{\zeta - z} d\zeta = f(z) - \sum_{k=1}^{n} \frac{B_k(z)}{B_k(z_k)} \frac{1 - |z_k|^2}{1 - \overline{z_k} z} f(z_k).$$

But because $|z| = 1$ and $|B(\zeta)| = 1$ on $\mathbb{T}$, the integral can be rewritten as

$$\frac{1}{2\pi i} B(z) \int_{\mathbb{T}} \frac{\overline{\zeta}\, \overline{B(\zeta)}}{1 - \overline{\zeta} z} |B(\zeta)|^p f(\zeta)\, d\zeta,$$

which is transformed by the Cauchy-Green formula (*cf.* Section 1.5) to the area integral

$$\begin{aligned} &B(z) \int_{\mathbb{D}} \frac{\partial}{\partial \overline{\zeta}} \left\{ \frac{\overline{\zeta}\, \overline{B(\zeta)}}{1 - \overline{\zeta} z} |B(\zeta)|^p \right\} f(\zeta)\, d\sigma(\zeta) \\ &= B(z) \int_{\mathbb{D}} \left\{ \frac{\overline{B(\zeta)}}{(1 - \overline{\zeta} z)^2} + \left(\frac{p}{2} + 1\right) \frac{\overline{\zeta}\, \overline{B'(\zeta)}}{1 - \overline{\zeta} z} \right\} |B(\zeta)|^p f(\zeta)\, d\sigma(\zeta). \end{aligned}$$

On the other hand, the reproducing property of the kernel function gives

$$\begin{aligned} &f(z) - \sum_{k=1}^{n} \frac{B_k(z)}{B_k(z_k)} \frac{1 - |z_k|^2}{1 - \overline{z_k} z} f(z_k) \\ &= \int_{\mathbb{D}} \left\{ J(z, \zeta) - \sum_{k=1}^{n} \frac{B_k(z)}{B_k(z_k)} \frac{1 - |z_k|^2}{1 - \overline{z_k} z} J(z_k, \zeta) \right\} |B(\zeta)|^p f(\zeta)\, d\sigma(\zeta). \end{aligned}$$

Since the two area integrals agree for every polynomial f, and the polynomials are dense in A^2_w because $w = |B|^p$ is bounded, the conclusion is that

$$J(z, \zeta) = \sum_{k=1}^{n} \frac{B_k(z)}{B_k(z_k)} \frac{1 - |z_k|^2}{1 - \overline{z_k} z} J(z_k, \zeta) + B(z) \left\{ \frac{\overline{B(\zeta)}}{(1 - \overline{\zeta} z)^2} + \left(\frac{p}{2} + 1\right) \frac{\overline{\zeta}\, \overline{B'(\zeta)}}{1 - \overline{\zeta} z} \right\}$$

for all $z, \zeta \in \mathbb{D}$. But the right-hand side is a rational function of z for each fixed $\zeta \in \mathbb{D}$, analytic in the whole plane except for poles at the points $1/\overline{\zeta}$ and $1/\overline{z_k}$. This proves that $J(z, \zeta)$ has an analytic continuation as described, at least in the case of simple zeros.

If B has multiple zeros, the residue calculation is more complicated but it yields a similar sum involving certain derivatives of f at points z_k.

These derivatives can be recovered from f by area integration against corresponding derivatives of the kernel function $J(z,\zeta)$, as is seen by a process of differentiation under the integral sign in the reproducing formula. Since the Blaschke product with multiple zeros is a uniform limit of Blaschke products with simple zeros, the resulting integral representation of the residue sum is justified by observing that the kernel function has uniformly continuous dependence on the weight function.

To be more precise, suppose $J_1(z,\zeta)$ and $J_2(z,\zeta)$ are the kernel functions of $A^2_{w_1}$ and $A^2_{w_2}$, respectively, where $w_k = |\psi_k|^p$ for functions ψ_k analytic in $\overline{\mathbb{D}}$. Let $S \subset \mathbb{D}$ be a compact subset. Then for each fixed $\zeta \in \mathbb{D}$,

$$|J_1(z,\zeta) - J_2(z,\zeta)| < \varepsilon \qquad \text{for all} \quad z \in S$$

whenever $|w_1(\zeta)-w_2(\zeta)| < \delta$ for all $\zeta \in \overline{\mathbb{D}}$. This can be proved by noting that the A^2_w norm of a function changes continuously with uniform perturbations of w, then appealing to the extremal description of the kernel function, as used in the previous section. In this manner, the result for multiple zeros is deduced from the result for simple zeros by a limiting argument. □

With the lemma in hand, it is now an easy matter to prove that canonical divisors of finite zero-sets have analytic continuations.

THEOREM 5. *In the space A^p for $0 < p < \infty$, let G be the canonical divisor of a finite zero-set. Then $G(z)$ has an analytic continuation to a disk $|z| < R$ for some $R > 1$.*

PROOF. By Theorem 4, the canonical divisor has the form

$$G(z) = J(0,0)^{-1/p}\, B(z)\, J(z,0)^{2/p},$$

where B is the finite Blaschke product corresponding to the prescribed zero-set, and $J(z,\zeta)$ is the kernel function of the space A^2_w with weight $w = |B|^p$. As shown in Lemma 9, $J(z,0) \neq 0$ for all $z \in \mathbb{D}$, so that the function $J(z,0)^{\frac{2}{p}}$ has a single-valued branch in $\mathbb{D}$ that is real and positive at the origin. According to Lemma 10, the kernel function $J(z,0)$ is a rational function with poles outside $\overline{\mathbb{D}}$, so it has an analytic continuation to a larger disk. To finish the proof, we need to show that $J(z,0) \neq 0$ on the unit circle. But in view of the structural formula for G, it follows from Lemma 10 that G has a continuous extension to $\overline{\mathbb{D}}$. Since G is an expansive multiplier of A^p, it then follows from Lemma 5 that $|G(z)| \geq 1$ on $\mathbb{T}$. In particular, $J(z,0) \neq 0$ on $\mathbb{T}$, which completes the proof. □

The proof of Lemma 10 is due to MacGregor and Stessin [1], at least in the case of simple zeros. They used the rational expression for $J(z,0)$ to obtain a structural formula for the canonical divisor of a finite zero-set, in terms of parameters $J(z_k,0)$ which can be determined numerically. Cima and Derrick [1] obtained a similar formula in the case $p = 2$, where the canonical divisor of a finite zero-set is a finite linear combination of ordinary kernel functions of A^2. For zero-sets consisting of two points, Hansbo [1] found an explicit expression for the canonical divisor in A^p.

The representation formula in Theorem 4 is the key to proving that the canonical divisor of a finite zero-set has an analytic continuation to a larger disk. Duren, Khavinson, and Shapiro [1] extended the argument to show that the canonical divisor of a *Blaschke* zero-set has an analytic continuation across each arc of the unit circle that is free of cluster points of the zeros. Sundberg [1] then generalized that result to canonical divisors of arbitrary A^p zero-sets.

§5.6. Contractive divisors.

It is now a short step to the main result of this chapter, the contractive property of canonical divisors. We state it first for finite zero-sets, then derive the general theorem by passage to a limit.

LEMMA 11. *Suppose $0 < p < \infty$, and let G be the canonical divisor in A^p of a finite zero-set $\{z_k\}$. Then G is a contractive divisor. If a function $f \in A^p$ vanishes on the zero-set $\{z_k\}$, then $f/G \in A^p$ and $\|f/G\|_p \le \|f\|_p$.*

PROOF. By the corollary to Theorem 4, the canonical divisor G is unique even for $0 < p < 1$. Theorem 3 says that G is an expansive multiplier in the sense that $\|Gf\| \ge \|f\|$ for all polynomials f. But Theorem 5 says that G is continuous in $\overline{\mathbb{D}}$. In particular, G is bounded, so the inequality $\|Gf\| \ge \|f\|$ extends to all functions $f \in A^p$, since the polynomials are dense. On the other hand, Theorem 2 asserts that G has no extraneous zeros. Thus for any function $f \in A^p$ that vanishes on the prescribed zero-set $\{z_k\}$, the quotient $g = f/G$ is analytic in $\mathbb{D}$. Moreover, $g \in A^p$ because G is continuous in $\overline{\mathbb{D}}$ and so $|G(z)| \ge 1$ on $\mathbb{T}$, by Lemma 5. The expansive multiplier property therefore implies

$$\|f\|_p = \|Gg\|_p \ge \|g\|_p = \|f/G\|_p\,,$$

as claimed. □

Our next objective is to show that the contractive property of Lemma 11 extends to canonical divisors of arbitrary A^p zero-sets. Conversely, it is easy to show that every contractive divisor is a solution to the extremal problem (1); it maximizes $|f(0)|$ among all functions $f \in A^p$ of unit norm that vanish on the prescribed zero-set. To see this, let g be a contractive divisor for a given A^p zero-set $\{z_k\}$. This is a function $g \in A^p$ of norm $\|g\|_p = 1$ with the property that $f/g \in A^p$ and $\|f/g\|_p \le \|f\|_p$ for all functions $f \in N^p$, the subspace of functions $f \in A^p$ that vanish on $\{z_k\}$. Then for any function $f \in N^p$,

$$|f(0)/g(0)| \le \|f/g\|_p \le \|f\|_p\,,$$

with $|f(0)/g(0)| = \|f/g\|_p$ only if f/g is constant. Therefore, if $f \in N^p$ and $\|f\|_p = 1$, we have $|f(0)| < |g(0)|$ unless f is a constant multiple of g. Consequently, any contractive divisor g with $g(0) > 0$ is the *unique* canonical divisor of the given A^p zero-set. In particular, there can be at most one normalized contractive divisor.

We are now ready prove that every canonical divisor is contractive.

THEOREM 6. *Suppose $0 < p < \infty$, and let G be a canonical divisor of an arbitrary A^p zero-set $\{z_k\}$. Then G is a contractive divisor. If a function $f \in A^p$ vanishes on $\{z_k\}$, then $f/G \in A^p$ and $\|f/G\|_p \le \|f\|_p$. Furthermore, the canonical divisor is unique.*

Note that when $\{z_k\}$ is an infinite zero-set, the uniqueness assertion for $0 < p < 1$ has not been proved until now.

PROOF OF THEOREM. As usual, let N^p denote the subspace of functions in A^p that vanish on the zero-set $\{z_k\}$. Let G_n be the canonical divisor in A^p of the finite zero-set $\{z_1, z_2, \ldots, z_n\}$, for $n = 1, 2, \ldots$. Then for each $f \in N^p$, it follows from Lemma 11 that $\|f/G_n\|_p \le \|f\|_p$. But $\|G_n\|_p = 1$, so the family $\{G_n\}$ is locally bounded and is therefore a normal family, by Montel's theorem. In other words, some subsequence of $\{G_n\}$ converges uniformly on compact subsets to a function H analytic in $\mathbb{D}$. According to Theorem 1 and Hurwitz' theorem, H has precisely the zeros z_k, with the prescribed multiplicities. By Fatou's lemma, $\|H\|_p \le 1$. Note also that $G_n(0) \ge G(0)$ by the extremal property of G_n, and so $H(0) \ge G(0) > 0$.

However, the inequality $\|f/G_n\|_p \le \|f\|_p$ shows that $\|f/H\|_p \le \|f\|_p$ for all $f \in N^p$, by another application of Fatou's lemma. Thus H is a contractive divisor and so, by the remarks preceding the theorem, it is a uniquely determined canonical divisor, a normalized solution to the original extremal problem (1). Thus $H = G$, and G is a contractive divisor. □

COROLLARY. *Let G be the canonical divisor of an A^p zero-set $\{z_k\}$, and let G_n be the canonical divisor in A^p of the finite zero-set $\{z_1, z_2, \ldots, z_n\}$. Then $\|G_n - G\|_p \to 0$ as $n \to \infty$.*

PROOF. This is really a corollary of the proof of Theorem 6. The argument shows that $\{G_n(z)\}$ converges locally uniformly to $G(z)$. Note also that $\|G_n\|_p = \|G\|_p = 1$. To infer that G_n tends in norm to G, one need only recall that in L^p of any measure space $(0 < p < \infty)$, the two conditions $f_n(x) \to f(x)$ a.e. and $\|f_n\|_p \to \|f\|_p$ imply that $\|f_n - f\|_p \to 0$. (See also Lemma 4 in Chapter 9.) □

Let us now observe that for $0 < p < \infty$, the Bergman space A^p has no *isometric* zero-divisors of unit norm. Since the contractive divisor is unique (under the normalization $G(0) > 0$), it is the only possible candidate. But the integral formula of Theorem 2 shows that $\|Gf\|_p > \|f\|_p$ for all nonconstant polynomials f, so the canonical divisor G is properly contractive when applied to any nonconstant polynomial multiple of itself. In Chapter 9 we will see that the polynomial multiples Gf are dense in N^p.

In fact, we can see directly that no function $g \in A^p$ of unit norm can be an isometric multiplier, unless g is constant. If $\|gf\|_p = \|f\|_p$ for every $f \in A^p$, then in particular $\|g^n\|_p = \|g\|_p = 1$ for $n = 1, 2, \dots$. But $|g(z)| > 1$ on a set of positive measure, since $\|g\|_p = 1$ and g is nonconstant, so this is impossible.

Thus the theorem of F. Riesz, asserting that Blaschke products are isometric zero-divisors for Hardy spaces, has no exact counterpart for Bergman spaces. Theorem 6 must be regarded as the best possible approximation.

Finally, it is interesting to ask under what circumstances the canonical divisor is bounded. That is certainly true for every finite zero-set. If $G \in H^\infty$ for an infinite zero-set, then the zero-set must be a Blaschke sequence. Is the converse true? In other words, if $\{z_k\}$ is a Blaschke sequence, must its canonical divisor G be bounded?

The answer is no. According Theorem 10 in Chapter 4, a Blaschke product B is a universal divisor of some Bergman space A^p, in the sense that $f/B \in A^p$ for every function $f \in N^p$, if and only if the zero-set of B is a finite union of uniformly separated sequences. But if $G \in H^\infty$, then $G/B \in H^\infty$ by the theorem of F. Riesz. On the other hand, $f/G \in A^p$ for each $f \in N^p$, so $f/B = (f/G)(G/B)$ also belongs to A^p; in other words, B is a universal divisor of A^p if $G \in H^\infty$. Thus a necessary condition for a canonical divisor to be bounded is that its zero-set be a finite union of uniformly separated sequences.

Conversely, Aleman and Richter [2] have shown that if $\{z_k\}$ is a Blaschke sequence and G is its canonical divisor in A^2, then

$$\log |G(z)| \leq \sum_{k=1}^{\infty} \frac{1 - |z_k|^2}{|1 - \overline{z_k} z|} .$$

Therefore, a sufficient condition for $G \in H^\infty$ is that

$$\sup_{z \in \mathbb{D}} \sum_{k=1}^{\infty} \frac{1 - |z_k|^2}{|1 - \overline{z_k} z|} < \infty .$$

§5.7. Invariant subspaces.

Much of the theory developed in this chapter extends readily to arbitrary invariant subspaces of A^p. A (closed) subspace M of A^p is said to be *invariant* if the function $zf(z)$ belongs to M whenever $f \in M$. In other words, M is invariant under multiplication by the independent variable z. It is equivalent to say that $f \in M$ implies $Qf \in M$ for every polynomial Q.

For any A^p zero-set $\{z_k\}$, the set N^p is an invariant subspace of A^p. Recall that N^p consists of all functions in A^p that vanish on the zero-set $\{z_k\}$ to at least the prescribed multiplicity. To see that N^p is *closed*, we have only to recall that norm convergence in A^p implies locally uniform convergence. Thus if $f_n \in N^p$ and $f_n \to f$ in A^p norm, it follows that $f \in N^p$.

Other examples of invariant subspaces arise by fixing a function $f \in A^p$ and constructing the A^p closure $[f]$ of the linear space of all polynomial multiples of f. We saw in Chapter 4 (Theorem 6) that if $\{z_k\}$ is a Blaschke sequence and B is the corresponding Blaschke product, then $[B] = N^p$. It is easy to see that $[G] = N^p$ if G is the canonical divisor of any finite zero-set. Indeed, Theorem 5 guarantees that $G \in H^\infty$. For any $f \in N^p$, Theorem 6 says that $f/G \in A^p$, so f/G can be approximated in A^p norm by polynomials. Since $G \in H^\infty$, it follows that f can be approximated by polynomial multiples of G, so that $f \in [G]$. Thus $N^p \subset [G]$. But $G \in N^p$, so it is clear that $[G] \subset N^p$. This proves that $[G] = N^p$. We will see in Chapter 9 that this last relation extends to arbitrary A^p zero-sets. In general, all functions in the subspace $[f]$ generated by a function $f \in A^p$ will vanish on the zero-set of f, but a nonvanishing function f need not generate the whole space A^p. A full description of the generators of A^p, also known as the cyclic elements, is a difficult problem to which we will return in Chapters 8 and 9.

For $0 < p < \infty$, let M be an arbitrary invariant subspace of A^p that contains some function f with $f(0) \neq 0$. Then we can generalize the extremal problem (1) to

$$\max_{f \in M,\, \|f\|_p = 1} |f(0)| \,. \tag{13}$$

As before, the problem has an equivalent formulation

$$\min_{f \in M,\, f(0) = 1} \|f\|_p \,. \tag{14}$$

For $1 < p < \infty$, the problem (14) again consists of finding a function of minimal norm in a closed convex subset of a uniformly convex Banach space, so it has a unique solution, by the standard theorem of functional analysis invoked earlier. For $p = 1$ there is at most one solution, since A^1 is strictly convex (*cf.* Section 2.2). For $0 < p < 1$, however, both the existence and the uniqueness of solutions are open problems for arbitrary invariant subspaces of A^p. It is also unknown whether an extremal function exists for every invariant subspace of A^1. We have proved existence and uniqueness of solutions in the zero-based subspaces N^p for $0 < p < \infty$, but for *general* invariant subspaces these questions remain unresolved when $0 < p < 1$.

A solution F to problem (13) with $F(0) > 0$ will be called a *canonical extremal function* of the invariant subspace M, under the assumption that the functions in M do not all vanish at the origin. A unique canonical extremal function always exists when $1 < p < \infty$.

Suppose now that F is a canonical extremal function for some invariant subspace $M \subset A^p$, where $0 < p < \infty$. The variational arguments of Section 5.1 apply with little change. Since $H = F/F(0)$ is a solution to the related extremal problem (14), it is clear that $H^* = H(1+\lambda f)$ is an admissible competitor for $\lambda \in \mathbb{C}$ and every polynomial f with $f(0)=0$, so $\|H^*\|_p \geq \|H\|_p$. The proof of Lemma 1 then gives

$$\int_{\mathbb{D}} |F(z)|^p \, z^n \, d\sigma(z) = 0\,, \qquad n = 1, 2, \ldots, \tag{15}$$

so that F is an A^p inner function. Consequently, the integral formula of Theorem 2 applies and shows that F is an expansive multiplier: $\|FQ\|_p \geq \|Q\|_p$ for every polynomial Q.

The same conclusions can be reached, with suitable interpretation, if the origin is a common zero of functions in an invariant subspace M. If all functions in M have a zero at the origin of order at least m, but $f^{(m)}(0) \neq 0$ for some $f \in M$, then the extremal problem (13) is replaced by that of maximizing $|f^{(m)}(0)|$ among all functions $f \in M$ with $\|f\|_p = 1$. A canonical extremal function of M is defined as a solution F to this modified problem with $F^{(m)}(0) > 0$. The variational proof of Lemma 1 can then be adapted to show that F is an A^p inner function, and is therefore an expansive multiplier.

The variational technique of Lemma 2 also extends to general invariant subspaces. Again let F be the canonical extremal function of an invariant subspace $M \subset A^p$, where $1 \leq p < \infty$. Then $H = F/F(0)$ solves the corresponding extremal problem (14), minimizing $\|f\|_p$ among all $f \in M$ with $f(0) = 1$. Now construct the variation $H^* = H + \lambda h$, where $\lambda \in \mathbb{C}$ and $h \in M$ with $h(0) = 0$. Then $\|H^*\|_p \geq \|H\|_p$, by the extremal property of H, so the proof of Lemma 2 gives

$$\int_{\mathbb{D}} |F(z)|^{p-2}\, \overline{F(z)} h(z) \, d\sigma = 0$$

for every $h \in M$ with $h(0) = 0$.

According to Theorem 4, the canonical divisor in A^p of a finite zero-set has the representation

$$G(z) = J(0,0)^{-1/p} \, B(z) \, J(z,0)^{2/p}\,,$$

where B is the associated Blaschke product and $J(z,\zeta)$ is the reproducing kernel of the space $\mathcal{P}^2_w$, the closure of the polynomials in A^2_w norm for the weight function $w = |B|^p$. This representation will now be generalized to arbitrary invariant subspaces of the form $[\psi]$, generated by any function $\psi \in A^p$. Subspaces of this form will be called *singly generated.* Recall that the space $\mathcal{P}^2_w$ with weight $w = |\psi|^p$ has a kernel function $J(z,\zeta)$, and that $J(z,0) \neq 0$ in $\mathbb{D}$, by Lemma 9. We will say that a function $\psi \in A^p$ is *normalized* if $\psi(0) > 0$, or more generally if its first nonvanishing derivative has the property $\psi^{(m)}(0) > 0$. With this terminology, we can state the following theorem.

THEOREM 7. *Let $0 < p < \infty$ and let $[\psi]$ be the invariant subspace of A^p generated by a normalized function $\psi \in A^p$. Let $J(z, \zeta)$ be the reproducing kernel of the space $\mathcal{P}_w^2$ with weight $w = |\psi|^p$. Then $[\psi]$ has a unique canonical extremal function, given by the formula*

$$F(z) = J(0,0)^{-1/p}\,\psi(z)\,J(z,0)^{2/p}\,. \tag{16}$$

As previously remarked, for $0 < p \le 1$ it is not known whether every invariant subspace of A^p has a canonical extremal function, and uniqueness is also an open question when $0 < p < 1$. For singly generated subspaces, however, we can assert the existence and uniqueness of a canonical extremal function even for $0 < p \le 1$, and this is part of the theorem.

Before embarking on the proof, we make some preliminary observations. By definition, the subspace $[\psi]$ consists of all functions $f \in A^p$ that can be approximated in norm by polynomial multiples of ψ. It follows that each function $f \in [\psi]$ must vanish wherever ψ does, to at least the same order, since norm convergence implies locally uniform convergence. In other words, f/ψ is analytic in $\mathbb{D}$, so each function $f \in [\psi]$ has the form $f = \psi h$ for some function h analytic in $\mathbb{D}$. It is easy to see that the admissible functions h are precisely those in the space $\mathcal{P}_w^p$, the closure of the polynomials in the A_w^p norm defined by

$$\|h\|_{p,w}^p = \int_{\mathbb{D}} |h|^p w\,d\sigma = \int_{\mathbb{D}} |\psi h|^p\,d\sigma\,.$$

Thus we can write $[\psi] = \psi\mathcal{P}_w^p$, where $w = |\psi|^p$. In Chapter 9 we will give another description of the invariant subspace $[\psi]$ when ψ is an A^p inner function.

For any function $\psi \in A^p$, it is clear that $[\psi]$ contains all functions of the form ψh where $h \in H^\infty$. Indeed, it follows from the Lebesgue dominated convergence theorem that for each function $h \in H^\infty$, the dilations $h_r(z) = h(rz)$ converge to h in A_w^p norm as $r \to 1$. On the other hand, each dilation h_r is analytic in $\overline{\mathbb{D}}$, so it can be approximated uniformly, hence in A_w^p norm, by polynomials. Thus $H^\infty \subset \mathcal{P}_w^p$. Since every polynomial is in H^∞, this shows that the H^∞ functions are dense in $\mathcal{P}_w^p$.

PROOF OF THEOREM 7. Suppose first that $\psi(0) > 0$. Recall that a canonical extremal function of $[\psi]$ is defined as a normalized function $F \in [\psi]$ of unit A^p norm that maximizes $|f(0)|$ among all functions $f \in [\psi]$ with norm $\|f\|_p = 1$. In view of the preceding remarks, this extremal problem reduces to finding the supremum of $|h(0)|$ among all polynomials h with $\|\psi h\|_p = 1$, or equivalently among all functions $h \in H^\infty$ with $\|\psi h\|_p = 1$. Our first goal is to show that no generality is lost if the extremal problem is posed over the special subset of *nonvanishing* functions $h \in H^\infty$ with $\|\psi h\|_p = 1$; in other words, the restricted problem will have the same solution. To see this, let

$$\mu = \sup\{|h(0)| \;:\; h \in H^\infty,\ \|\psi h\|_p = 1\}\,,$$

and let $\{Q_n\}$ be an extremal sequence of polynomials, with $\|\psi Q_n\|_p = 1$ and $|Q_n(0)| \to \mu$ as $n \to \infty$. Since $\mu > 0$, we may assume that $Q_n(0) \neq 0$ for all n. If Q_n has zeros in $\mathbb{D}$, let G_n be the canonical divisor in A^p of the (finite) subset of its zeros that lie in $\mathbb{D}$. Then Q_n/G_n belongs to H^∞ and has no zeros in $\mathbb{D}$, by Theorems 1 and 5. By the contractive divisor property of G_n (see Lemma 11),

$$\|\psi Q_n/G_n\|_p \leq \|\psi Q_n\|_p = 1\,,$$

and so $(Q_n/G_n)(0) \leq \mu$ by definition of μ. However, $0 < G_n(0) < 1$, so

$$|Q_n(0)| < |(Q_n/G_n)(0)| \leq \mu$$

and therefore $|(Q_n/G_n)(0)| \to \mu$ as $n \to \infty$. This shows that the canonical extremal function F of the subspace $[\psi]$ is a solution to the problem of maximizing $h(0)$ among all nonvanishing functions $h \in H^\infty$ with $\|\psi h\|_p = 1$ and $h(0) > 0$. Equivalently, it is enough to consider functions $f = \psi k^{2/p}$ with $k \in H^\infty$, $k(z) \neq 0$ in $\mathbb{D}$, and $k(0) > 0$, such that

$$\|f\|_p^p = \|\psi k^{2/p}\|_p^p = \int_{\mathbb{D}} |k(z)|^2 |\psi(z)|^p \, d\sigma = \|k\|_{2,w}^2 = 1\,, \qquad w = |\psi|^p\,,$$

and to pose the extremal problem of maximizing $k(0)$.

On the other hand, the function $J(0,0)^{-1/2} J(z,0)$ uniquely maximizes $\mathrm{Re}\{k(0)\}$ among all functions $k \in \mathcal{P}_w^2$ of unit norm. The same function maximizes $\mathrm{Re}\{k(0)\}$ among all functions $k \in H^\infty$ with $\|k\|_{2,w} = 1$, since these functions can be approximated by polynomials and are therefore dense in the space of all functions in $\mathcal{P}_w^2$ of unit norm. But by Lemma 9, the function $J(z,0) \neq 0$ in $\mathbb{D}$, so the function $J(0,0)^{-1/2} J(z,0)$ uniquely maximizes $k(0)$ among all functions $k \in H^\infty$ with $\|k\|_{2,w} = 1$, $k(0) > 0$, and $k(z) \neq 0$ in $\mathbb{D}$. However, this is precisely the problem that leads to a canonical extremal function $F = \psi k^{2/p}$ for the subspace $[\psi]$. Consequently, a canonical extremal function exists and necessarily has the form (16). In particular, the canonical extremal function is uniquely determined.

The same argument applies, with minor adjustments, if $\psi(0) = 0$. Suppose ψ has a zero of order m at the origin, and $\psi^{(m)}(0) > 0$. Then a canonical extremal function is determined by maximizing $|(\psi h)^{(m)}(0)| = |\psi^{(m)}(0) h(0)|$ among all functions $\psi h \in [\psi]$ of unit A^p norm. This problem reduces in the same way to finding the supremum of $|h(0)|$ among all nonvanishing functions $h \in H^\infty$ with $\|\psi h\|_p = 1$, so the canonical extremal function is again expressed by (16). □

It will be proved in Chapter 9 that the extremal function F of the subspace $[\psi]$ is itself a generator: $[F] = [\psi]$. Thus we can infer from Theorem 7 that

$$F(z) = \widetilde{J}(0,0)^{-1/p} F(z) \, \widetilde{J}(z,0)^{2/p}\,,$$

where $\widetilde{J}(z,\zeta)$ is the kernel function of $\mathcal{P}^2_\omega$ with weight $\omega = |F|^p$. This says that $\widetilde{J}(z,0) \equiv 1$, and the reproducing property gives

$$\int_{\mathbb{D}} Q(z)|F(z)|^p \, d\sigma = \int_{\mathbb{D}} Q(z)\widetilde{J}(0,z)\omega(z)\, d\sigma = Q(0) = \int_{\mathbb{D}} Q(z)\, d\sigma\,,$$

or

$$\int_{\mathbb{D}} \big(|F(z)|^p - 1\big)Q(z)\, d\sigma = 0$$

for all polynomials Q, in agreement with (15).

Theorem 7 was proved by Duren, Khavinson, and Shapiro [1] for the special case where ψ is a classical inner function. As a specific example they considered the invariant subspace $[S]$ generated in A^p by the atomic singular inner function

$$S(z) = \exp\left\{-\frac{1+z}{1-z}\right\}.$$

This is a *proper* subspace of A^p, according to a general theorem to be proved in Chapter 8, but this fact can also be gleaned from the following considerations.

We propose to show that the canonical extremal function of $[S]$ has the form

$$F(z) = (1+p)^{-1/p}\, S(z)\left(1 + \frac{p}{1-z}\right)^{2/p}. \tag{17}$$

Since $F(z) \not\equiv 1$, it follows that $[S] \neq A^p$. Comparing (17) with (16), we can deduce information about the reproducing kernel $J(z,\zeta)$ of the Hilbert space $\mathcal{P}^2_w$ with weight $w = |S|^p$. When $\zeta = 0$, it takes the simple form

$$J(z,0) = 1 + \frac{p}{1-z}\,.$$

The proof of (17) depends on the observation that the finite Blaschke products

$$B_n(z) = \left\{\frac{(1-\frac{1}{n}) - z}{1-(1-\frac{1}{n})z}\right\}^n = \{b_n(z)\}^n$$

converge to $S(z)$ as $n \to \infty$, uniformly on compact subsets of $\mathbb{D}$. The subspace $[B_n]$ consists of all functions in A^p with a zero of order at least n at the point $1 - \frac{1}{n}$, so its canonical extremal function (or canonical divisor) is

$$G_n(z) = C\, B_n(z)\big\{1 + \tfrac{np}{2}\big(1 - (1-\tfrac{1}{n})b_n(z)\big)\big\}^{2/p},$$

where

$$C = \big\{1 + \tfrac{np}{2}\big(1 - (1-\tfrac{1}{n})^2\big)\big\}^{-1/p},$$

as shown at the end of Section 5.2. After algebraic manipulation, the formula reduces to

$$G_n(z) = \big\{1 + p(1-\tfrac{1}{2n})\big\}^{-1/p}\, B_n(z)\left\{1 + \frac{p(1-\frac{1}{2n})}{1-(1-\frac{1}{n})z}\right\}^{2/p}.$$

Since $B_n(z) \to S(z)$ and $G_n(z)$ tends locally uniformly to the function $F(z)$ given by (16), it is reasonable to expect F to be the canonical extremal function of $[S]$.

To prove this, we will show first that $[F] = [S]$. Since the function $(1 + p/(1-z))^{-1}$ is bounded in $\mathbb{D}$, it is clear that $S \in [F]$ and hence that $[S] \subset [F]$. To see that $F \in [S]$, observe first that because $\|G_n\|_p = 1$ and $G_n(z) \to F(z)$ pointwise in $\mathbb{D}$, Fatou's lemma gives $\|F\|_p \le 1$. This implies that

$$\int_{\mathbb{D}} |S(z)|^p \, |1-z|^{-2} \, d\sigma < \infty . \tag{18}$$

Now let $Q_n(z) = 1 + z + \cdots + z^n$ and observe that $1 + p\,Q_n(z) \neq 0$ in $\mathbb{D}$, since $|z + pz^{n+1}| < 1 + p$. Thus $(1 + p\,Q_n)^{2/p}$ is analytic in $\overline{\mathbb{D}}$, so that $(1 + p\,Q_n)^{2/p} S \in [S]$. But

$$(1+p)^{-1/p}\, S(z)(1 + p\,Q_n(z))^{2/p} \to F(z)$$

pointwise in $\mathbb{D}$, so (18) allows us to apply the Lebesgue dominated convergence theorem to conclude that the convergence holds in the A^p norm. Thus $F \in [S]$, and we have shown that $[F] = [S]$.

In order to conclude that the function F is the canonical extremal function of $[S]$, we have to show that $\|F\|_p = 1$ and $F(0) = \lambda$, where

$$\lambda = \sup_Q |Q(0)S(0)| ,$$

the supremum extending over all polynomials Q with $\|QS\|_p = 1$. But G_n is the canonical divisor of $[B_n]$, so by definition,

$$G_n(0) = \sup_Q |Q(0)B_n(0)| ,$$

for all polynomials Q with $\|QB_n\|_p = 1$. Since $B_n(z) \to S(z)$ pointwise in $\mathbb{D}$, it follows from the Lebesgue dominated convergence theorem that $\|QB_n - QS\|_p \to 0$ for each polynomial Q. Thus $\|QB_n\|_p \to \|QS\|_p$ as $n \to \infty$. Now let Q be a polynomial with $\|QS\|_p = 1$. Then for each $\varepsilon > 0$, there is an N such that $\|QB_n\|_p \le 1 + \varepsilon$ for all $n \ge N$, and so

$$|Q(0)B_n(0)| \le (1+\varepsilon)G_n(0) , \qquad n \ge N.$$

But $B_n(0) \to S(0)$ and $G_n(0) \to F(0)$, so this implies that $|Q(0)S(0)| \le F(0)$, showing that $\lambda \le F(0)$. On the other hand, $F \in [S]$ implies that $F(0) \le \lambda \, \|F\|_p$, hence $\|F\|_p \ge 1$. However, we saw above that $\|F\|_p \le 1$, so this shows $\|F\|_p = 1$ and $F(0) = \lambda$. This completes the proof that F is the canonical extremal function of $[S]$.

CHAPTER 6

Sampling and Interpolation

In Chapter 4 we studied zero-sets of the Bergman spaces A^p. We turn now to a closely related topic, the sampling and interpolation sequences. After some preliminary discussion, a special family of sequences is examined directly by explicit constructions. Then comes a definition of density and statements of the two main theorems characterizing sampling and interpolation sequences for A^p. The proofs are deferred to Chapter 7, but the present chapter concludes by exploring some consequences of the main theorems and giving further examples.

§6.1. Definitions and motivations.

A sequence $\{z_k\}$ of distinct points in the unit disk is called a *sampling sequence for* A^p, where $0 < p < \infty$, if there exist positive constants K_1 and K_2 such that

$$K_1 \|f\|_p^p \le \sum_{k=1}^{\infty} (1 - |z_k|^2)^2 |f(z_k)|^p \le K_2 \|f\|_p^p$$

for all $f \in A^p$. This pair of inequalities will be called the *sampling inequalities.*

Sampling sequences may be viewed as those for which the integral $\int_{\mathbb{D}} |f(z)|^p d\sigma$ is comparable to a Riemann sum. If the disk is partitioned into nonoverlapping regions S_k with normalized areas ΔA_k, and if $z_k \in S_k$, then the corresponding Riemann sum is $\sum |f(z_k)|^p \Delta A_k$. But if the subregions S_k can be chosen to have equal *hyperbolic* area, then ΔA_k has the approximate form $C(1 - |z_k|^2)^2$, and we can say that $\{z_k\}$ is a sampling sequence if and only if the Riemann sum $\sum |f(z_k)|^p \Delta A_k$ is essentially a constant multiple of the integral $\int_{\mathbb{D}} |f(z)|^p \, d\sigma$ as f ranges over A^p.

Another motivation for sampling sequences comes from the notion of a frame, which plays a role in the theory of wavelets. If $\mathcal{H}$ is any separable Hilbert space and $\{u_n\}$ is an orthonormal basis, Parseval's formula states that

$$\|x\|^2 = \sum_{n=1}^{\infty} |\langle x, u_n \rangle|^2 , \qquad x \in \mathcal{H} .$$

More generally, a sequence $\{v_n\}$ is said to be a *frame for* $\mathcal{H}$ if there are positive constants A and B such that

$$A\,\|x\|^2 \le \sum_{n=1}^{\infty} |\langle x, v_n\rangle|^2 \le B\,\|x\|^2\,, \qquad x \in \mathcal{H}\,.$$

Every orthonormal basis is a frame.

Recall that the Bergman kernel function $k_\zeta(z) = (1 - \overline{\zeta}z)^{-2}$ has the reproducing property $\langle f, k_\zeta\rangle = f(\zeta)$ for all $f \in A^2$. These kernel functions are not mutually orthogonal, so for no sequence of distinct points $z_n \in \mathbb{D}$ do the normalized kernel functions $k_{z_n}/\|k_{z_n}\|_2$ form an orthonormal basis. However, for certain choices of points z_n, these functions will constitute a frame for A^2. Since

$$\|k_\zeta\|_2 = \sqrt{k_\zeta(\zeta)} = \frac{1}{1 - |\zeta|^2}\,,$$

we see by the reproducing property $\langle f, k_{z_n}\rangle = f(z_n)$ that this occurs precisely when $\{z_n\}$ is a sampling sequence for A^2.

There is no counterpart of sampling sequences for Hardy spaces. The analogous requirement for the Hardy space H^2 is that the normalized Szegő kernels form a frame. Generalizing the resulting condition to H^p, we are led to say that $\{z_k\}$ is a sampling sequence for H^p if

$$A\,\|f\|_{H^p}^p \le \sum_{k=1}^{\infty} (1 - |z_k|^2)|f(z_k)|^p \le B\,\|f\|_{H^p}^p\,, \qquad f \in H^p\,,$$

for some constants A and B. However, an application of the upper inequality to the function $f = 1$ shows that $\{z_k\}$ must be a Blaschke sequence, which contradicts the lower inequality since the Blaschke sequences are zero-sets for H^p. Thus H^p has no sampling sequences.

Nevertheless, the Bergman spaces A^p do have sampling sequences. Their existence is not obvious, and actual examples are not so easy to produce, but some constructions will be carried out in this chapter.

The upper sampling inequality is satisfied whenever the sequence $\{z_k\}$ is uniformly discrete. Indeed, we know from results in Section 2.11 (*cf.* Theorem 15) that $\sum (1 - |z_k|^2)^2|f(z_k)|^p \le C\,\|f\|_p^p$ for every $f \in A^p$ if and only if $\{z_k\}$ is a finite union of uniformly discrete sequences. On the other hand, we define $K_1(\Gamma)$ to be the largest constant K_1 for which the lower sampling inequality holds, and we set $K_1(\Gamma) = 0$ if the inequality fails to hold. A uniformly discrete sequence Γ is then a sampling sequence if and only if $K_1(\Gamma) > 0$. The number $K_1(\Gamma)$ will be called the *sampling constant* of Γ for the space A^p.

An important property of sampling sequences is their conformal invariance. In other words, if $\Gamma = \{z_k\}$ is a sampling sequence for A^p, then so is $\varphi(\Gamma) = \{\zeta_k\} = \{\varphi(z_k)\}$ for every conformal mapping

$$\zeta = \varphi(z) = \frac{\alpha - z}{1 - \overline{\alpha}z}\,, \qquad \alpha \in \mathbb{D}\,,$$

of the disk onto itself. In fact, $K_1(\varphi(\Gamma)) = K_1(\Gamma)$. To see this, recall that the mapping $g \mapsto f$ defined by $f(z) = g(\varphi(z))\varphi'(z)^{2/p}$ is a surjective isometry of A^p, while a calculation shows that

$$(1 - |\zeta_k|^2)^2 |g(\zeta_k)|^p = (1 - |z_k|^2)^2 |f(z_k)|^p ,$$

in view of the identity

$$1 - |\varphi(z)|^2 = (1 - |z|^2)|\varphi'(z)| , \tag{1}$$

which expresses the conformal invariance of the hyperbolic metric $\frac{|dz|}{1-|z|^2}$.

Closely related to the concept of sampling, and in some sense dual to it, is that of interpolation. A sequence $\{z_k\}$ of distinct points in $\mathbb{D}$ is called an *interpolation sequence for* A^p if the interpolation problem $f(z_k) = w_k$ for $k = 1, 2, \ldots$ has a solution $f \in A^p$ whenever

$$\sum_{k=1}^{\infty} (1 - |z_k|^2)^2 |w_k|^p < \infty .$$

We know that $M_\infty(r, f) = o\big((1 - r)^{-2/p}\big)$ for every function $f \in A^p$ (see Section 3.2), so the general term of the infinite series

$$\sum_{k=1}^{\infty} (1 - |z_k|^2)^2 |f(z_k)|^p$$

tends to zero whenever $|z_k| \to 1$. As noted above, however, the series converges for every $f \in A^p$ if and only if $\{z_k\}$ is a finite union of uniformly discrete sequences.

Every interpolation sequence is a zero-set. Indeed, if $\{z_k\}$ is an A^p interpolation sequence, then there is a function $f \in A^p$ with $f(z_1) = 1$ and $f(z_k) = 0$ for $k = 2, 3, \ldots$. Then the function $g(z) = (z - z_1)f(z)$ belongs to A^p and vanishes on $\{z_k\}$, but is not the zero function. This says that $\{z_k\}$ is a subset of an A^p zero-set. But by the theorem of Horowitz (Section 4.4, Theorem 4), every subset of an A^p zero-set is an A^p zero-set.

On the other hand, an A^p sampling sequence is *never* an A^p zero-set. This follows directly from the lower sampling inequality; if a function $f \in A^p$ vanishes on a sampling sequence $\{z_k\}$, then $\|f\|_p = 0$ and f is the zero function. Thus no sequence can be both sampling and interpolating.

As for sampling sequences, one can associate a constant with each interpolation sequence. Given a sequence Γ, let $\ell^p(\Gamma)$ be the space consisting of sequences $\{w_k\}$ for which

$$\|\{w_k\}\|_p^p = \sum_{k=1}^{\infty} (1 - |z_k|^2)^2 |w_k|^p < \infty .$$

If Γ is an interpolation sequence, one can define the linear operator R_p that maps an $\ell^p(\Gamma)$ sequence $\{w_k\}$ to the coset $f+N^p$ in the space A^p/N^p, where f is any function performing the interpolation, and N^p is the subspace of A^p consisting of functions vanishing on Γ. It is clear from the definition that R_p is a well-defined closed mapping, which by the closed graph theorem is bounded. Equivalently, there is a constant M such that for each $\{w_k\}$ satisfying $\sum(1-|z_k|^2)^2|w_k|^p < \infty$, the inequality

$$\|f\|_p^p \leq M \sum_{k=1}^{\infty}(1-|z_k|^2)^2|w_k|^p$$

holds for some $f \in A^p$ with $f(z_k) = w_k$. We define $M(\Gamma)$ to be the smallest such constant M. This number $M(\Gamma)$ will be called the *interpolation constant* of Γ for the space A^p.

It may also be noted that A^p interpolation sequences, like sampling sequences, are conformally invariant and moreover that $M(\varphi(\Gamma)) = M(\Gamma)$, where

$$\varphi(z) = \frac{\alpha - z}{1-\overline{\alpha}z}$$

is a conformal self-mapping of the disk. The proof uses the identity (1) and the isometry defined by $f(z) = g(\varphi(z))\varphi'(z)^{2/p}$. To be more precise, suppose $\{z_k\}$ is an A^p interpolation sequence, and let $\zeta_k = \varphi(z_k)$. If

$$\sum_{k=1}^{\infty}(1-|\zeta_k|^2)^2|v_k|^p < \infty\,,$$

take $w_k = v_k\,\varphi'(z_k)^{2/p}$ and observe that

$$\sum_{k=1}^{\infty}(1-|z_k|^2)^2|w_k|^p = \sum_{k=1}^{\infty}(1-|\zeta_k|^2)^2|v_k|^p < \infty\,.$$

Therefore, because $\{z_k\}$ is an interpolation sequence, $f(z_k) = w_k$ for some $f \in A^p$. But if g is related to f as above, then $\|g\|_p = \|f\|_p$ and $g(\zeta_k) = v_k$, which shows that $\{\zeta_k\}$ is also an A^p interpolation sequence.

One trivial observation is often useful. If Γ is an interpolation sequence for A^p, then every subsequence has the same property. Similarly, if Γ is a finite union of uniformly discrete sequences and some subsequence of Γ is a sampling sequence for A^p, then Γ is a sampling sequence for A^p.

The sampling and interpolation sequences for A^p can be completely described in terms of certain densities introduced by Kristian Seip [3]. The precise statements require some preparation and will be given later in this chapter. Meanwhile, by way of orientation, we propose to discuss interpolation sequences for Hardy spaces and to show by direct construction of an interpolation series that they coincide with the uniformly separated sequences. We then classify a special family of sequences as interpolating or sampling for the Bergman space A^p.

§6.2. Interpolation in Hardy spaces.

The interpolation sequences for the Hardy space H^p were discussed briefly in Section 4.6. Recall that a Blaschke sequence $\{z_k\}$ is said to be *uniformly separated* if

$$\prod_{j\neq k}\left|\frac{z_j - z_k}{1-\overline{z_j}z_k}\right| \geq \delta\,, \qquad k = 1, 2, \ldots\,,$$

for some constant $\delta > 0$ independent of k. A sequence $\{z_k\}$ is called a *universal interpolation sequence* if for each complex sequence $\{w_k\} \in \ell^\infty$, there is a function $f \in H^\infty$ with $f(z_k) = w_k$ for $n = 1, 2, \ldots$. It is called an *interpolation sequence for H^p*, where $0 < p < \infty$, if for each sequence $\{w_k\}$ with $\sum(1 - |z_k|^2)|w_k|^p < \infty$, there is a function $f \in H^p$ for which $f(z_k) = w_k$. Carleson [2] proved that $\{z_k\}$ is a universal interpolation sequence if and only if it is uniformly separated. Shapiro and Shields [1] then showed for $1 \leq p < \infty$ that $\{z_k\}$ is an interpolation sequence for H^p if and only if it is uniformly separated. For $0 < p < \infty$, the *necessity* of the uniform separation condition is proved rather easily by judicious choice of the sequence $\{w_k\}$. To prove the *sufficiency*, Shapiro and Shields devised a clever argument that appeals to the theory of dual extremal problems and is therefore applicable only for $1 \leq p < \infty$. Kabaĭla [1,2] proved the sufficiency for $0 < p \leq 1$ by an explicit interpolation series, but his construction fails for $1 < p < \infty$. (Further details can be found in Duren [5], Chapter 9.)

More recently, Schuster and Seip [1] found a simpler proof of sufficiency in the case $1 < p < \infty$, giving an H^p interpolation series in the spirit of Kabaĭla's proof for $0 < p \leq 1$. Their construction, which is reproduced below, relies on the fact that the Szegő projection

$$(Tf)(z) = \frac{1}{2\pi}\int_0^{2\pi} \frac{f(e^{i\theta})e^{i\theta}}{e^{i\theta} - z}\, d\theta$$

is a bounded operator from L^p to H^p for $1 < p < \infty$. (See Section 2.1.) It also uses the fact (contained in Theorem 10 of Section 4.6) that if $\{z_k\}$ is uniformly separated, then

$$\sum_{k=1}^{\infty}(1 - |z_k|^2)|f(z_k)|^p < \infty$$

for every $f \in H^p$ and so, by the closed graph theorem,

$$\left\{\sum_{k=1}^{\infty}(1 - |z_k|^2)|f(z_k)|^p\right\}^{1/p} \leq C\,\|f\|_{H^p}\,, \qquad f \in H^p\,,$$

for some constant C.

Let $\{z_k\}$ be a uniformly separated sequence, and let $B(z)$ be its corresponding Blaschke product. An equivalent expression of the uniform separation condition is that

$$(1-|z_k|^2)|B'(z_k)| \geq \delta > 0\,, \qquad k=1,2,\dots.$$

Suppose that $\sum(1-|z_k|^2)|w_k|^p < \infty$ for some $p>1$, and consider the infinite series

$$f(z) = \sum_{k=1}^{\infty} w_k \frac{B(z)}{(z-z_k)B'(z_k)}\,. \tag{2}$$

By the uniform separation of $\{z_k\}$, the estimate

$$\left| w_k \frac{B(z)}{(z-z_k)B'(z_k)} \right| \leq \frac{2}{\delta(1-R)}(1-|z_k|^2)|w_k|\,, \qquad |z| \leq R < 1\,,$$

holds for all k sufficiently large. But the condition $\sum(1-|z_k|^2)|w_k|^p < \infty$ implies $\sum(1-|z_k|^2)|w_k| < \infty$, since $\{z_k\}$ is a Blaschke sequence. Thus the above estimate shows that the series (2) converges locally uniformly in $\mathbb{D}$ and therefore represents an analytic function $f(z)$. Observe now that the partial sums

$$s_n(z) = \sum_{k=1}^{n} w_k \frac{B(z)}{(z-z_k)B'(z_k)}$$

satisfy $s_n(z_k) = w_k$ for $k=1,2,\dots,n$. This implies that $f(z_k)=w_k$ for $k=1,2,\dots$, since $s_n(z)\to f(z)$ pointwise in $\mathbb{D}$.

To complete the proof, we need to show that $f\in H^p$. This will follow from Fatou's lemma if we can show that the H^p norms of the functions s_n are bounded. To prove this, let $h\in L^q$, where $q=\frac{p}{p-1}$ is the conjugate exponent, and write

$$\begin{aligned}\frac{1}{2\pi}\int_0^{2\pi} s_n(e^{i\theta})h(e^{i\theta})e^{i\theta}\,d\theta &= \frac{1}{2\pi}\sum_{k=1}^{n}\frac{w_k}{B'(z_k)}\int_0^{2\pi}\frac{B(e^{i\theta})h(e^{i\theta})e^{i\theta}}{(e^{i\theta}-z_k)}\,d\theta \\ &= \sum_{k=1}^{n}\frac{w_k}{B'(z_k)}T(Bh)(z_k)\,,\end{aligned}$$

where T is the Szegő projection. Thus by Hölder's inequality and the boundedness of the Szegő projection,

$$\left|\frac{1}{2\pi}\int_0^{2\pi} s_n(e^{i\theta})h(e^{i\theta})e^{i\theta}\,d\theta\right| \leq \frac{1}{\delta}\sum_{k=1}^{n}(1-|z_k|^2)|w_k||T(Bh)(z_k)|$$

$$\leq \frac{1}{\delta}\left\{\sum_{k=1}^{n}(1-|z_k|^2)|w_k|^p\right\}^{1/p}\left\{\sum_{k=1}^{n}(1-|z_k|^2)|T(Bh)(z_k)|^q\right\}^{1/q}$$

$$\leq C\,\|T(Bh)\|_{H^q} \leq C\,\|Bh\|_{L^q} = C\,\|h\|_{L^q}\,,$$

since $\{z_k\}$ is uniformly separated. This shows that

$$\|s_n\|_{H^p} = \sup_{\|h\|_{L^q}=1} \left| \frac{1}{2\pi} \int_0^{2\pi} s_n(e^{i\theta})h(e^{i\theta})e^{i\theta}\, d\theta \right| \le C\,, \qquad n = 1, 2, \dots,$$

as claimed. Consequently, the function f defined by (2) is in H^p, and we have proved that every uniformly separated sequence is an interpolation sequence for H^p, where $1 < p < \infty$.

Another constructive proof was given earlier by Eric Amar [2].

§6.3. A family of sampling and interpolation sequences.

In this section we return to the main topic of the chapter, sampling and interpolation sequences for Bergman spaces. We develop a criterion that classifies certain kinds of sequences as sampling or interpolating. The results are then applied to a special parametrized family of sequences, where they determine precisely which values of the parameters give sampling or interpolation sequences for the space A^p. Seip [2] carried out a similar analysis for A^2 as a forerunner to his general density description of sampling and interpolation sequences. The techniques developed here, as well as the results themselves, will play a role in the proofs of those general theorems.

Recall that the pseudohyperbolic metric ρ is defined by

$$\rho(z, \zeta) = |\varphi_\zeta(z)|\,, \qquad \text{where } \varphi_\zeta(z) = \frac{\zeta - z}{1 - \bar{\zeta} z}\,,$$

and that a sequence of points $\Gamma = \{z_k\}$ is uniformly discrete if

$$\delta(\Gamma) = \inf_{j \ne k} \rho(z_j, z_k) > 0\,.$$

The number $\delta = \delta(\Gamma)$ is called the *separation constant* of Γ. Let $\rho(z, \Gamma) = \inf_k \rho(z, z_k)$ denote the pseudohyperbolic distance from a point $z \in \mathbb{D}$ to the sequence Γ. If $F(z)$ and $G(z)$ are nonnegative functions in $\mathbb{D}$, we will write $F(z) \simeq G(z)$ if there are positive constants C_1 and C_2 such that

$$C_1\, F(z) \le G(z) \le C_2\, F(z) \qquad \text{for all } z \in \mathbb{D}\,.$$

Suppose now that Γ is a uniformly discrete sequence that admits an analytic function g with the property

$$|g(z)| \simeq \rho(z, \Gamma)(1 - |z|^2)^{-\alpha}\,, \qquad z \in \mathbb{D}\,, \tag{3}$$

for some $\alpha > 0$. Note that there can be at most one number α associated in this way with a given sequence Γ. To see this, observe first that the zero-set of g is precisely Γ. If there is another analytic function h with the property $h(z) \simeq \rho(z, \Gamma)(1 - |z|^2)^{-\beta}$, where $\beta < \alpha$, then h/g is analytic in $\mathbb{D}$ and vanishes at the boundary, so it must vanish everywhere in $\mathbb{D}$, by the maximum modulus principle. But this is clearly impossible.

In the next section we will discuss a specific class of examples with the property (3). In terms of the parameter α specified by (3), we now prove the following two theorems for $1 \le p < \infty$.

THEOREM 1. *For $0 < p < \infty$, a uniformly discrete sequence Γ with the property (3) is an interpolation sequence for A^p if and only if $\alpha < \frac{1}{p}$.*

THEOREM 2. *For $0 < p < \infty$, a uniformly discrete sequence Γ with the property (3) is a sampling sequence for A^p if and only if $\alpha > \frac{1}{p}$.*

These theorems, when applied to a specific class of sequences to be discussed in the next section, will provide examples of sampling and interpolation sequences for A^p. In fact, the theorems show that in some sense, sampling and interpolation sequences can be arbitrarily close to each other. They also give the first hint of a geometric quantity associated with a sequence that determines whether it is sampling or interpolating or neither.

Observe at the outset that the property (3) implies $g \in A^p$ when $\alpha < \frac{1}{p}$. This shows directly that Γ is an A^p zero-set if $\alpha < \frac{1}{p}$. Theorem 1 makes the stronger assertion that Γ is an A^p interpolation set if $\alpha < \frac{1}{p}$, and Theorem 2 tells us that Γ is not an A^p zero-set when $\alpha > \frac{1}{p}$. The proof of Theorem 1 will also show that Γ is not an A^p zero-set when $\alpha = \frac{1}{p}$.

The proofs will make use of several lemmas, which are stated below. Lemma 1 was proved in Section 2.3, but is transcribed here for easy reference. Lemma 2 is a generalization of a result in Section 2.11.

LEMMA 1. *For $1 < t < s$, there is a constant $C = C(t, s) > 0$ such that*

$$\int_{\mathbb{D}} \frac{(1-|z|^2)^{t-2}}{|1-\bar{\zeta}z|^s}\, d\sigma(z) \le C\,(1-|\zeta|^2)^{t-s}, \qquad \zeta \in \mathbb{D}.$$

LEMMA 2. *Let $\Gamma = \{z_k\}$ be a uniformly discrete sequence with separation constant $\delta = \delta(\Gamma)$. Then for $t \in \mathbb{R}$ and $0 < p < \infty$ there is a constant $C = C(t, \delta) > 0$ such that*

$$\sum_{z_k \in \Omega} (1-|z_k|^2)^t |f(z_k)|^p \le C \int_{\rho(z,\Omega)<\delta} (1-|z|^2)^{t-2} |f(z)|^p\, d\sigma$$

for every measurable set $\Omega \subset \mathbb{D}$ and every function f analytic in $\mathbb{D}$.

LEMMA 3. *Let $\Gamma = \{z_k\}$ be a uniformly discrete sequence with separation constant $\delta = \delta(\Gamma)$. Then for $1 < t < s$, there is a constant $C = C(t, s, \delta) > 0$ such that*

$$\sum_{k=1}^{\infty} \frac{(1-|z_k|^2)^t}{|1-\bar{\zeta}z_k|^s} \le C\,(1-|\zeta|^2)^{t-s}, \qquad \zeta \in \mathbb{D}.$$

LEMMA 4. *Let $0 < p < \infty$, $0 < \varepsilon < 1$, and $w \in \mathbb{D}$. Then*

$$\int_{\Delta(w,\frac{\varepsilon}{2})} \frac{|h(z)|^p}{1-|z|^2}\, d\sigma \le \int_{\Omega(w,\frac{\varepsilon}{2},\varepsilon)} \frac{|h(z)|^p}{1-|z|^2}\, d\sigma$$

for each function h analytic in $\mathbb{D}$, where $\Omega(w, r, s)$ denotes the pseudohyperbolic annulus $\{z : r < \rho(z, w) < s\}$.

PROOF OF LEMMA 2. The proof is essentially the same as that of Lemma 14 in Chapter 2. The pseudohyperbolic disks $\Delta(z_j, \frac{\delta}{2})$ and $\Delta(z_k, \frac{\delta}{2})$ are disjoint for $j \neq k$. For $z_k \in \Omega$, the disk $\Delta(z_k, \frac{\delta}{2})$ is contained in the set where $\rho(z, \Omega) < \delta$. Thus

$$\int_{\rho(z,\Omega)<\delta} (1-|z|^2)^{t-2}|f(z)|^p\, d\sigma \geq \sum_{z_k\in\Omega} \int_{\Delta(z_k,\frac{\delta}{2})} (1-|z|^2)^{t-2}|f(z)|^p\, d\sigma$$

$$= \sum_{z_k\in\Omega} \int_{\Delta(0,\frac{\delta}{2})} (1-|\varphi_{z_k}(\zeta)|^2)^{t-2}|f(\varphi_{z_k}(\zeta))|^p|\varphi'_{z_k}(\zeta)|^2\, d\sigma$$

$$= \sum_{z_k\in\Omega} \int_{\Delta(0,\frac{\delta}{2})} (1-|\zeta|^2)^{t-2}|f(\varphi_{z_k}(\zeta))|^p|\varphi'_{z_k}(\zeta)|^t\, d\sigma\,,$$

in view of the identity $(1-|\zeta|^2)|\varphi'_w(\zeta)| = 1-|\varphi_w(\zeta)|^2$. But the factor $(1-|\zeta|^2)^{t-2}$ is bounded below by a positive constant, so the last sum is bounded below by a constant multiple of

$$\sum_{z_k\in\Omega} \int_{\Delta(0,\frac{\delta}{2})} |f(\varphi_{z_k}(\zeta))|^p|\varphi'_{z_k}(\zeta)|^t\, d\sigma \geq \frac{\delta^2}{4} \sum_{z_k\in\Omega} |f(\varphi_{z_k}(0))|^p|\varphi'_{z_k}(0)|^t$$

$$= \frac{\delta^2}{4} \sum_{z_k\in\Omega} (1-|z_k|^2)^t|f(z_k)|^p\,,$$

by the sub-mean-value property of each integrand. □

PROOF OF LEMMA 3. Apply Lemmas 2 and 1 to the function $f(z) = (1-\overline{\zeta}z)^{-s/p}$, with $\Omega = \mathbb{D}$. □

PROOF OF LEMMA 4. By changing variables, we see that

$$\int_{\Delta(w,\frac{\varepsilon}{2})} \frac{|h(z)|^p}{1-|z|^2}\, d\sigma = \int_{\Delta(0,\frac{\varepsilon}{2})} \frac{|h(\varphi_w(\zeta))|^p}{1-|\varphi_w(\zeta)|^2}\, |\varphi'_w(\zeta)|^2\, d\sigma\,.$$

By the standard identity

$$1-|\varphi_w(\zeta)|^2 = \frac{(1-|w|^2)(1-|\zeta|^2)}{|1-\overline{w}\zeta|^2}\,,$$

the last integrand equals $g_w(\zeta)/(1-|\zeta|^2)$, where

$$g_w(\zeta) = \frac{|h(\varphi_w(\zeta))|^p|1-\overline{w}\zeta|^2|\varphi'_w(\zeta)|^2}{1-|w|^2}\,.$$

But for any subharmonic function $u(\zeta)$, the integral means $\int_0^{2\pi} u(re^{i\theta})d\theta$ are nondecreasing functions of r. Since $g_w(\zeta)$ is subharmonic,

$$\begin{aligned}
&\int_{\Delta(0,\frac{\varepsilon}{2})} \frac{g_w(\zeta)}{1-|\zeta|^2}\, d\sigma = \frac{1}{\pi}\int_0^{\frac{\varepsilon}{2}} \int_0^{2\pi} \frac{g_w(re^{i\theta})}{1-r^2}\, r\, d\theta\, dr \\
&\leq \frac{1}{\pi}\int_{\frac{\varepsilon}{2}}^{\varepsilon} \int_0^{2\pi} \frac{g_w(re^{i\theta})}{1-r^2}\, r\, d\theta\, dr = \int_{\Omega(0,\frac{\varepsilon}{2},\varepsilon)} \frac{g_w(\zeta)}{1-|\zeta|^2}\, d\sigma \\
&= \int_{\Omega(w,\frac{\varepsilon}{2},\varepsilon)} \frac{|h(z)|^p}{1-|z|^2}\, d\sigma\,. \qquad \square
\end{aligned}$$

One immediate consequence of Lemma 2 is that every uniformly discrete sequence is almost a Blaschke sequence in the sense that $\sum (1-|z_k|)^{1+\varepsilon} < \infty$ for every $\varepsilon > 0$. To see this, simply choose $f(z) \equiv 1$ and let $\Omega = \mathbb{D}$. In fact, it can be shown that

$$\sum_{k=1}^{\infty} (1-|z_k|) \left\{ \log \frac{1}{1-|z_k|} \right\}^{-1-\varepsilon} < \infty$$

for every $\varepsilon > 0$. This follows from the estimate $n(r) = O\big(1/(1-r)\big)$ for the counting function (Chapter 2, Lemma 15), by the technique of integration by parts used in Section 4.1. Moreover, the series may diverge if $\varepsilon = 0$. Examples are constructed in Duren, Schuster, and Vukotić [1]. Therefore, uniformly discrete sequences have exactly the same maximum growth property as do zero-sets of functions in a Bergman space A^p. (See Sections 4.1 and 4.3.)

PROOF OF THEOREM 1. Let us consider first the proof of sufficiency. Assuming that $\alpha < 1/p$, we wish to show that $\Gamma = \{z_k\}$ is an interpolating sequence for A^p. Let $\{w_k\}$ be a sequence of complex numbers for which $\sum (1-|z_k|^2)^2 |w_k|^p < \infty$. Using the analytic function g associated with Γ by the relation (3), we construct the infinite series

$$f(z) = \sum_{k=1}^{\infty} \frac{w_k}{g'(z_k)} \frac{g(z)}{z-z_k} \left(\frac{1-|z_k|^2}{1-\overline{z_k} z} \right)^{\gamma}, \tag{4}$$

where γ is a positive number to be specified later.

It is clear formally that $f(z_k) = w_k$, but as in the corresponding proof of the Hardy space analogue (in Section 6.2), we need to show that the series converges to a function in A^p. To estimate the summands, observe that by (3),

$$\left| \frac{g(z)-g(z_k)}{z-z_k} \right| = \frac{|g(z)|}{\rho(z,z_k)|1-\overline{z_k}z|} \geq C\,(1-|z|^2)^{-\alpha} \frac{1}{|1-\overline{z_k}z|}\,,$$

since $\rho(z, z_k) = \rho(z, \Gamma)$ for z sufficiently close to z_k. Therefore,

$$|g'(z_k)| = \lim_{z \to z_k} \left| \frac{g(z) - g(z_k)}{z - z_k} \right| \geq C\,(1 - |z_k|^2)^{-\alpha - 1}. \tag{5}$$

Another consequence of (3) is that

$$\left| \frac{g(z)}{z - z_k} \right| = \frac{|g(z)|}{|1 - \overline{z_k} z|\rho(z, z_k)} \leq \frac{C}{|1 - \overline{z_k} z|(1 - |z|^2)^{\alpha}}, \qquad z \in \mathbb{D}. \tag{6}$$

If $\gamma \geq 1 - \alpha$, the estimates (5) and (6) show that

$$\begin{aligned} \left| \frac{w_k}{g'(z_k)} \frac{g(z)}{z - z_k} \left(\frac{1 - |z_k|^2}{1 - \overline{z_k} z} \right)^{\gamma} \right| &\leq C\, \frac{(1 - |z_k|^2)^{\gamma + \alpha + 1}\, |w_k|}{|1 - \overline{z_k} z|^{\gamma + 1} (1 - |z|^2)^{\alpha}} \\ &\leq C\, \frac{(1 - |z_k|^2)^2\, |w_k|}{(1 - R)^{\gamma + \alpha + 1}} \end{aligned}$$

for $|z| \leq R < 1$. But the series $\sum (1 - |z_k|^2)^2 |w_k|$ converges by the assumption that $\sum (1 - |z_k|^2)^2 |w_k|^p < \infty$ and the fact that $\sum (1 - |z_k|^2)^2 < \infty$ for every uniformly discrete sequence. (Refer to the remarks following the proof of Lemma 4.) Thus the series (4) converges locally uniformly if $\gamma \geq 1 - \alpha$, and its sum f is analytic in $\mathbb{D}$. Since the partial sums

$$s_n(z) = \sum_{k=1}^{n} \frac{w_k}{g'(z_k)} \frac{g(z)}{z - z_k} \left(\frac{1 - |z_k|^2}{1 - \overline{z_k} z} \right)^{\gamma}$$

satisfy $s_n(z_k) = w_k$ for $k = 1, 2, \ldots, n$, it follows that $f(z_k) = w_k$ for $k = 1, 2, \ldots$.

To show that $f \in A^p$, it will be enough to prove that the norms $\|s_n\|_p$ are bounded. Consider first the case $p = 1$. Estimating the terms of the series as above, we have

$$|s_n(z)| \leq C\,(1 - |z|^2)^{-\alpha} \sum_{k=1}^{n} \frac{(1 - |z_k|^2)^{\gamma + \alpha + 1}\, |w_k|}{|1 - \overline{z_k} z|^{\gamma + 1}}. \tag{7}$$

Since $\alpha < \frac{1}{p} = 1$, we can choose $\gamma > 1 - \alpha$ and apply Lemma 1 to infer that

$$\|s_n\|_1 \leq C \sum_{k=1}^{n} (1 - |z_k|^2)^2\, |w_k|,$$

which shows that $f \in A^1$ and completes the proof of sufficiency in the case $p = 1$.

If $1 < p < \infty$, let q be the conjugate index and apply Hölder's inequality to obtain

$$\sum_{k=1}^{n} \frac{(1-|z_k|^2)^{\gamma+\alpha+1}\,|w_k|}{|1-\overline{z_k}z|^{\gamma+1}}$$

$$\leq \left\{ \sum_{k=1}^{n} \frac{(1-|z_k|^2)^{\gamma+\frac{1}{2}(\alpha p+3)}|w_k|^p}{|1-\overline{z_k}z|^{\gamma+1}} \right\}^{1/p} \left\{ \sum_{k=1}^{n} \frac{(1-|z_k|^2)^{\gamma+\frac{q}{2}(\alpha-\frac{1}{p})+1}}{|1-\overline{z_k}z|^{\gamma+1}} \right\}^{1/q}.$$

Recalling that $\alpha < \frac{1}{p}$ and choosing $\gamma > \frac{q}{2}(\frac{1}{p}-\alpha)$, we may apply Lemma 3 to estimate the last sum by

$$\sum_{k=1}^{n} \frac{(1-|z_k|^2)^{\gamma+\frac{q}{2}(\alpha-\frac{1}{p})+1}}{|1-\overline{z_k}z|^{\gamma+1}} \leq C\,(1-|z|^2)^{\frac{q}{2}(\alpha-\frac{1}{p})}.$$

Combining these estimates with (7) and applying Lemma 1, we find that

$$\begin{aligned} \|s_n\|_p^p &\leq C \sum_{k=1}^{n} (1-|z_k|^2)^{\gamma+\frac{1}{2}(\alpha p+3)}\,|w_k|^p \int_{\mathbb{D}} \frac{(1-|z|^2)^{-\frac{1}{2}(\alpha p+1)}}{|1-\overline{z_k}z|^{\gamma+1}}\,d\sigma \\ &\leq C \sum_{k=1}^{n} (1-|z_k|^2)^2\,|w_k|^p. \end{aligned}$$

The last application of Lemma 1 requires that $\gamma > \frac{1}{2}(1-\alpha p) > 0$. Thus the norms $\|s_n\|_p$ are bounded if γ is chosen sufficiently large, which implies that $f \in A^p$ and concludes the proof of sufficiency.

Conversely, suppose $\Gamma = \{z_k\}$ is an interpolation sequence for A^p. To prove that this implies $\alpha < \frac{1}{p}$, it will suffice to show that Γ is not an A^p zero-set when $\alpha \geq \frac{1}{p}$, since every interpolation sequence is a zero-set. Suppose on the contrary that some function $f \in A^p$ vanishes on Γ. Then $h = f/g$ is analytic in $\mathbb{D}$, where g is an analytic function with the property (3). Let

$$\mathbb{D}_\delta = \mathbb{D} \setminus \left\{ \bigcup_{k=1}^{\infty} \Delta(z_k, \tfrac{\delta}{4}) \right\}$$

denote the "Swiss cheese" consisting of $\mathbb{D}$ with holes of pseudohyperbolic radius $\delta/4$ about each of the points z_k. Recall that the disks $\Delta(z_k, \frac{\delta}{2})$ are disjoint, by the triangle inequality for the pseudohyperbolic metric, since $\delta = \delta(\Gamma)$. Then

$$\int_{\mathbb{D}} \frac{|h(z)|^p}{1-|z|^2}\,d\sigma = \int_{\mathbb{D}_\delta} \frac{|h(z)|^p}{1-|z|^2}\,d\sigma + \sum_{k=1}^{\infty} \int_{\Delta(z_k,\frac{\delta}{4})} \frac{|h(z)|^p}{1-|z|^2}\,d\sigma$$

$$\leq \int_{\mathbb{D}_\delta} \frac{|h(z)|^p}{1-|z|^2}\,d\sigma + \sum_{k=1}^{\infty} \int_{\Omega(z_k,\frac{\delta}{4},\frac{\delta}{2})} \frac{|h(z)|^p}{1-|z|^2}\,d\sigma \leq 2 \int_{\mathbb{D}_\delta} \frac{|h(z)|^p}{1-|z|^2}\,d\sigma,$$

by Lemma 4 and the fact that the annuli $\Omega(z_k, \frac{\delta}{4}, \frac{\delta}{2})$ are disjoint. But $\rho(z,\Gamma) \geq \frac{\delta}{4}$ for $z \in \mathbb{D}_\delta$, so the property (3) gives

$$\begin{aligned}\int_{\mathbb{D}_\delta} \frac{|h(z)|^p}{1-|z|^2}\, d\sigma &= \int_{\mathbb{D}_\delta} \frac{|f(z)|^p}{1-|z|^2} \frac{1}{|g(z)|^p}\, d\sigma \leq C \int_{\mathbb{D}_\delta} \frac{|f(z)|^p}{1-|z|^2} \frac{(1-|z|^2)^{\alpha p}}{\rho(z,\Gamma)^p}\, d\sigma \\ &\leq C\,\delta^{-p} \int_{\mathbb{D}_\delta} |f(z)|^p (1-|z|^2)^{\alpha p - 1}\, d\sigma \\ &\leq C\,\delta^{-p} \int_{\mathbb{D}} |f(z)|^p\, d\sigma = C\,\delta^{-p}\, \|f\|_p^p < \infty\,, \cdot\end{aligned}$$

since $\alpha p - 1 \geq 0$ and $f \in A^p$. This shows that

$$\int_{\mathbb{D}} \frac{|h(z)|^p}{1-|z|^2}\, d\sigma < \infty\,,$$

which is impossible unless $h = 0$, since the integral means of h are nondecreasing and $\int_0^1 \frac{dr}{1-r^2} = \infty$. But $h = 0$ implies $f = 0$, which shows that Γ is not an A^p zero-set when $\alpha \geq \frac{1}{p}$. This concludes the proof. $\square$

Proof of Theorem 2. If Γ is a sampling sequence for A^p, then $\alpha \geq \frac{1}{p}$, since we know directly from (3) that Γ is an A^p zero-set if $\alpha < \frac{1}{p}$. We will show later, by appeal to the main theorem, that $\alpha \neq \frac{1}{p}$ if Γ is a sampling sequence. Thus it remains to prove that Γ is a sampling sequence for A^p if $\alpha > \frac{1}{p}$. We have already remarked that the upper sampling inequality is satisfied whenever Γ is uniformly discrete (Chapter 2, Lemma 14), so it will be enough to establish the lower inequality.

The proof will make essential use of the representation formula

$$f(z) = \sum_{k=1}^{\infty} \frac{f(z_k)}{g'(z_k)} \frac{g(z)}{z - z_k} \left(\frac{1-|z|^2}{1-\overline{z}z_k} \right)^{\gamma} \tag{8}$$

for arbitrary functions $f \in A^p$, valid when $\alpha > \frac{1}{p}$ and γ is an arbitrary real number. This formula has some remarkable features. The sum is independent of γ, and the sum is analytic although the partial sums are not. Moreover, the sum is in A^p although the function g is not when $\alpha \geq \frac{1}{p}$. If we could show directly that the series converges pointwise in $\mathbb{D}$ to a function $F \in A^p$, then it would be clear that $F(z_k) = f(z_k)$ for $k = 1, 2, \ldots,$ and we could conclude that $F = f$, since $\Gamma = \{z_k\}$ is not an A^p zero-set when $\alpha \geq \frac{1}{p}$. However, the proof of (8) will follow a different route.

To show that the series (8) converges uniformly on compact subsets of $\mathbb{D}$, we may apply (5) and (6) to obtain

$$\begin{aligned}\left| \frac{f(z_k)}{g'(z_k)} \frac{g(z)}{z - z_k} \left(\frac{1-|z|^2}{1-\overline{z}z_k} \right)^{\gamma} \right| &\leq C(1-|z|^2)^{\gamma-\alpha} \frac{(1-|z_k|^2)^{\alpha+1}\, |f(z_k)|}{|1-\overline{z_k}z|^{\gamma+1}} \\ &\leq C(1-|z|^2)^{-\alpha-1}\, (1-|z_k|^2)^{\alpha+1}\, |f(z_k)|\,.\end{aligned} \tag{9}$$

If $p = 1$, then $\alpha > \frac{1}{p} = 1$ and

$$\sum_{k=1}^{\infty} (1 - |z_k|^2)^{\alpha+1} |f(z_k)| \leq \sum_{k=1}^{\infty} (1 - |z_k|^2)^2 |f(z_k)| \leq C \|f\|_1 .$$

If $1 < p < \infty$, Hölder's inequality gives

$$\sum_{k=1}^{\infty} (1 - |z_k|^2)^{\alpha+1} |f(z_k)|$$

$$\leq \left\{ \sum_{k=1}^{\infty} (1 - |z_k|^2)^2 |f(z_k)|^p \right\}^{1/p} \left\{ \sum_{k=1}^{\infty} (1 - |z_k|^2)^{q(\alpha+1-\frac{2}{p})} \right\}^{1/q} \leq C \|f\|_p ,$$

since Γ is uniformly discrete and $q(\alpha + 1 - \frac{2}{p}) > 1$ if $\alpha > \frac{1}{p}$. Thus the series (8) converges uniformly on compact subsets of $\mathbb{D}$ to a sum $F(z)$.

The next step is to establish the formula (8) when f is a polynomial. For $r < 1$, let S_r be the loop consisting of those z on the circle $|z| = r$ for which $\rho(z, \Gamma) \geq \delta/2$ and of the shorter arcs of the circles $\rho(z, z_k) = \delta/2$ that intersect the circle $|z| = r$. Then $\rho(z, \Gamma) \geq \delta/2$ for all $z \in S_r$. Denote by Ω_r the domain bounded by S_r, and suppose P is a polynomial. For fixed $z \in \Omega_r \setminus \Gamma$, the residue theorem gives

$$\frac{1}{2\pi i} \int_{S_r} \frac{P(\zeta)}{(1 - \overline{z}\zeta)^\gamma g(\zeta)} \frac{d\zeta}{\zeta - z} = \frac{P(z)}{(1 - |z|^2)^\gamma g(z)} - \sum_{z_k \in \Omega_r} \frac{P(z_k)}{(1 - \overline{z} z_k)^\gamma g'(z_k)} \frac{1}{z - z_k} .$$

But by (3), the integral on the left-hand side of this equation is bounded by

$$C \|P\|_\infty \int_{S_r} \frac{(1 - |\zeta|^2)^\alpha}{|1 - \overline{z}\zeta|^\gamma \rho(\zeta, \Gamma)} \frac{|d\zeta|}{|\zeta - z|} .$$

Since $\rho(\zeta, \Gamma) \geq \delta/2$ and the length of S_r stays bounded, the integral tends to 0 as $r \to 1$. This proves the formula (8) for polynomials.

Now let f be an arbitrary A^p function, and let P be any polynomial. Let $F(z)$ denote the sum of the series (8). By the validity of (8) for polynomials and the estimates following (9), we have for each fixed $z \in \mathbb{D}$

$$\begin{aligned} |f(z) - F(z)| &\leq |f(z) - P(z)| + \left| \sum_{k=1}^{\infty} \frac{f(z_k) - P(z_k)}{g'(z_k)} \frac{g(z)}{z - z_k} \left(\frac{1 - |z|^2}{1 - \overline{z} z_k} \right)^\gamma \right| \\ &\leq (1 - |z|^2)^{-\frac{2}{p}} \|f - P\|_p + C (1 - |z|^2)^{-\alpha-1} \|f - P\|_p , \end{aligned}$$

where we have applied the formula $\|\phi\| = (1 - |z|^2)^{-\frac{2}{p}}$ for the norm of a point-evaluation functional on A^p (see Section 2.4), although only the boundedness of ϕ (*cf.* Section 1.1) is actually needed. But the last estimate can be made arbitrarily small by choosing P close to f in norm. This proves that $f(z) = F(z)$, which verifies the representation formula (8).

Combining (8) with the estimate (9), we see that

$$|f(z)| \le C\,(1-|z|^2)^{\gamma-\alpha} \sum_{k=1}^{\infty} \frac{(1-|z_k|^2)^{\alpha+1}\,|f(z_k)|}{|1-\overline{z_k}z|^{\gamma+1}}\,, \tag{10}$$

for $f \in A^p$ and $\alpha > \frac{1}{p}$. With $p = 1$ and $\gamma > \alpha - 1 > 0$, this implies by Lemma 1 that

$$\begin{aligned}\int_{\mathbb{D}} |f(z)|\,d\sigma &\le C \sum_{k=1}^{\infty} (1-|z_k|^2)^{\alpha+1}\,|f(z_k)| \int_{\mathbb{D}} \frac{(1-|z|^2)^{\gamma-\alpha}}{|1-\overline{z_k}z|^{\gamma+1}}\,d\sigma \\ &\le C \sum_{k=1}^{\infty} (1-|z_k|^2)^2\,|f(z_k)|\,,\end{aligned}$$

which is the lower sampling inequality for $p = 1$. For $1 < p < \infty$ and $\alpha > \frac{1}{p}$, an application of Hölder's inequality to (10) yields

$$\begin{aligned}|f(z)|^p \le C\,(1-|z|^2)^{p(\gamma-\alpha)} &\left\{\sum_{k=1}^{\infty} \frac{(1-|z_k|^2)^{\frac{1}{2}(\alpha p+3)}|f(z_k)|^p}{|1-\overline{z_k}z|^{\gamma+1}}\right\} \\ &\times \left\{\sum_{k=1}^{\infty} \frac{(1-|z_k|^2)^{\frac{q}{2}(\alpha-\frac{1}{p})+1}}{|1-\overline{z_k}z|^{\gamma+1}}\right\}^{p/q}.\end{aligned}$$

But for $\gamma > \frac{q}{2}(\alpha - \frac{1}{p})$, Lemma 3 can be applied to the last sum, showing that

$$\sum_{k=1}^{\infty} \frac{(1-|z_k|^2)^{\frac{q}{2}(\alpha-\frac{1}{p})+1}}{|1-\overline{z_k}z|^{\gamma+1}} \le C\,(1-|z|^2)^{\frac{q}{2}(\alpha-\frac{1}{p})-\gamma}.$$

Thus

$$|f(z)|^p \le C\,(1-|z|^2)^{\gamma-\frac{1}{2}(\alpha p+1)} \sum_{k=1}^{\infty} \frac{(1-|z_k|^2)^{\frac{1}{2}(\alpha p+3)}|f(z_k)|^p}{|1-\overline{z_k}z|^{\gamma+1}}\,.$$

Finally, take $\gamma > \frac{1}{2}(\alpha p - 1)$ and apply Lemma 1 to arrive at the conclusion

$$\begin{aligned}\int_{\mathbb{D}} |f(z)|^p\,d\sigma &\le C \sum_{k=1}^{\infty} (1-|z_k|^2)^{\frac{1}{2}(\alpha p+3)}|f(z_k)|^p \int_{\mathbb{D}} \frac{(1-|z|^2)^{\gamma-\frac{1}{2}(\alpha p+1)}}{|1-\overline{z_k}z|^{\gamma+1}}\,d\sigma \\ &\le C \sum_{k=1}^{\infty} (1-|z_k|^2)^2\,|f(z_k)|^p\,,\end{aligned}$$

which is the lower sampling inequality. □

§6.4. Some explicit examples.

Theorems 1 and 2 will now be applied to produce concrete examples of sampling and interpolation sequences for Bergman spaces. With respect to parameters $a > 1$ and $b > 0$, define the lattice

$$\Lambda(a,b) = \{a^m(bn+i) : m \in \mathbb{Z}, n \in \mathbb{Z}\}$$

of points in the upper half-plane $\mathbb{H}^+$. Let ψ be a fixed conformal mapping of $\mathbb{D}$ onto $\mathbb{H}^+$, defined by

$$\psi(z) = i\,\frac{1+z}{1-z}, \qquad \psi^{-1}(\zeta) = \frac{\zeta - i}{\zeta + i},$$

for $z \in \mathbb{D}$ and $\zeta \in \mathbb{H}^+$. Then $\Gamma(a,b) = \psi^{-1}(\Lambda(a,b))$ is a lattice of distinct points in $\mathbb{D}$. The following theorem is due to Seip [2] for $p = 2$.

THEOREM 3. *For $a > 1$ and $b > 0$, let $\Gamma(a,b)$ be the set described above, and let $\beta = \frac{2\pi}{b\log a}$. Let $0 < p < \infty$. Then $\Gamma(a,b)$ is a sampling set for A^p if and only if $\beta > \frac{1}{p}$, and it is an interpolation set for A^p if and only if $\beta < \frac{1}{p}$. Also, $\Gamma(a,b)$ is not an A^p zero-set when $\beta = \frac{1}{p}$.*

The pseudohyperbolic metric of the upper half-plane has the form

$$\rho^+(z,\zeta) = \left|\frac{z-\zeta}{z-\overline{\zeta}}\right|, \qquad z,\zeta \in \mathbb{H}^+.$$

The invariance property

$$\rho(z,\zeta) = \rho^+(\psi(z),\psi(\zeta))$$

is easily checked. A straightforward calculation shows that

$$\rho^+(z,\zeta) \geq \min\left\{\frac{a-1}{a+1}, \frac{b}{\sqrt{b^2+4}}\right\} > 0, \qquad z,\zeta \in \Lambda(a,b),\ z \neq \zeta.$$

Thus by the invariance property, $\Gamma(a,b)$ is uniformly discrete. Consequently, in view of Theorems 1 and 2, the proof of Theorem 3 reduces to the construction of a function g analytic in $\mathbb{D}$ that has $\Gamma = \Gamma(a,b)$ as its zero-set and satisfies the condition (3) with $\alpha = \beta$.

It is more convenient to carry out the construction in the upper half-plane. Consider the function

$$h(z) = \prod_{k=0}^{\infty} \frac{\sin(\frac{\pi}{b}a^{-k}(ia^k - z))}{\sin(\frac{\pi}{b}a^{-k}(ia^k + z))} \prod_{m=1}^{\infty} e^{\frac{2\pi}{b}} \frac{\sin(\frac{\pi}{b}a^m(z - ia^{-m}))}{\sin(\frac{\pi}{b}a^m(z + ia^{-m}))}$$

for $z \in \mathbb{H}^+$. It will be shown that the infinite products converge, and that h is analytic in $\mathbb{H}^+$ with zero-set $\Lambda(a,b)$. The desired function is then

$$g(z) = (1-z)^{-2\beta} h(\psi(z)). \tag{11}$$

LEMMA 5. *The function g defined by (11) is analytic in $\mathbb{D}$, and its zero-set is $\Gamma = \Gamma(a,b)$. Furthermore, it has the property (3) with $\alpha = \beta$.*

PROOF OF LEMMA. Denote the factors by

$$f_k(z) = \frac{\sin(\frac{\pi}{b}a^{-k}(ia^k - z))}{\sin(\frac{\pi}{b}a^{-k}(ia^k + z))}, \qquad g_m(z) = e^{\frac{2\pi}{b}} \frac{\sin(\frac{\pi}{b}a^m(z - ia^{-m}))}{\sin(\frac{\pi}{b}a^m(z + ia^{-m}))}.$$

A calculation shows that $f_k(z) = -\varphi_\zeta(w)$ with $\zeta = e^{-\frac{2\pi}{b}}$ and $w = e^{\frac{2\pi}{b}ia^{-k}z} \in \mathbb{D}$, where

$$\varphi_\zeta(w) = \frac{\zeta - w}{1 - \bar{\zeta}w}$$

is the standard Möbius self-mapping of the disk. Thus $|f_k(z)| \le 1$ for all $z \in \mathbb{H}^+$, and the identities

$$1 + \varphi_\zeta(w) = \frac{(1+\zeta)(1-w)}{1 - \zeta w},$$

$$1 - |\varphi_\zeta(w)|^2 = \frac{(1-\zeta^2)(1-|w|^2)}{|1 - \zeta w|^2},$$

which hold for w and ζ defined above, give the estimates

$$|1 - f_k(z)| \le \frac{1 + e^{-\frac{2\pi}{b}}}{1 - e^{-\frac{2\pi}{b}}} \left|1 - e^{\frac{2\pi}{b}ia^{-k}z}\right| \le \frac{4\pi|z|}{b(1 - e^{-\frac{2\pi}{b}})} a^{-k}$$

and

$$1 - |f_k(z)| \le \frac{1 + e^{-\frac{2\pi}{b}}}{1 - e^{-\frac{2\pi}{b}}} \left|1 - e^{-\frac{4\pi}{b}a^{-k}y}\right| \le \frac{8\pi y}{b(1 - e^{-\frac{2\pi}{b}})} a^{-k}$$

for $z = x + iy \in \mathbb{H}^+$ and k sufficiently large. A similar calculation yields

$$|1 - g_m(z)| \le \frac{e^{\frac{4\pi}{b}} - 1}{e^{\frac{2\pi}{b}(1 + a^m y)} - 1}.$$

These estimates show that both infinite products in the definition of h converge uniformly on compact subsets of $\mathbb{H}^+$ and therefore represent functions analytic in $\mathbb{H}^+$. Note that $f_k(z) = 0$ if and only if $z = a^k(bn + i)$ for some integer n, and that each of these zeros is simple. Similarly, $g_m(z) = 0$ if and only if $z = a^{-m}(bn + i)$ for some integer n. This proves that the function h is analytic in $\mathbb{H}^+$ and has zero-set $\Lambda(a,b)$.

We see by direct calculation that h has the property

$$h(az) = -e^{-\frac{2\pi}{b}} h(z). \tag{12}$$

This relation allows us to deduce global estimates for $h(z)$ from estimates over the horizontal strip

$$S_a = \{z = x + iy : a^{-\frac{1}{2}} \le y \le a^{\frac{1}{2}}\}.$$

Note that all zeros of $h(z)$ in the strip S_a occur on the line $y = 1$ and are due to the factor $f_0(z)$. The other zeros are bounded away from S_a, so the above estimates for $1 - |f_k(z)|$ and $|1 - |g_m(z)|| \le |1 - g_m(z)|$ show that $|h(z)/f_0(z)|$ is bounded above and below by positive constants for all $z \in S_a$. To estimate the factor

$$f_0(z) = \frac{\sin(\frac{\pi}{b}(i - z))}{\sin(\frac{\pi}{b}(i + z))},$$

we need only observe that

$$\left| \frac{\sin(\frac{\pi}{b}(i - z))}{\sin(\frac{\pi}{b}(i + z))} \frac{z - bn + i}{z - bn - i} \right|$$

is bounded above and below by positive constants near the point $z = bn + i$, and the bounds are independent of the integer n. This shows that $|f_0(z)|$ is bounded above and below by positive constant multiples of $\rho^+(z, \Lambda(a, b))$ for all $z \in S_a$. Putting everything together, we conclude that

$$C_1 \, \rho^+(z, \Lambda(a, b)) \le |h(z)| \le C_2 \, \rho^+(z, \Lambda(a, b)), \qquad z \in S_a, \tag{13}$$

for some positive constants C_1 and C_2.

The next step is to extend the estimate of $h(z)$ to the entire half-plane $\mathbb{H}^+$. Given a point $z = x + iy \in \mathbb{H}^+$, one can choose an integer m such that

$$m - \tfrac{1}{2} \le \frac{\log y}{\log a} \le m + \tfrac{1}{2},$$

so that $a^{-m} z \in S_a$. Then by (12) and (13),

$$\begin{aligned} |h(z)| = |h(a^m a^{-m} z)| = e^{-\frac{2\pi}{b} m} |h(a^{-m} z)| &\le C \, e^{-\frac{2\pi}{b} \frac{\log y}{\log a}} \rho^+(a^{-m} z, \Lambda(a, b)) \\ &= C \, y^{-\beta} \rho^+(z, \Lambda(a, b)), \end{aligned}$$

where $\beta = \frac{2\pi}{b \log a}$, since it is clear that

$$\rho^+(az, \Lambda(a, b)) = \rho^+(z, \Lambda(a, b)).$$

Similar estimates lead to the lower bound

$$|h(z)| \ge C \, y^{-\beta} \rho^+(z, \Lambda(a, b)).$$

The final step is to translate these estimates of h to estimates of the function g defined by (11). In view of the identity

$$\operatorname{Im}\{\psi(z)\} = \operatorname{Im}\left\{ i \, \frac{1 + z}{1 - z} \right\} = \frac{1 - |z|^2}{|1 - z|^2}$$

and the invariance property of the pseudohyperbolic metric, we have

$$\begin{aligned}|g(z)| &= |1-z|^{-2\beta}\,|h(\psi(z))| \le C\,|1-z|^{-2\beta}\,(\operatorname{Im}\{\psi(z)\})^{-\beta}\,\rho^{+}(\psi(z),\Lambda(a,b))\\ &= C\,|1-z|^{-2\beta}\left(\frac{1-|z|^2}{|1-z|^2}\right)^{-\beta}\rho(z,\Gamma(a,b))\\ &= C(1-|z|^2)^{-\beta}\,\rho(z,\Gamma(a,b))\,.\end{aligned}$$

The lower bound is derived in a similar way. Thus we have shown that the function g defined by (11) has the property (3), and the proof of Lemma 5 is complete. □

§6.5. The density theorems.

We turn now to the main theorems, the characterizations of sampling and interpolation sequences in terms of densities introduced by Kristian Seip [3]. Before the theorems can be stated, we need to define the densities. Recall that a sequence $\Gamma = \{z_k\}$ is uniformly discrete if

$$\delta(\Gamma) = \inf_{j\neq k} \rho(z_j, z_k) > 0\,,$$

where $\rho(z,\zeta) = |(\zeta - z)/(1-\overline{\zeta}z)|$ is the pseudohyperbolic metric. The pseudohyperbolic disk $\rho(z,\zeta) < r$ is denoted by $\Delta(\zeta, r)$. The hyperbolic area of a measurable set $\Omega \subset \mathbb{D}$ is

$$a(\Omega) = \int_\Omega \frac{d\sigma(z)}{(1-|z|^2)^2}\,.$$

Let Γ be a uniformly discrete sequence. For $0 < s < 1$, let $n(\Gamma,\zeta,s)$ be its pseudohyperbolic counting function, the number of points of Γ that lie in the disk $\Delta(\zeta, s)$. Then the quantity

$$E(\Gamma,\zeta,r) = \frac{\int_0^r n(\Gamma,\zeta,s)\,ds}{2\int_0^r a(\Delta(\zeta,s))\,ds}$$

is, roughly speaking, a measure of the average number of points of Γ contained in the disks $\Delta(\zeta, s)$, per unit of hyperbolic area, for $0 < s < r$.

The *lower uniform density* of Γ is defined to be

$$D^-(\Gamma) = \liminf_{r\to 1}\ \inf_{\zeta\in\mathbb{D}} E(\Gamma,\zeta,r)\,,$$

and the *upper uniform density* is

$$D^+(\Gamma) = \limsup_{r\to 1}\ \sup_{\zeta\in\mathbb{D}} E(\Gamma,\zeta,r)\,.$$

It is clear from the definitions that $0 \le D^-(\Gamma) \le D^+(\Gamma)$. We will see that $D^+(\Gamma)$ is finite whenever Γ is uniformly discrete.

These notions of density allow a complete description of sampling and interpolation sequences. The main theorems can be stated as follows.

THEOREM 4. *For $0 < p < \infty$, a sequence Γ of distinct points in the unit disk is an interpolation sequence for A^p if and only if it is uniformly discrete and $D^+(\Gamma) < \frac{1}{p}$.*

THEOREM 5. *For $0 < p < \infty$, a sequence Γ of distinct points in the disk is a sampling sequence for A^p if and only if it is a finite union of uniformly discrete sequences and it has a uniformly discrete subsequence Γ' for which $D^-(\Gamma') > \frac{1}{p}$.*

Both theorems are due to Seip [3], who considered only the case $p = 2$. The extensions to general values of p were carried out by Schuster [2,3] and Schuster and Varolin [1], using the methods of Seip [3] and Berndtsson and Ortega-Cerdà [1], respectively. Jevtić, Massaneda, and Thomas [1] provided some details regarding Theorem 4. The theorems are rather deep and their proofs require considerable effort. In the next chapter, we will give complete proofs for $1 \le p < \infty$. Meanwhile, we propose to make a few general remarks and to calculate the densities for some specific sets.

For a uniformly discrete set Γ, the theorems say that Γ is interpolating for A^p if $D^+(\Gamma) < \frac{1}{p}$ and is sampling for A^p if $D^-(\Gamma) > \frac{1}{p}$. Thus the points of an interpolation set are not too densely packed in any part of the disk, while those of a sampling set are not too sparse in any part of the disk. Since $D^-(\Gamma) \le D^+(\Gamma)$, the theorems reflect the familiar fact that a set Γ cannot be both sampling and interpolating for the same A^p space.

An immediate corollary is worthy of note. It follows from Theorem 4 that if Γ is an interpolation set for some space A^p, then it is an interpolation set for all spaces A^q with $0 < q < p$. Similarly, Theorem 5 shows that an A^p sampling set is an A^q sampling set for $p < q < \infty$. These implications are not apparent from the definitions.

Because the sampling and interpolation sets for A^p are both Möbius invariant, as are the uniformly discrete sets, it is to be expected that the upper and lower densities are also Möbius invariant. This can be deduced from the theorems, but a direct confirmation is of interest. First recall (*cf.* Section 2.5) that hyperbolic area is Möbius invariant, and that

$$a(\Delta(\zeta, r)) = \frac{r^2}{1 - r^2},$$

so that

$$2\int_0^r a(\Delta(\zeta, s))\, ds = 2\int_0^r \frac{s^2}{1 - s^2}\, ds = \log\frac{1+r}{1-r} - 2r.$$

Consequently, the densities can be expressed as

$$D^-(\Gamma) = \liminf_{r\to 1}\left(\log\frac{1}{1-r}\right)^{-1} \inf_{\zeta\in\mathbb{D}} \int_0^r n(\Gamma, \zeta, s)\, ds,$$

$$D^+(\Gamma) = \limsup_{r\to 1}\left(\log\frac{1}{1-r}\right)^{-1} \sup_{\zeta\in\mathbb{D}} \int_0^r n(\Gamma, \zeta, s)\, ds.$$

Now let $\widetilde{\Gamma} = \varphi_w(\Gamma)$ be the image of Γ under a Möbius transformation

$$\varphi_w(z) = \frac{w - z}{1 - \overline{w}z},$$

for some point $w \in \mathbb{D}$, and let $\widetilde{\zeta} = \varphi_w(\zeta)$. Then the counting function has the property $n(\widetilde{\Gamma}, \widetilde{\zeta}, r) = n(\Gamma, \zeta, r)$, so it is clear from the above formulas that $D^-(\widetilde{\Gamma}) = D^-(\Gamma)$ and $D^+(\widetilde{\Gamma}) = D^+(\Gamma)$.

Recall next that for any uniformly discrete set Γ with separation constant $\delta > 0$, we have shown (Chapter 2, Lemma 15) that

$$n(\Gamma, \zeta, r) \leq \left(\tfrac{2}{\delta} + 1\right)^2 \frac{1}{1 - r^2},$$

which implies that

$$\int_0^r n(\Gamma, \zeta, s)\, ds \leq C \log \frac{1}{1 - r},$$

where C is a constant independent of ζ. Therefore, $D^+(\Gamma) < \infty$ for every uniformly discrete set Γ. The same argument shows that $D^+(\Gamma) = 0$ if

$$\int_0^r n(\Gamma, \zeta, s)\, ds = o\left(\log \frac{1}{1 - r}\right), \qquad r \to 1,$$

uniformly for $\zeta \in \mathbb{D}$. In particular, $D^+(\Gamma) = 0$ if Γ is a finite set.

There is an equivalent version of the density formulas that is often easier to work with. For $\frac{1}{2} < r < 1$, let

$$D(\Gamma, \zeta, r) = \left(\log \frac{1}{1 - r}\right)^{-1} \sum_{z_k \in \Omega(\zeta, \frac{1}{2}, r)} \log \frac{1}{|\varphi_\zeta(z_k)|},$$

where $\Omega(\zeta, r_1, r_2)$ denotes the pseudohyperbolic annulus $r_1 < \rho(\zeta, z) < r_2$, or $r_1 < |\varphi_\zeta(z)| < r_2$. Then the densities can be reformulated as

$$D^-(\Gamma) = \liminf_{r \to 1} \inf_{\zeta \in \mathbb{D}} D(\Gamma, \zeta, r), \qquad D^+(\Gamma) = \limsup_{r \to 1} \sup_{\zeta \in \mathbb{D}} D(\Gamma, \zeta, r).$$

To see this, use integration by parts to write

$$\begin{aligned}
\int_0^r n(\Gamma, \zeta, s)\, ds &= \int_0^r n(\varphi_\zeta(\Gamma), 0, s)\, ds \\
&= r\, n(\varphi_\zeta(\Gamma), 0, r) - \int_0^r s\, dn(\varphi_\zeta(\Gamma), 0, s) \\
&= r\, n(\varphi_\zeta(\Gamma), 0, r) - \sum_{z_k \in \Delta(\zeta, r)} |\varphi_\zeta(z_k)| \\
&= \sum_{z_k \in \Delta(\zeta, r)} (1 - |\varphi_\zeta(z_k)|) - (1 - r)\, n(\varphi_\zeta(\Gamma), 0, r) \\
&\leq 2\left(\tfrac{2}{\delta} + 1\right)^2 + \sum_{z_k \in \Omega(\zeta, \frac{1}{2}, r)} \log \frac{1}{|\varphi_\zeta(z_k)|},
\end{aligned}$$

since $1-x \le \log\frac{1}{x}$ for $x>0$ and $n(\Gamma,\zeta,\frac{1}{2}) \le 2\left(\frac{2}{\delta}+1\right)^2$, by Lemma 15 in Chapter 2. In the other direction, we can use the inequality

$$\log\frac{1}{x} \le 1-x+(1-x)^2\,, \qquad \tfrac{1}{2}\le x\le 1\,,$$

to infer that

$$\int_0^r n(\Gamma,\zeta,s)\,ds \ge -\left(\tfrac{2}{\delta}+1\right)^2 + \sum_{z_k\in\Omega(\zeta,\frac{1}{2},r)} (1-|\varphi_\zeta(z_k)|)$$

$$\ge -\left(\tfrac{2}{\delta}+1\right)^2 + \sum_{z_k\in\Omega(\zeta,\frac{1}{2},r)} \log\frac{1}{|\varphi_\zeta(z_k)|} - \sum_{z_k\in\Omega(\zeta,\frac{1}{2},r)} (1-|\varphi_\zeta(z_k)|)^2\,.$$

But by the corollary to Lemma 14 in Chapter 2,

$$\sum_{k=1}^{\infty}(1-|\varphi_\zeta(z_k)|)^2 \le \left(\tfrac{2}{\delta}\right)^2\,,$$

where $\delta=\delta(\Gamma)=\delta(\varphi_\zeta(\Gamma))$ is the separation constant of the uniformly discrete set Γ. Consequently,

$$\left|\int_0^r n(\Gamma,\zeta,s)\,ds - \sum_{z_k\in\Omega(\zeta,\frac{1}{2},r)} \log\frac{1}{|\varphi_\zeta(z_k)|}\right| \le 2\left(\tfrac{2}{\delta}+1\right)^2\,,$$

which implies that the two formulations of the densities are equivalent.

Of course, the choice of inner radius $\frac{1}{2}$ is not essential. Equivalent definitions result if the annuli $\Omega(\zeta,\frac{1}{2},r)$ are replaced by $\Omega(\zeta,c,r)$ for any constant c in the interval $0<c<1$. In particular, c may be chosen arbitrarily close to 1.

We can now deduce that every uniformly separated sequence has zero density. Recall from Chapter 4 that a Blaschke sequence $\Gamma=\{z_k\}$ is said to be uniformly separated if

$$\prod_{j\ne k}\left|\frac{z_j-z_k}{1-\overline{z_j}z_k}\right| \ge \delta\,, \qquad k=1,2,\dots\,,$$

for some $\delta>0$ independent of k. It is clear from the definitions that every uniformly separated sequence is uniformly discrete. We have seen that if Γ is uniformly separated, then

$$\sup_{\zeta\in\mathbb{D}}\sum_{k=1}^{\infty}(1-|\varphi_\zeta(z_k)|) < \infty$$

(*cf.* Section 4.6, Theorem 10). Since $\log \frac{1}{x} \le 1 - x^2$ for $\frac{1}{2} \le x \le 1$, it follows that

$$\left(\log \frac{1}{1-r}\right) D(\Gamma, \zeta, r) = \sum_{z_k \in \Omega(\zeta, \frac{1}{2}, r)} \log \frac{1}{|\varphi_\zeta(z_k)|} \le \sum_{k=1}^{\infty} (1 - |\varphi_\zeta(z_k)|^2) \le C\,,$$

where C is a constant independent of ζ. This shows that $D^+(\Gamma) = 0$ if Γ is uniformly separated.

As an immediate consequence, we conclude from Theorem 4 that uniformly separated sequences are interpolating for every A^p space. To put it another way, every H^p interpolation sequence is an A^p interpolation sequence. The converse is certainly false, but if a sequence is situated on a ray and is interpolating for some A^p space, then it is uniformly separated and therefore interpolating for H^p. To see this, note first that an A^p interpolation sequence is uniformly discrete, by Theorem 4. But it can be shown that a uniformly discrete set on a ray is exponential and therefore uniformly separated. (See for instance Duren [5], Chapter 9.)

Suppose now that Γ is a set of the type studied in Section 6.3; that is, a uniformly discrete set admitting an analytic function g with the property (3). Comparing Theorem 1 with Theorem 4, we see that $D^+(\Gamma) = \alpha$. Recall that Theorem 2 was not proved in full strength, since it was shown only that $\alpha \ge \frac{1}{p}$ when Γ is a sampling set for A^p, while the theorem claimed $\alpha > \frac{1}{p}$. Suppose, however, that some set Γ of the given type is a sampling set for A^p and has $\alpha = \frac{1}{p}$. Then $D^-(\Gamma) > \alpha$ by Theorem 5, and so $D^-(\Gamma) > \frac{1}{q} > \alpha$ for some q, which implies that Γ is a sampling set for A^q. But this contradicts Theorem 1, which says that Γ is an interpolation set for A^q when $\alpha < \frac{1}{q}$. Thus $\alpha > \frac{1}{p}$ whenever Γ is a sampling set for A^p. This completes the proof of Theorem 2, with help from Theorem 5.

A comparison of Theorems 1 and 2 with Theorems 4 and 5 shows that

$$D^-(\Gamma) = D^+(\Gamma) = \alpha$$

for uniformly discrete sets Γ that admit an analytic function g with the property (3). Actually, our proof applies only for $0 < \alpha < 1$, but in Chapter 7 we will give a direct proof for arbitrary $\alpha > 0$. In particular, for the set $\Gamma(a, b)$ of Theorem 3, we see that

$$D^-(\Gamma(a,b)) = D^+(\Gamma(a,b)) = \frac{2\pi}{b \log a}\,.$$

Instead of applying the main theorems to calculate the densities of sets, one would hope to use the theorems to identify given sets as sampling or interpolating for A^p spaces. The difficulty is to calculate the densities directly from their definitions, but this can be done for certain regularly distributed sets, as we shall see in the next section.

§6.6. Direct calculation of densities.

Theorems 4 and 5 will be effective in describing specific sampling and interpolation sets only insofar as the densities can be computed. In the previous section, the theorems were applied in the opposite way, to calculate densities of certain families of sets that had been classified independently as sampling or interpolating for A^p. In Chapter 7 we will calculate the densities of those sets in a different manner, but our purpose is now to consider another natural family of sets in the disk and to calculate their densities directly. As a corollary of the methods, we will also find the densities of some sets investigated in Chapter 4.

For a fixed constant $a > 0$, let

$$d\mu(z) = \frac{a\, d\sigma(z)}{(1-|z|^2)^2}$$

denote the corresponding multiple of hyperbolic area measure. Partition the disk into disjoint annuli

$$R_n = \{z : t_{n-1} \le |z| < t_n\}\,, \qquad n = 1, 2, 3, \dots\,,$$

where the radii t_n are defined inductively by $t_0 = 0$ and $\mu(R_n) = 2^{n-1}$. A calculation shows that

$$\frac{1}{1-t_n^2} = \frac{2^n + a - 1}{a}\,,$$

so that

$$t_n = 1 - 2^{-n-1}a + O(2^{-2n})\,. \tag{14}$$

In particular, $t_n \to 1$ as $n \to \infty$.

Next divide each annulus R_n into 2^{n-1} cells Q_{nj} by placing radial segments at angles $j2^{-n+2}\pi$, where $j = 1, 2, \dots, 2^{n-1}$. Then all cells have equal area $\mu(Q_{nj}) = 1$. Let

$$\zeta_{nj} = \int_{Q_{nj}} z\,d\mu(z)$$

denote the hyperbolic center of mass of the cell Q_{nj}. A calculation shows that

$$|\zeta_{nj}| = 1 - 2^{-n}(a \log 2) + O(2^{-2n})\,, \tag{15}$$

so that $\zeta_{nj} \in Q_{nj}$ at least for all n sufficiently large. Finally, we define our sequence $\Gamma = \{z_k\}$ to be an enumeration of the points ζ_{nj}. We will prove

THEOREM 6. *The sequence Γ is uniformly discrete and*

$$D^-(\Gamma) = D^+(\Gamma) = \frac{a}{2}.$$

It is not difficult to see that Γ is uniformly discrete. This assertion is part of the following preliminary result. The *pseudohyperbolic diameter* of a set $E \subset \mathbb{D}$ is defined by

$$d(E) = \sup_{z,\zeta \in E} \rho(z, \zeta)\,.$$

PROPOSITION 1.

(a) *The set* Γ *is uniformly discrete.*

(b) *The pseudohyperbolic diameters of the cells* Q_{nj} *are bounded away from 1; there is a constant* $K < 1$ *such that* $d(Q_{nj}) \leq K$ *for all indices* n, j.

(c) *There is a constant* C *such that if* α *and* β *belong to the same cell* Q_{nj}, *then*

$$\left|\frac{1-\overline{\alpha}z}{1-\overline{\beta}z}\right| \leq C \qquad \text{for all} \quad z \in \mathbb{D}.$$

PROOF. Calculations based on the asymptotic formulas (14) and (15) for t_n and $|\zeta_{nj}|$ show that there exist radii R_1 and R_2 for which

$$\Delta(\zeta_{nj}, R_1) \subset Q_{nj} \subset \Delta(\zeta_{nj}, R_2),$$

at least for all n sufficiently large. From this (a) and (b) follow immediately.

For the proof of (c), note first that

$$\left|\frac{1-\overline{\alpha}z}{1-\overline{\beta}z}\right|^2 = \frac{1-|\alpha|^2}{1-|\beta|^2}\,\frac{1-\rho(\beta,z)^2}{1-\rho(\alpha,z)^2}.$$

The first factor can be bounded by appeal to Lemma 3 of Chapter 2, since $Q_{nj} \subset \Delta(\zeta_{nj}, R_2)$. To estimate the second factor, we apply the strong form of the triangle inequality for the pseudohyperbolic metric to write

$$\begin{aligned}\frac{1-\rho(\beta,z)}{1-\rho(\alpha,z)} &\leq \frac{1-\rho(\beta,z)}{1-\frac{\rho(\alpha,\beta)+\rho(\beta,z)}{1+\rho(\alpha,\beta)\rho(\beta,z)}} = \frac{(1-\rho(\beta,z))(1+\rho(\alpha,\beta)\rho(\beta,z))}{(1-\rho(\alpha,\beta))(1-\rho(\beta,z))} \\ &\leq \frac{2}{1-\rho(\alpha,\beta)} \leq \frac{2}{1-K},\end{aligned}$$

because $\rho(\alpha,\beta) \leq d(Q_{nj}) \leq K < 1$. □

To calculate the densities of $\Gamma = \{z_k\}$, we introduce the discrete integer-valued counting measure $\nu = \sum \delta_{z_k}$, where δ_z is the Dirac-delta measure concentrated at z. For $\alpha \in \mathbb{D}$ and $0 < c < r < 1$, recall that

$$\Omega(\alpha, c, r) = \{z \in \mathbb{D} : c < \rho(\alpha, z) < r\}$$

is the pseudohyperbolic annulus with center α and radii c and r. The main step in the proof of Theorem 6 is to establish the following fact.

PROPOSITION 2.

$$\sup_{\alpha\in\mathbb{D},\frac{1}{2}<r<1}\left|\int_{\Omega(\alpha,\frac{1}{2},r)} \log|\varphi_\alpha(z)|\, d(\mu(z)-\nu(z))\right| < \infty. \tag{16}$$

Deferring the proof, we observe first that this proposition implies the theorem.

DEDUCTION OF THEOREM 6. By a change of variables,

$$\begin{aligned}\int_{\Omega(\alpha,\frac{1}{2},r)} \log|\varphi_\alpha(z)|\,d\mu(z) &= a\int_{\Omega(0,\frac{1}{2},r)} \log|w|\,\frac{d\sigma(w)}{(1-|w|^2)^2}\\ &= 2a\int_{\frac{1}{2}}^{r} \frac{s\log s}{(1-s^2)^2}\,ds\\ &= -\frac{a}{2}\log\frac{1}{1-r} + o\left(\log\frac{1}{1-r}\right).\end{aligned}$$

On the other hand,

$$\begin{aligned}\int_{\Omega(\alpha,\frac{1}{2},r)} \log|\varphi_\alpha(z)|\,d\nu(z) &= \sum_{z_k\in\Omega(\alpha,\frac{1}{2},r)} \log|\varphi_\alpha(z_k)|\\ &= -\left(\log\frac{1}{1-r}\right)D(\Gamma,\alpha,r).\end{aligned}$$

These two formulas are now combined to give

$$\begin{aligned}\int_{\Omega(\alpha,\frac{1}{2},r)} \log|\varphi_\alpha(z)|\,d(\mu(z)-\nu(z)) &= \left(D(\Gamma,\alpha,r)-\frac{a}{2}\right)\log\frac{1}{1-r}\\ &\quad + o\left(\log\frac{1}{1-r}\right).\end{aligned}$$

Thus the boundedness of the integral implies that $D^-(\Gamma) = D^+(\Gamma) = \frac{a}{2}$. $\square$

PROOF OF PROPOSITION 2. Denote by $\Omega_1 = \Omega_1(\alpha,\frac{1}{2},r)$ the union of the closures of the cells Q_{nj} that are entirely contained in $\Omega(\alpha,\frac{1}{2},r)$, and let

$$\Omega_2 = \Omega_2(\alpha,\tfrac{1}{2},r) = \Omega(\alpha,\tfrac{1}{2},r)\setminus\Omega_1.$$

Our first task is to estimate the part of the integral in (16) taken over Ω_1. To this end, define

$$L(z) = \log\left(\frac{z-\alpha}{1-\overline{\alpha}z}\right).$$

Integration by parts yields

$$L(z) = L(z_k) + L'(z_k)(z-z_k) + \int_{z_k}^{z} L''(w)(z-w)\,dw.$$

Denoting by Q_k the cell containing z_k, we apply this equation to obtain

$$\int_{Q_k} \log\left|\frac{z-\alpha}{1-\overline{\alpha}z}\right| d(\mu(z)-\nu(z)) = \mathrm{Re}\left\{\int_{Q_k}\int_{z_k}^{z} L''(w)(z-w)\,dw\,d\mu(z)\right\}. \tag{17}$$

Here we have used the facts that

$$\int_{Q_k} d(\mu(z)-\nu(z)) = 1-1=0\,, \qquad \int_{Q_k} z\,d(\mu(z)-\nu(z)) = z_k - z_k = 0\,,$$

and

$$\operatorname{Re}\left\{\int_{Q_k}\int_{z_k}^{z} L''(w)(z-w)\,dw\,d\nu(z)\right\} = \operatorname{Re}\left\{\int_{z_k}^{z_k} L''(w)(z_k-w)\,dw\right\} = 0\,.$$

To estimate the right-hand side of (17), we make the explicit calculation

$$L''(w) = (1-|\alpha|^2)\,\frac{\overline{\alpha}(w-\alpha)-(1-\overline{\alpha}w)}{(1-\overline{\alpha}w)^2(w-\alpha)^2}\,.$$

But if $Q_k \subset \Omega(\alpha, \frac{1}{2}, r)$, then $\frac{1}{2} < \rho(\alpha, w) < r$ for all $w \in Q_k$. This implies that

$$\begin{aligned}
|L''(w)| &\le (1-|\alpha|^2)\left(\frac{1}{|w-\alpha||1-\overline{\alpha}w|^2} + \frac{1}{|w-\alpha|^2|1-\overline{\alpha}w|}\right)\\
&\le (1-|\alpha|^2)\left(\frac{2}{|1-\overline{\alpha}w|^3} + \frac{4}{|1-\overline{\alpha}w|^3}\right) = \frac{6(1-|\alpha|^2)}{|1-\overline{\alpha}w|^3}
\end{aligned}$$

for all $w \in Q_k \subset \Omega_1$. Therefore, applying Proposition 1(c) with $\alpha = w$ and $\beta = z$, we find

$$\begin{aligned}
\left|\int_{z_k}^{z} L''(w)(z-w)\,dw\right| &\le C(1-|\alpha|^2)\int_{z_k}^{z}\frac{|z-w|}{|1-\overline{\alpha}w|^3}\,|dw|\\
&\le C\frac{(1-|\alpha|^2)}{|1-\overline{\alpha}z|^3}\int_{z_k}^{z}|z-w|\,|dw|\\
&\le C\frac{(1-|\alpha|^2)}{|1-\overline{\alpha}z|^3}\sup_{w\in Q_k}|w-z|^2\,.
\end{aligned}$$

But if $z, w \in Q_k$, another application of Proposition 1(c) shows that

$$|w-z| \le |1-\overline{w}z| \le C(1-|z|^2)\,.$$

Therefore,

$$\left|\int_{z_k}^{z} L''(w)(z-w)\,dw\right| \le C\,\frac{(1-|\alpha|^2)(1-|z|^2)^2}{|1-\overline{\alpha}z|^3}\,.$$

Thus it follows from (17) that

$$\left|\sum_{Q_k\subset\Omega_1}\int_{Q_k}\log|\varphi_\alpha(z)|(d\mu(z)-d\nu(z))\right| \le C(1-|\alpha|^2)\int_{\mathbb{D}}\frac{d\sigma(z)}{|1-\overline{\alpha}z|^3}\,.$$

But

$$\int_{\mathbb{D}} \frac{d\sigma(z)}{|1-\overline{\alpha}z|^3} \leq \frac{C}{1-|\alpha|^2}, \qquad \alpha \in \mathbb{D},$$

by Lemma 1. This shows that the part of the integral in (16) taken over $\Omega_1 = \Omega_1(\alpha, \frac{1}{2}, r)$ is bounded for $\alpha \in \mathbb{D}$ and $\frac{1}{2} < r < 1$.

It remains to show that the part of the integral over $\Omega_2 = \Omega_2(\alpha, \frac{1}{2}, r)$ is bounded. First choose a number B with $K < B < 1$, where K is the constant in Proposition 1. Next observe that

$$B^* = \frac{1+4B+B^2}{2(1+B+B^2)} < 1,$$

and take $B^* < r < 1$. Let

$$c = \frac{1+2B}{2+B} \qquad \text{and} \qquad h(r) = \frac{r-B}{1-rB}.$$

Then it can be checked that

$$\tfrac{1}{2} < c < h(r) < r < 1.$$

We claim now that $\Omega(\alpha, c, h(r)) \subset \Omega_1(\alpha, \frac{1}{2}, r)$, or equivalently that

$$\Omega_2(\alpha, \tfrac{1}{2}, r) \cap \Omega(\alpha, c, h(r)) = \emptyset.$$

To see this, it suffices to take $\alpha = 0$ and to verify that

$$\rho(\tfrac{1}{2}, c) = \rho(h(r), r) = B > K.$$

Then by Proposition 1(b) and the Möbius invariance of the pseudohyperbolic metric, any cell Q_{nj} that meets the circle $\rho(\alpha, z) = r$ must lie entirely outside the circle $\rho(\alpha, z) = h(r)$. Similarly, any cell that meets the inner circle $\rho(\alpha, z) = \frac{1}{2}$ must lie inside the circle $\rho(\alpha, z) = c$. (See the remark about pseudohyperbolically concentric circles in Section 2.5.) This proves that

$$\Omega_2(\alpha, \tfrac{1}{2}, r) \subset \Omega(\alpha, \tfrac{1}{2}, c) \cup \Omega(\alpha, h(r), r),$$

which is equivalent to our claim that $\Omega(\alpha, c, h(r)) \subset \Omega_1(\alpha, \frac{1}{2}, r)$.

Consequently, the part of the integral in (16) taken over $\Omega_2(\alpha, \frac{1}{2}, r)$ is bounded by $I_1 + I_2$, where

$$I_1 = \int_{\Omega(\alpha, \frac{1}{2}, c)} \log \frac{1}{|\varphi_\alpha(z)|}\, d(\mu(z) + \nu(z)),$$

$$I_2 = \int_{\Omega(\alpha, h(r), r)} \log \frac{1}{|\varphi_\alpha(z)|}\, d(\mu(z) + \nu(z)).$$

It is easy to see that I_1 is bounded. Indeed,

$$\begin{aligned} I_1 &= a \int_{\Omega(\alpha,\frac{1}{2},c)} \log \frac{1}{|\varphi_\alpha(z)|} \frac{d\sigma(z)}{(1-|z|^2)^2} + \sum_{z_k\in\Omega(\alpha,\frac{1}{2},c)} \log \frac{1}{|\varphi_\alpha(z_k)|} \\ &\le \frac{ac^2 \log 2}{1-c^2} + n(\Gamma,\alpha,c)\log 2 \le \frac{\log 2}{1-c^2}\left(ac^2 + \left(\tfrac{2}{\delta}+1\right)^2\right), \end{aligned}$$

by Lemma 15 of Chapter 2, where δ is the separation constant of the uniformly discrete set Γ.

To show that I_2 is bounded, write $I_2 = I_3 + I_4$, where

$$\begin{aligned} I_3 &= a \int_{\Omega(\alpha,h(r),r)} \log \frac{1}{|\varphi_\alpha(z)|} \frac{d\sigma(z)}{(1-|z|^2)^2}, \\ I_4 &= \sum_{z_k\in\Omega(\alpha,h(r),r)} \log \frac{1}{|\varphi_\alpha(z_k)|}. \end{aligned}$$

Since

$$-\frac{s\log s}{(1-s^2)^2} \le \frac{s}{1-s^2}$$

for $\frac{1}{2} \le s < 1$,

$$I_3 = -2a \int_{h(r)}^{r} \frac{s\log s}{(1-s^2)^2}\, ds \le a \int_{h(r)}^{r} \frac{s}{1-s^2}\, ds = a \log \frac{1-h(r)^2}{1-r^2}.$$

But

$$\frac{1-h(r)^2}{1-r^2} = \frac{1-B^2}{(1-rB)^2} \le \frac{1+B}{1-B},$$

showing that I_3 stays bounded as $r \to 1$.

Finally, with the notation $\zeta_k = \varphi_\alpha(z_k)$, observe that

$$\begin{aligned} I_4 &= \sum_{z_k\in\Omega(\alpha,h(r),r)} \log \frac{1}{|\varphi_\alpha(z_k)|} = \sum_{\zeta_k\in\Omega(0,h(r),r)} \log \frac{1}{|\zeta_k|} \\ &\le \sum_{\zeta_k\in\Omega(0,h(r),r)} (1-|\zeta_k|^2), \end{aligned}$$

so that Lemma 2 gives

$$I_4 \le C \int_{\rho(z,\Omega)<\delta} \frac{d\sigma(z)}{1-|z|^2},$$

where $\Omega = \Omega(0,h(r),r)$ and $\delta = \delta(\Gamma) = \delta(\varphi_\alpha(\Gamma))$ is the separation constant of the uniformly discrete set Γ. But $\rho(z,\Omega) < \delta$ if and only if $r_1 < |z| < r_2$, where

$$r_1 = \frac{h(r)-\delta}{1-\delta h(r)}, \qquad r_2 = \frac{r+\delta}{1+\delta r}.$$

Therefore,

$$\begin{aligned}\int_{\rho(z,\Omega)<\delta} \frac{d\sigma(z)}{1-|z|^2} &= 2\int_{r_1}^{r_2} \frac{s}{1-s^2}\,ds = \log\frac{1-{r_1}^2}{1-{r_2}^2} \\ &= \log\frac{(1+\delta r)^2}{(1-\delta h(r))^2} + \log\frac{1-h(r)^2}{1-r^2} \\ &\le 2\log\frac{1+\delta}{1-\delta} + \log\frac{1+B}{1-B}\,.\end{aligned}$$

This shows that I_4 is bounded, which completes the proof of Proposition 2, hence of Theorem 6. □

Theorem 6 is due to Duren, Schuster, and Seip [1]. Some of the ideas had appeared earlier in a paper of Seip [5].

§6.7. Sharpened forms of Horowitz' theorems.

With Theorem 6 in hand, we are now in position to compute the density of the set of zeros of the infinite product

$$f(z) = \prod_{n=1}^{\infty}(1 - bz^{2^n})\,, \qquad b > 1\,,$$

introduced by Horowitz [2] and discussed in Chapter 4. This zero-set $\mathcal{Z}$ has 2^n equally spaced points on the circle $|z| = b_n = (1/b)^{2^{-n}}$, $n = 1, 2, \dots$. Luecking [4] constructed a similar set Λ by placing 2^n equally spaced points on the circle $|z| = r_n = 1 - 2^{-n}\gamma$, for $n = N, N+1, \dots$, where γ is a positive constant and $2^N > \gamma$. Here each group is aligned to include the point $z = r_n$.

The densities of these sets will be computed by regarding them as small pseudohyperbolic perturbations of the set Γ treated in the previous section and applying the following lemma.

Lemma 6. *If $\Gamma = \{z_k\}$ and $\widetilde{\Gamma} = \{\zeta_k\}$ are uniformly discrete sets satisfying $\rho(z_k, \zeta_k) \le \varepsilon < 1$ for all k, then*

$$\frac{1-\varepsilon}{1+\varepsilon} D^+(\Gamma) \le D^+(\widetilde{\Gamma}) \le \frac{1+\varepsilon}{1-\varepsilon} D^+(\Gamma) \tag{18}$$

and

$$\frac{1-\varepsilon}{1+\varepsilon} D^-(\Gamma) \le D^-(\widetilde{\Gamma}) \le \frac{1+\varepsilon}{1-\varepsilon} D^-(\Gamma)\,. \tag{19}$$

PROOF. It suffices to prove the upper inequality in (18), since the lower inequality is then deduced by reversing the roles of Γ and $\widetilde{\Gamma}$. For any point $\alpha \in \mathbb{D}$, it follows from the strong form of the triangle inequality (*cf.* Section 2.5) that if $\rho(\alpha, \zeta_k) < r < 1$, then

$$\rho(\alpha, z_k) \leq \frac{\rho(\alpha, \zeta_k) + \rho(\zeta_k, z_k)}{1 + \rho(\alpha, \zeta_k)\rho(\zeta_k, z_k)} \leq \frac{r + \varepsilon}{1 + r\varepsilon}.$$

With the notation $g(r) = \frac{r+\varepsilon}{1+r\varepsilon}$, we conclude that $n(\widetilde{\Gamma}, \alpha, r) \leq n(\Gamma, \alpha, g(r))$. Therefore, by a change of variables,

$$\begin{aligned}\int_0^r n(\widetilde{\Gamma}, \alpha, s)\, ds \leq \int_0^r n(\Gamma, \alpha, g(s))\, ds &\leq \frac{1+\varepsilon}{1-\varepsilon} \int_0^{g(r)} n(\Gamma, \alpha, u)\, du \\ &= \frac{1+\varepsilon}{1-\varepsilon} \int_0^r n(\Gamma, \alpha, u)\, du + \frac{1+\varepsilon}{1-\varepsilon} \int_r^{g(r)} n(\Gamma, \alpha, u)\, du.\end{aligned}$$

Since Γ is uniformly discrete, we know from Lemma 15 in Chapter 2 that $n(\Gamma, \alpha, u) \leq C/(1-u)$, and so

$$\int_r^{g(r)} n(\Gamma, \alpha, u)\, du \leq C \int_r^{g(r)} \frac{du}{1-u} \leq C \log\left(\frac{1+\varepsilon}{1-\varepsilon}\right).$$

Consequently,

$$\begin{aligned}D^+(\widetilde{\Gamma}) &= \limsup_{r\to 1} \left(\log \frac{1}{1-r}\right)^{-1} \sup_{\alpha\in\mathbb{D}} \int_0^r n(\widetilde{\Gamma}, \alpha, s)\, ds \\ &\leq \frac{1+\varepsilon}{1-\varepsilon} \limsup_{r\to 1} \left(\log \frac{1}{1-r}\right)^{-1} \sup_{\alpha\in\mathbb{D}} \int_0^r n(\Gamma, \alpha, s)\, ds = \frac{1+\varepsilon}{1-\varepsilon} D^+(\Gamma).\end{aligned}$$

The proof of (19) is essentially the same. □

COROLLARY 1. *The Horowitz set $\mathcal{Z}$ has density*

$$D^-(\mathcal{Z}) = D^+(\mathcal{Z}) = \frac{\log b}{\log 2}.$$

Luecking's set Λ has density

$$D^-(\Lambda) = D^+(\Lambda) = \frac{\gamma}{\log 2}.$$

PROOF. Let $\Gamma = \{\zeta_{nj}\}$ be the set constructed in the previous section. The estimate (15) for $|\zeta_{nj}|$ and the symmetry of the construction show that Γ is obtained by placing 2^{n-1} equally spaced points ζ_{nj} on a circle of radius

$$v_n = 1 - 2^{-n}(a \log 2) + O(2^{-2n}).$$

Suppose now that a set $\widetilde{\Gamma}$ is constructed by placing 2^{n-1} equally spaced points on the circle of radius

$$s_n = 1 - 2^{-n}(a \log 2)\,,$$

where the arguments of the points agree with those of ζ_{nj}. Since $\rho(s_n, v_n) \to 0$, the points z_k of Γ and $\widetilde{z}_k$ of $\widetilde{\Gamma}$ can be paired so that $\rho(z_k, \widetilde{z}_k) \to 0$ as $k \to \infty$. It is then clear that $\widetilde{\Gamma}$ is uniformly discrete, since the triangle inequality gives

$$\rho(\widetilde{z}_j, \widetilde{z}_k) \geq \rho(z_j, z_k) - \rho(z_j, \widetilde{z}_j) - \rho(z_k, \widetilde{z}_k)\,,$$

and we have already shown that Γ is uniformly discrete. But the densities are unaffected by adding or removing a finite number of points, so it follows from Lemma 6 that

$$D^-(\widetilde{\Gamma}) = D^+(\widetilde{\Gamma}) = D^-(\Gamma) = D^+(\Gamma) = \frac{a}{2}\,.$$

Now observe that Luecking's set Λ is constructed by placing 2^{n-1} equally spaced points on the circle of radius $u_n = r_{n-1}$, where $r_n = 1 - 2^{-n}\gamma$; that is, $u_n = 1 - 2^{-n}(2\gamma)$. Thus Λ coincides with a rotation of $\widetilde{\Gamma}$, at least for large n, if $2\gamma = a \log 2$. This implies that

$$D^-(\Lambda) = D^+(\Lambda) = \frac{\gamma}{\log 2}\,.$$

If we choose $\gamma = \log b$, then Horowitz' zero-set $\mathcal{Z}$ has 2^n equally spaced points on the circle of radius

$$b_n = (1/b)^{2^{-n}} = e^{-2^{-n}\gamma} = 1 - 2^{-n}\gamma + O(2^{-2n})\,,$$

and these points are aligned with those of Λ. Since $\rho(b_n, r_n) \to 0$, Lemma 6 shows that

$$D^-(\mathcal{Z}) = D^+(\mathcal{Z}) = \frac{\log b}{\log 2}\,. \qquad \square$$

Luecking [4] found that Λ is an A^p zero-set if and only if $p\gamma < \log 2$. Horowitz [2] showed that the infinite product belongs to A^p, hence that $\mathcal{Z}$ is an A^p zero-set, if $p \log b < \log 2$. For $p \log b > \log 2$, he proved that $\mathcal{Z}$ is not an A^p zero-set. (See Chapter 4.) Luecking sharpened Horowitz' result by showing that $\mathcal{Z}$ is not an A^p zero-set if $p \log b = \log 2$. Appealing to Theorems 4 and 5, we can use the above calculations of densities to round out the picture as follows.

COROLLARY 2. (a) *The Horowitz set $\mathcal{Z}$ is an interpolation set for A^p if and only if $p \log b < \log 2$. It is a sampling set for A^p if and only if $p \log b > \log 2$. When $p \log b = \log 2$, it is a set of uniqueness for A^p.*

(b) *Luecking's set Λ is an interpolation set for A^p if and only if $p\gamma < \log 2$. It is a sampling set for A^p if and only if $p\gamma > \log 2$. When $p\gamma = \log 2$, it is a set of uniqueness for A^p.*

These results can be combined with the main sampling and interpolation theorems to recapture some of Horowitz' theorems on zero-sets, as given in Chapter 4, in sharper form. For instance, Theorem 2 of Chapter 4 states that the A^p zero-sets depend on p. More precisely, for $0 < p < q < \infty$, there is an A^p zero-set that is not an A^q zero-set. The following theorem makes a stronger assertion.

THEOREM 7. *If $0 < p < q < \infty$, then there is an interpolation set for A^p that is a sampling set for A^q.*

PROOF. In the set Γ of Theorem 6, for instance, choose the parameter a so that $\frac{1}{q} < \frac{a}{2} < \frac{1}{p}$. Or in Horowitz' set $\mathcal{Z}$, choose b with $\frac{1}{q} < \frac{\log 2}{\log b} < \frac{1}{p}$. Then apply Theorems 4 and 5. □

Similar examples can be used to sharpen Theorem 3 of Chapter 4, which asserts that the union of two A^p zero-sets need not be an A^q zero-set for any $q > \frac{p}{2}$.

THEOREM 8. *For $0 < \frac{p}{2} < q < \infty$, there exist two A^p interpolation sets whose union is an A^q sampling set.*

PROOF. Choose γ such that

$$\frac{1}{q} < \frac{\gamma}{\log 2} < \frac{2}{p},$$

and consider Luecking's set $\Lambda = \Lambda(\gamma)$, written in this way to emphasize the dependence on γ. Let $\Lambda_1 = \Lambda(\frac{\gamma}{2})$. Note that Λ_1 is constructed by placing 2^n equally spaced points on the circle of radius $1 - \frac{\gamma}{2}2^{-n} = r_{n+1}$, while the set $\Lambda(\gamma)$ places 2^{n+1} equally spaced points on the same circle. The set $\Lambda_2 = \Lambda(\gamma) \setminus \Lambda_1$ is simply a rotation of Λ_1, so it has the same densities. Thus

$$D^-(\Lambda_1) = D^+(\Lambda_1) = \frac{\gamma}{2\log 2} = D^-(\Lambda_2) = D^+(\Lambda_2).$$

However, $\Lambda = \Lambda(\gamma) = \Lambda_1 \cup \Lambda_2$ has densities

$$D^-(\Lambda) = D^+(\Lambda) = \frac{\gamma}{\log 2}.$$

By Theorem 4, the inequalities $D^+(\Lambda_1) < \frac{1}{p}$ and $D^+(\Lambda_2) < \frac{1}{p}$ show that Λ_1 and Λ_2 are A^p interpolation sets. By Theorem 5, the inequality $D^-(\Lambda) > \frac{1}{q}$ implies that Λ is an A^q sampling set. □

The theorem has a natural generalization that will become an important tool in our discussion of invariant subspaces in Chapter 8.

THEOREM 9. *For $0 < p < \infty$ and for each integer $n \geq 2$, there is an A^p sampling set that is a disjoint union of the form $\Gamma = \Gamma_1 \cup \Gamma_2 \cup \cdots \cup \Gamma_n$, where $\Gamma \setminus \Gamma_j$ is an A^p interpolation set for each $j = 1, 2, \ldots, n$.*

The proof will require some general inequalities describing the behavior of upper and lower uniform densities under unions of sets.

LEMMA 7. *Let Γ_1 and Γ_2 be disjoint uniformly discrete sets in the unit disk. Then*

$$D^-(\Gamma_1) + D^-(\Gamma_2) \leq D^-(\Gamma_1 \cup \Gamma_2) \leq D^-(\Gamma_1) + D^+(\Gamma_2) \\ \leq D^+(\Gamma_1 \cup \Gamma_2) \leq D^+(\Gamma_1) + D^+(\Gamma_2) .$$

PROOF. The counting functions satisfy

$$n(\Gamma_1, \zeta, r) + n(\Gamma_2, \zeta, r) = n(\Gamma_1 \cup \Gamma_2, \zeta, r) ,$$

so it is clear that

$$E(\Gamma_1, \zeta, r) + E(\Gamma_2, \zeta, r) = E(\Gamma_1 \cup \Gamma_2, \zeta, r) \tag{20}$$

for every $\zeta \in \mathbb{D}$ and $r < 1$. Therefore, it follows directly from the definitions that

$$D^-(\Gamma_1) + D^-(\Gamma_2) \leq D^-(\Gamma_1 \cup \Gamma_2) \quad \text{and} \quad D^+(\Gamma_1 \cup \Gamma_2) \leq D^+(\Gamma_1) + D^+(\Gamma_2) .$$

By definition of the densities, for each $\varepsilon > 0$ there is a number $r_0 < 1$ such that $r_0 < r < 1$ implies

$$E(\Gamma_1 \cup \Gamma_2, \zeta, r) > D^-(\Gamma_1 \cup \Gamma_2) - \varepsilon \qquad \text{and} \qquad E(\Gamma_2, \zeta, r) < D^+(\Gamma_2) + \varepsilon$$

for every $\zeta \in \mathbb{D}$. Then by (20) we conclude that

$$E(\Gamma_1, \zeta, r) > D^-(\Gamma_1 \cup \Gamma_2) - D^+(\Gamma_2) - 2\varepsilon ,$$

for $r_0 < r < 1$ and $\zeta \in \mathbb{D}$, so that $D^-(\Gamma_1) \geq D^-(\Gamma_1 \cup \Gamma_2) - D^+(\Gamma_2)$. The remaining inequality is proved in a similar way. □

PROOF OF THEOREM 9. We shall revert to the construction in Section 6.4 of the lattice

$$\Lambda(a, b) = \{a^m(bk + i) : m \in \mathbb{Z}, \ k \in \mathbb{Z}\} , \qquad a > 1, \ b > 0 ,$$

in the upper half-plane and its image $\Gamma(a, b) = \psi^{-1}(\Lambda(a, b))$ in the unit disk. It was observed in Section 6.5 that

$$D^-(\Gamma(a, b)) = D^+(\Gamma(a, b)) = \frac{2\pi}{b \log a} .$$

For $j = 1, 2, \ldots, n$, consider now the subsets

$$\Lambda_j^n(a,b) = \{a^m(bk+i) : m \in \mathbb{Z},\ k \equiv j \pmod{n}\},$$

and define $\Gamma_j^n(a,b) = \psi^{-1}(\Lambda_j^n(a,b))$. Since $\Lambda_n^n(a,b) = \Lambda(a, nb)$, we find that

$$\begin{aligned} D^-(\Gamma_j^n(a,b)) = D^+(\Gamma_j^n(a,b)) &= D^-(\Gamma_n^n(a,b)) = D^+(\Gamma_n^n(a,b)) \\ &= \frac{2\pi}{nb\log a}, \qquad j = 1, 2, \ldots, n. \end{aligned}$$

This statement can be verified by the method of Section 6.4. (See also Hedenmalm, Richter, and Seip [1].) Taking $\Gamma = \Gamma(a,b)$ and $\Gamma_j = \Gamma_j^n(a,b)$, we have $\Gamma = \Gamma_1 \cup \Gamma_2 \cup \ldots \Gamma_n$. By Lemma 7,

$$D^-(\Gamma \setminus \Gamma_j) \geq D^-(\Gamma) - D^+(\Gamma_j) = \frac{n-1}{n}\,\frac{2\pi}{b\log a}$$

and

$$D^+(\Gamma \setminus \Gamma_j) \leq D^+(\Gamma) - D^-(\Gamma_j) = \frac{n-1}{n}\,\frac{2\pi}{b\log a}.$$

This shows that

$$D^-(\Gamma \setminus \Gamma_j) = D^+(\Gamma \setminus \Gamma_j) = \frac{n-1}{n}\,\frac{2\pi}{b\log a}.$$

Now choose $a > 1$ and $b > 0$ such that

$$\frac{1}{p} < \frac{2\pi}{b\log a} < \frac{n}{n-1}\,\frac{1}{p}.$$

Then Theorem 5 shows that Γ is a sampling set for A^p, while Theorem 4 shows that each set $\Gamma \setminus \Gamma_j$ is interpolating for A^p. $\square$

With suitable choice of parameters, the proof actually shows that Γ can be chosen to be a sampling set for A^q, where q is any specified index with $q > \frac{n-1}{n}p$. For $n = 2$ the result then reduces to Theorem 7.

§6.8. Sufficient conditions with the pseudohyperbolic metric.

Theorems 4 and 5 say that interpolation sequences are sparse everywhere in the unit disk, whereas sampling sequences must be sufficiently dense. The precise conditions are expressed by upper and lower uniform densities, but these densities can be quite difficult to compute. Our goal is now to develop sufficient conditions for sampling and interpolation, directly based on the pseudohyperbolic metric, that are relatively easy to verify.

For $0 < \varepsilon < 1$, a sequence $\Gamma = \{z_k\}$ of points in the unit disk $\mathbb{D}$ is said to be an *ε-net* if each point $z \in \mathbb{D}$ has the property $\rho(z, z_k) < \varepsilon$ for some z_k in Γ. An equivalent statement is that

$$\mathbb{D} = \bigcup_{k=1}^{\infty} \Delta(z_k, \varepsilon),$$

where $\Delta(z_k, \varepsilon)$ denotes a pseudohyperbolic disk. Luecking [3] showed that every uniformly discrete ε-net with sufficiently small ε is a sampling set for A^p. With the aid of Theorem 5, this principle can be expressed in quantitative form, as shown by Duren, Schuster, and Vukotić [1].

THEOREM 10. *Let $0 < p < \infty$. If Γ is a uniformly discrete ε-net with*

$$\varepsilon < \frac{1}{1 + \sqrt{\frac{2}{p}}},$$

then Γ is a sampling set for A^p.

In particular, all uniformly discrete ε-nets with $\varepsilon < \frac{1}{2}$ are sampling sets for A^2. Another consequence of the theorem is that every ε-net is a sampling set for some Bergman space A^p.

Theorem 10 is a direct consequence of Theorem 5 and the following lemma.

LEMMA 8. *Let $0 < \varepsilon < 1$. If Γ is an ε-net, then*

$$D^-(\Gamma) \geq \frac{(1-\varepsilon)^2}{2\varepsilon^2}.$$

DEDUCTION OF THEOREM 10. Since the function

$$f(x) = \frac{(1-x)^2}{2x^2}$$

is decreasing for $0 < x < 1$, Lemma 8 implies that

$$D^-(\Gamma) \geq f(\varepsilon) > f\left(\frac{1}{1 + \sqrt{\frac{2}{p}}}\right) = \frac{1}{p},$$

and the result follows from Theorem 5. □

Before passing to the proof of Lemma 8, we note the following corollary.

COROLLARY. *A sequence Γ has lower density $D^-(\Gamma) > 0$ if and only if Γ is an ε-net for some $\varepsilon < 1$.*

In light of Theorem 5, the corollary shows that a uniformly discrete set Γ is a sampling set for some A^p space if and only if Γ is an ε-net for some ε. This is analogous to the fact that Γ is an interpolation set for some A^p space if and only if Γ is uniformly discrete. (Recall that $D^+(\Gamma) < \infty$ if Γ is uniformly discrete.)

PROOF OF COROLLARY. It need only be observed that Γ is an ε-net if $D^-(\Gamma) > 0$. Suppose, on the contrary, that Γ is *not* an ε-net for any $\varepsilon < 1$. Then for each radius $r < 1$, the pseudohyperbolic counting function $n(\Gamma, \alpha, r) = 0$ for some $\alpha \in \mathbb{D}$ depending on r. Consequently, $n(\Gamma, \alpha, s) = 0$ for all $s < r$, so that $\int_0^r n(\Gamma, \alpha, s)\, ds = 0$. In particular,

$$\inf_{\zeta \in \mathbb{D}} \int_0^r n(\Gamma, \zeta, s)\, ds = 0$$

for each $r < 1$, which implies that $D^-(\Gamma) = 0$ by the definition of lower density (*cf.* Section 6.5) . □

PROOF OF LEMMA 8. Let $\zeta \in \mathbb{D}$ and define

$$h(s) = \frac{s - \varepsilon}{1 - s\varepsilon}, \qquad \varepsilon < s < 1 .$$

It follows from the strong form of the triangle inequality for the pseudohyperbolic metric (see Section 2.5) that

$$\Delta(\zeta, h(s)) \subset \bigcup_{z_k \in \Delta(\zeta, s)} \Delta(z_k, \varepsilon) . \tag{21}$$

Indeed, since Γ is an ε-net, each point $w \in \mathbb{D}$ belongs to some disk $\Delta(z_k, \varepsilon)$. If $w \in \Delta(\zeta, h(s))$, then

$$\rho(z_k, \zeta) \le \frac{\rho(z_k, w) + \rho(w, \zeta)}{1 + \rho(z_k, w)\rho(w, \zeta)} .$$

But $\rho(z_k, w) < \varepsilon$ and $\rho(w, \zeta) < h(s)$, so

$$\rho(z_k, \zeta) < \frac{\varepsilon + h(s)}{1 + \varepsilon h(s)} = s .$$

This shows that $z_k \in \Delta(\zeta, s)$, which proves (21). Recall now that a pseudohyperbolic disk $\Delta(\alpha, r)$ has hyperbolic area

$$a(\Delta(\alpha, r)) = \frac{r^2}{1 - r^2} .$$

Thus it follows from (21) that

$$n(\Gamma, \zeta, s) \ge \frac{a(\Delta(\zeta, h(s)))}{a(\Delta(z_k, \varepsilon))} = \frac{h(s)^2}{1 - h(s)^2} \frac{1 - \varepsilon^2}{\varepsilon^2} = \frac{(s - \varepsilon)^2}{\varepsilon^2 (1 - s^2)} .$$

Consequently, for $r > \varepsilon$,

$$\int_0^r n(\Gamma, \zeta, s)\, ds \ge \int_\varepsilon^r n(\Gamma, \zeta, s)\, ds \ge \frac{1}{\varepsilon^2} \int_\varepsilon^r \frac{(s - \varepsilon)^2}{1 - s^2}\, ds .$$

But a straightforward calculation gives

$$\int_{\varepsilon}^{r} \frac{(s-\varepsilon)^2}{1-s^2}\,ds = -r + \tfrac{1}{2}(1+\varepsilon)^2 \log(1+r) - \tfrac{1}{2}(1-\varepsilon)^2 \log(1-r) + C(\varepsilon)\,,$$

where $C(\varepsilon)$ denotes a constant depending only on ε. Therefore,

$$D^-(\Gamma) \geq \lim_{r\to 1} \left(\log \frac{1}{1-r}\right)^{-1} \frac{1}{\varepsilon^2} \int_{\varepsilon}^{r} \frac{(s-\varepsilon)^2}{1-s^2}\,ds = \frac{(1-\varepsilon)^2}{2\varepsilon^2}\,,$$

which is the desired inequality. □

There is a similar criterion for interpolation. If a sequence $\Gamma = \{z_k\}$ is uniformly discrete with separation constant

$$\delta = \delta(\Gamma) = \inf_{j\neq k} \rho(z_j, z_k)$$

sufficiently close to 1, then Γ is an interpolation set for A^p. This principle was discovered by Amar [1] and Rochberg [1]. The following lemma, when combined with Theorem 4, gives a quantitative version for $0 < p < \infty$.

LEMMA 9. *If Γ is uniformly discrete with separation constant δ, where $0 < \delta < 1$, then*

$$D^+(\Gamma) \leq (2\pi + 1)\frac{\sqrt{1-\delta}}{(1-\sqrt{1-\delta})^2}\,.$$

PROOF. It is convenient to define

$$\varepsilon = 1 - \sqrt{1-\delta}\,, \qquad \text{so that} \quad \delta = \frac{2\varepsilon}{1+\varepsilon^2} \quad \text{and} \quad 0 < \varepsilon < \delta\,.$$

Begin by writing

$$\int_0^r n(\Gamma,\zeta,s)\,ds = \int_0^{\varepsilon} n(\Gamma,\zeta,s)\,ds + \int_{\varepsilon}^r n(\Gamma,\zeta,s)\,ds$$

for $\varepsilon < r < 1$. Note that

$$\int_0^{\varepsilon} n(\Gamma,\zeta,s)\,ds \leq n(\Gamma,\zeta,\varepsilon) \leq \left(\frac{2}{\delta}+1\right)^2 \frac{1}{1-\varepsilon^2}\,,$$

by Lemma 15 of Chapter 2. Next write

$$\int_{\varepsilon}^r n(\Gamma,\zeta,s)\,ds = \int_{\varepsilon}^r n(\Gamma,\zeta,h(s))\,ds + \int_{\varepsilon}^r n(\Gamma,\zeta,h(s),s)\,ds\,,$$

where

$$h(s) = \frac{s-\varepsilon}{1-s\varepsilon} < s$$

and $n(\Gamma, \zeta, r_1, r_2)$ is the number of points of Γ lying in the pseudohyperbolic annulus $\Omega(\zeta, r_1, r_2)$.

To deal with the first integral on the right-hand side above, apply the strong form of the triangle inequality to see that $\Delta(z_j, \varepsilon) \cap \Delta(z_k, \varepsilon) = \emptyset$ for $j \neq k$, since $\rho(z_j, z_k) \geq \delta$. Moreover, the triangle inequality shows that if $z_k \in \Delta(\zeta, h(s))$, then $\Delta(z_k, \varepsilon) \subset \Delta(\zeta, s)$. Therefore,

$$n(\Gamma, \zeta, h(s)) \leq \frac{a(\Delta(\zeta, s))}{a(\Delta(z_k, \varepsilon))} = \frac{(1-\varepsilon^2)s^2}{\varepsilon^2(1-s^2)},$$

and so

$$\begin{aligned} \int_\varepsilon^r n(\Gamma, \zeta, h(s))\, ds &\leq \frac{1-\varepsilon^2}{\varepsilon^2} \int_0^r \frac{s^2}{1-s^2}\, ds \\ &= \frac{1-\varepsilon^2}{2\varepsilon^2} \left(\log \frac{1+r}{1-r} - 2r \right). \end{aligned}$$

Our next task is to estimate $\int_\varepsilon^r n(\Gamma, \zeta, h(s), s)\, ds$. Since the pseudohyperbolic metric is invariant under Möbius self-mappings of the disk, there is no loss of generality in taking $\zeta = 0$. Then $\Omega(0, h(s), s)$ is a Euclidean annulus bounded by the circles $|z| = s$ and $|z| = h(s)$. Use radial segments of angular separation θ to divide this annulus into a minimum number of congruent cells of pseudohyperbolic diameter smaller than δ. Then there can be at most one point of Γ in each cell. To determine this minimum number of cells, suppose first that θ is chosen to make cells of precise diameter δ, so that $\rho(h(s), se^{i\theta}) = \delta$. Using a standard identity for the pseudohyperbolic metric, one computes

$$\begin{aligned} 1 - \rho(h(s), se^{i\theta})^2 &= \frac{(1-s^2)(1-h(s)^2)}{|1-h(s)se^{i\theta}|^2} \\ &= \frac{(1-s^2)^2(1-\varepsilon^2)}{(1-s^2)^2 + 2s(s-\varepsilon)(1-s\varepsilon)(1-\cos\theta)}. \end{aligned}$$

Setting this expression equal to $1-\delta^2$, we find after some manipulation

$$\theta^2 \geq 2(1-\cos\theta) = \frac{\delta^2-\varepsilon^2}{1-\delta^2} \frac{(1-s^2)^2}{s(s-\varepsilon)(1-s\varepsilon)} \geq \frac{\delta^2-\varepsilon^2}{1-\delta^2} \frac{(1-s^2)^2}{s^2(1-\varepsilon^2)}.$$

Further calculation gives

$$\frac{\delta^2-\varepsilon^2}{(1-\delta^2)(1-\varepsilon^2)} = \frac{\varepsilon^2(\varepsilon^2+3)}{(1-\varepsilon^2)^2} \geq \frac{\varepsilon^4}{(1-\varepsilon^2)^2},$$

so we have

$$\theta \geq \frac{\varepsilon^2}{1-\varepsilon^2} \frac{1-s^2}{s}.$$

Consequently,

$$n(\Gamma, \zeta, h(s), s) \leq \frac{2\pi}{\theta} \leq \frac{2\pi(1-\varepsilon^2)}{\varepsilon^2} \frac{s}{1-s^2},$$

and

$$\int_\varepsilon^r n(\Gamma, \zeta, h(s), s)\, ds \leq \frac{2\pi(1-\varepsilon^2)}{\varepsilon^2} \int_0^r \frac{s}{1-s^2}\, ds = \frac{\pi(1-\varepsilon^2)}{\varepsilon^2} \log \frac{1}{1-r^2}.$$

Putting everything together, we conclude that

$$\begin{aligned} D^+(\Gamma) &\leq \lim_{r\to 1} \left(\log \frac{1}{1-r} \right)^{-1} \left\{ \frac{1-\varepsilon^2}{2\varepsilon^2} \log \frac{1+r}{1-r} + \frac{\pi(1-\varepsilon^2)}{\varepsilon^2} \log \frac{1}{1-r^2} \right\} \\ &= (\pi + \tfrac{1}{2}) \frac{1-\varepsilon^2}{\varepsilon^2} < (2\pi+1) \frac{\sqrt{1-\delta}}{(1-\sqrt{1-\delta})^2}. \qquad \square \end{aligned}$$

Lemma 9 was found by Duren, Schuster, and Vukotić [1]. With an appeal to Theorem 4, it yields the following theorem.

THEOREM 11. *Let $0 < p < \infty$. If Γ is a uniformly discrete sequence with separation constant δ satisfying*

$$(2\pi+1) \frac{\sqrt{1-\delta}}{(1-\sqrt{1-\delta})^2} < \frac{1}{p},$$

then Γ is an interpolation set for A^p.

§6.9. Duality relations.

Sampling and interpolation sequences are in some sense dual to each other. While this is evident in the statements of Theorems 4 and 5, it can also be deduced from the definitions. To see this, we fix a sequence $\Gamma = \{z_n\}$ in the disk and define the linear map T_p that takes an analytic function f to the sequence $\{f(z_n)(1-|z_n|^2)^{\frac{2}{p}}\}$. With this notation the sampling inequalities become

$$K_1 \|f\|_p \leq \|T_p(f)\|_{\ell^p} \leq K_2 \|f\|_p,$$

and the interpolation condition is $\ell^p \subset T_p(A^p)$.

Let us now suppose that Γ is uniformly discrete. By results in Section 2.11, this implies that the upper sampling inequality holds and therefore $T_p(A^p) \subset \ell^p$. Thus Γ is a sampling sequence for A^p if and only if T_p, viewed as a map from A^p to ℓ^p, is bounded below, while Γ is an interpolation sequence for A^p if and only if T_p is surjective.

It was shown in Chapter 2 that with respect to the pairing

$$\langle f, g\rangle_{\mathbb{D}} = \lim_{r\to 1}\int_{r\mathbb{D}} f(z)\overline{g(z)}\,d\sigma(z)\,,$$

the dual of A^p $(1 < p < \infty)$ can be identified with A^q, where p and q are conjugate indices, and the dual of A^1 is the Bloch space $\mathcal{B}$. Also, by the Riesz representation theorem, the dual of ℓ^p $(1 < p < \infty)$ can be identified with ℓ^q, and the dual of ℓ^1 is ℓ^∞. Here the pairing is given by

$$\langle \mathbf{a}, \mathbf{b}\rangle = \sum_{n=1}^{\infty} a_n \overline{b_n}\,, \qquad \text{where } \mathbf{a} = \{a_n\},\ \mathbf{b} = \{b_n\}\,.$$

Under these pairings, the adjoint of T_p is identified with the bounded map T_p^* from ℓ^q (resp. ℓ^∞) into A^q (resp. $\mathcal{B}$) defined by the equation

$$\langle T_p(f), \mathbf{a}\rangle = \langle f, T_p^*(\mathbf{a})\rangle_{\mathbb{D}}$$

for all $\mathbf{a} \in \ell^q$ (resp. ℓ^∞) and $f \in A^p$ (resp. A^1).

To determine T_p^*, we denote by $\mathbf{e}_n$ the sequence with 1 in the nth position and 0 in all the other positions. For each $f \in A^p$,

$$\langle f, T_p^*(\mathbf{e}_n)\rangle_{\mathbb{D}} = \langle T_p(f), \mathbf{e}_n\rangle = (1-|z_n|^2)^{\frac{2}{p}} f(z_n) = \langle f, (1-|z_n|^2)^{\frac{2}{p}} k_{z_n}\rangle_{\mathbb{D}}\,,$$

where $k_\zeta(z) = (1-\overline{\zeta}z)^{-2}$ is the Bergman kernel function. This implies

$$T_p^*(\mathbf{e}_n)(z) = (1-|z_n|^2)^{\frac{2}{p}} k_{z_n}(z) = \frac{(1-|z_n|^2)^{\frac{2}{p}}}{(1-\overline{z_n}z)^2}\,,$$

and so by linearity

$$T_p^*(\mathbf{a})(z) = \sum_{n=1}^{\infty} a_n \frac{(1-|z_n|^2)^{\frac{2}{p}}}{(1-\overline{z_n}z)^2} \tag{22}$$

if $a_n = 0$ for all but a finite number of indices. We claim now that (22) holds for arbitrary $\mathbf{a} \in \ell^q$. Since the uniformly discrete sequence $\{z_n\}$ has the property $\sum (1-|z_n|^2)^2 < \infty$, an application of Hölder's inequality shows that for each $\mathbf{a} \in \ell^q$ the infinite series (22) converges uniformly on compact subsets of $\mathbb{D}$ and therefore represents an analytic function. On the other hand, if $1 < p < \infty$ the series converges in A^q norm because T_p^* is continuous and the truncated sequence $\{a_1, \dots, a_m, 0, 0, \dots\}$ converges to $\mathbf{a}$ in ℓ^q norm as $m \to \infty$. Since norm convergence implies pointwise convergence, this proves (22) for arbitrary $\mathbf{a} \in \ell^q$ when $1 < p < \infty$. When $p = 1$, the series converges in the weak-star topology for the Bloch space, because T_1 is weak-star continuous and $\{a_1, \dots, a_m, 0, \dots\}$ converges to $\mathbf{a}$ in the weak-star topology of ℓ^∞. However, weak-star convergence in $\mathcal{B}$ implies pointwise convergence (by pairing with the kernel function), so (22) holds for arbitrary $\mathbf{a} \in \ell^\infty$ when $p = 1$.

Dual characterizations of interpolation and sampling sequences can be deduced from the following general theorem of functional analysis (*cf.* Rudin [3], p. 97).

THEOREM A. *Suppose X and Y are Banach spaces. Let $L : X \to Y$ be a bounded linear operator, and let $L^* : Y^* \to X^*$ denote its adjoint. Then L is surjective if and only if L^* is bounded below, and L is bounded below if and only if L^* is surjective.*

Actually, Rudin's book gives an explicit proof only of the first statement, that L is surjective if and only if L^* is bounded below. The two statements are equivalent if X is reflexive. In the general case, let $\widehat{X} \subset X^{**}$ be the image of X under the natural isometric embedding. Then we can deduce the second statement from the first by the following argument. If L^* maps Y^* onto X^*, then L^{**} is bounded below on X^{**}, hence on $\widehat{X}$, which implies that L is bounded below on X. Conversely, if L is bounded below, then its range $L(X)$ is a closed subspace of Y, and its inverse L^{-1} is bounded on $L(X)$. Then for any $\psi \in X^*$, the linear functional ϕ defined on $L(X)$ by $\phi(y) = \psi(L^{-1}(y))$ is bounded. In other words, $\phi \in L(X)^*$. By the Hahn–Banach theorem, ϕ can be extended to a functional $\Phi \in Y^*$. Then

$$L^*(\Phi)(x) = \Phi(L(x)) = \phi(L(x)) = \psi(x), \qquad x \in X.$$

Hence $L^*(\Phi) = \psi$, so L^* maps Y^* onto X^*.

Recall now that a uniformly discrete sequence $\{z_n\}$ is an interpolation sequence for A^p if and only if T_p maps A^p onto ℓ^p, and is a sampling sequence if and only if T_p is bounded below. Therefore, Theorem A can be combined with the representation (22) of the adjoint operator to characterize interpolation and sampling sequences for A^p spaces in the two cases $1 < p < \infty$ and $p = 1$, as follows.

THEOREM 12. *Let $\Gamma = \{z_n\}$ be uniformly discrete. Suppose $1 < p < \infty$ and let q be the conjugate index. Then Γ is an interpolating sequence for A^p if and only if*

$$\int_{\mathbb{D}} \left| \sum_{n=1}^{\infty} a_n \frac{(1-|z_n|^2)^{\frac{2}{p}}}{(1-\overline{z_n}z)^2} \right|^q d\sigma(z) \ge C \sum_{n=1}^{\infty} |a_n|^q$$

for some constant $C > 0$ and all sequences $\{a_n\} \in \ell^q$. On the other hand, Γ is a sampling sequence for A^p if and only if for each $f \in A^q$, there is a sequence $\{a_n\} \in \ell^q$ such that

$$f(z) = \sum_{n=1}^{\infty} a_n \frac{(1-|z_n|^2)^{\frac{2}{p}}}{(1-\overline{z_n}z)^2},$$

where the series converges in A^q norm.

THEOREM 13. *Let $\Gamma = \{z_n\}$ be uniformly discrete. Then Γ is an interpolation sequence for A^1 if and only if*

$$\left\| \sum_{n=1}^{\infty} a_n \frac{(1-|z_n|^2)^2}{(1-\overline{z_n}z)^2} \right\| \ge C \sup_n |a_n|$$

for some constant $C > 0$ and all sequences $\{a_n\} \in \ell^\infty$, where $\|\cdot\|$ denotes the norm in the Bloch space $\mathcal{B}$. On the other hand, Γ is a sampling sequence for A^1 if and only if for each $f \in \mathcal{B}$, there is a sequence $\{a_n\} \in \ell^\infty$ such that

$$f(z) = \sum_{n=1}^{\infty} a_n \frac{(1-|z_n|^2)^2}{(1-\overline{z_n}z)^2},$$

where the series converges in the weak-star topology of $\mathcal{B}$.

It was shown in Section 2.4 that $\|k_\zeta\|_p \simeq (1-|\zeta|^2)^{-\frac{2}{q}}$. Because of this relationship, the representation of A^q functions in Theorem 12 is often called an *atomic decomposition*, since it states that every function in the space can be expressed as an infinite linear combination of normalized reproducing kernels or *atoms*. A similar formula can also be obtained for A^1 functions; details are given in Zhu [1], p. 70. The existence of such decompositions was first proved by Coifman and Rochberg [1].

CHAPTER 7

Proofs of Sampling and Interpolation Theorems

Chapter 6 was devoted to a broad exposition of sampling and interpolation sequences, with a variety of examples and applications. Much of the discussion was focused on the two central theorems that characterize a sequence Γ as interpolating or sampling for A^p according to its upper and lower density $D^+(\Gamma)$ and $D^-(\Gamma)$. The main purpose of the present chapter is to supply proofs of those theorems, which are now restated for convenience of reference.

THEOREM 1. *Let $0 < p < \infty$. A sequence Γ of distinct points in the disk is an interpolation sequence for A^p if and only if it is uniformly discrete and $D^+(\Gamma) < \frac{1}{p}$.*

THEOREM 2. *Let $0 < p < \infty$. A sequence Γ of distinct points in the disk is a sampling sequence for A^p if and only if it is a finite union of uniformly discrete sequences and it contains a uniformly discrete subsequence Γ' for which $D^-(\Gamma') > \frac{1}{p}$.*

Both theorems are essentially due to Kristian Seip [3]. Other contributions were cited in Chapter 6. Here we shall carry out the proofs only for $1 \leq p < \infty$. Most of the techniques are outlined in Seip's papers [3] and [6].

§7.1. Perturbation of sampling sequences.

In this section we will see that if $\Gamma = \{z_k\}$ is a uniformly discrete sampling sequence for the Bergman space A^p, then every sufficiently small perturbation of Γ is itself a sampling sequence for A^p. In fact, with an appeal to Theorem 2 (yet to be proved), it follows from Lemma 6 of Chapter 6 that if $\varepsilon > 0$ is sufficiently small, then every sequence $\widetilde{\Gamma} = \{\zeta_k\}$ with $\rho(z_k, \zeta_k) < \varepsilon$ for all k is an A^p sampling sequence. An important ingredient in the *proof* of Theorem 2 is that sampling sequences actually have this property of stability under small perturbations.

Recall that the upper sampling inequality is satisfied whenever Γ is uniformly discrete. The sampling constant $K_1(\Gamma)$ was defined as the largest constant K_1 for which the lower sampling inequality

$$K_1 \|f\|_p^p \leq \sum_{k=1}^{\infty} (1 - |z_k|^2)^2 |f(z_k)|^p$$

holds for all functions $f \in A^p$. Thus a uniformly discrete sequence Γ is a sampling sequence if and only if $K_1(\Gamma) > 0$.

A natural notion of distance between subsets of the unit disk is the Hausdorff distance with respect to the pseudohyperbolic metric. For a relatively closed set $A \subset \mathbb{D}$, let

$$A_t = \{z \in \mathbb{D} \ : \ \rho(z, A) < t\}, \qquad 0 < t < 1,$$

where $\rho(z, A) = \inf_{\zeta \in A} \rho(z, \zeta)$. The *pseudohyperbolic Hausdorff distance* between relatively closed sets A and B in $\mathbb{D}$ is defined by

$$[A, B] = \inf \{t \ : \ A \subset B_t \quad \text{and} \quad B \subset A_t\}.$$

It can be verified that the Hausdorff distance is a true metric. In terms of the asymmetric "distance"

$$\rho^*(A, B) = \sup_{z \in A} \rho(z, B) = \sup_{z \in A} \inf_{\zeta \in B} \rho(z, \zeta)$$

between the two sets, the Hausdorff distance is

$$[A, B] = \max \{\rho^*(A, B), \ \rho^*(B, A)\}.$$

The main goal of this section is to prove the following lemma.

LEMMA 1. *Let $1 \le p < \infty$ and let $\Gamma = \{z_k\}$ be a uniformly discrete sequence with separation constant*

$$\delta = \delta(\Gamma) = \inf_{j \ne k} \rho(z_j, z_k).$$

Then there is a constant C depending only on p and δ, such that

$$|K_1(\Gamma)^{\frac{1}{p}} - K_1(\widetilde{\Gamma})^{\frac{1}{p}}| \le C\,[\Gamma, \widetilde{\Gamma}]$$

for every sequence $\widetilde{\Gamma}$ with $[\Gamma, \widetilde{\Gamma}] < \delta/8$.

Note that $[\Gamma, \widetilde{\Gamma}] \le \sup_k \rho(z_k, \zeta_k)$, so that Lemma 1 implies the preservation of sampling sequences under perturbations with $\rho(z_k, \zeta_k) < \varepsilon$ for all k, when ε is sufficiently small. The key to the proof of Lemma 1 is a uniform continuity property of Bergman space functions with respect to the pseudohyperbolic metric. This is the content of the next lemma. Recall that $\Delta(\alpha, r)$ denotes a pseudohyperbolic disk.

LEMMA 2. *Let $0 < p < \infty$ and $0 < r < 1$. Then there is a constant C depending only on p and r such that*

$$\left|(1 - |z|^2)^{\frac{2}{p}} |f(z)| - (1 - |\zeta|^2)^{\frac{2}{p}} |f(\zeta)|\right| \le C\,\rho(z, \zeta) \left\{ \int_{\Delta(z,r)} |f(w)|^p \, d\sigma \right\}^{1/p}$$

for all $f \in A^p$, provided that $\rho(z, \zeta) < r/4$.

PROOF OF LEMMA 2. Suppose that $\rho(z,\zeta) < r/4$ and let

$$S(z) = (1-|z|^2)^{\frac{2}{p}} f(z) \qquad \text{and} \qquad S_\zeta(z) = (1-|z|^2)^{\frac{2}{p}} f_\zeta(z),$$

where

$$f_\zeta(z) = f(\varphi_\zeta(z))\varphi_\zeta'(z)^{\frac{2}{p}}, \qquad \varphi_\zeta(z) = \frac{\zeta - z}{1-\bar{\zeta}z}.$$

Since $\varphi_\zeta(\varphi_\zeta(z)) = z$ and

$$|\varphi_\zeta'(z)| = \frac{1-|\varphi_\zeta(z)|^2}{1-|z|^2}, \tag{1}$$

we see that $|S_\zeta(\varphi_\zeta(z))| = |S(z)|$ and $|S_\zeta(0)| = |S(\zeta)|$. Therefore,

$$\begin{aligned}
\big||S(z)| - |S(\zeta)|\big| &= \big||S_\zeta(\varphi_\zeta(z))| - |S_\zeta(0)|\big| \le \big|S_\zeta(\varphi_\zeta(z)) - S_\zeta(0)\big| \\
&= \big|(1-|\varphi_\zeta(z)|^2)^{\frac{2}{p}} f_\zeta(\varphi_\zeta(z)) - f_\zeta(0)\big| \\
&\le \big|(1-|\varphi_\zeta(z)|^2)^{\frac{2}{p}} f_\zeta(\varphi_\zeta(z)) - f_\zeta(\varphi_\zeta(z))\big| + \big|f_\zeta(\varphi_\zeta(z)) - f_\zeta(0)\big| \\
&= (1-(1-|\varphi_\zeta(z)|^2)^{\frac{2}{p}})|f_\zeta(\varphi_\zeta(z))| + |f_\zeta(\varphi_\zeta(z)) - f_\zeta(0)| \\
&= \Phi_1 + \Phi_2.
\end{aligned}$$

Now recall the inequality

$$(1-|z|^2)^2|f(z)|^p \le \frac{1}{r^2}\int_{\Delta(z,r)} |f(w)|^p\,d\sigma, \tag{2}$$

which was derived in Chapter 2 as Corollary 1 to Lemma 14. Note that $f_\zeta(\varphi_\zeta(z)) = f(z)\varphi_\zeta'(z)^{-\frac{2}{p}}$, since $\varphi_\zeta'(\varphi_\zeta(z))\varphi_\zeta'(z) = 1$, and apply (1) to infer from (2) that

$$\begin{aligned}
\Phi_1 &\le \frac{1-(1-|\varphi_\zeta(z)|^2)^{\frac{2}{p}}}{(1-|\varphi_\zeta(z)|^2)^{\frac{2}{p}}} \left\{\frac{1}{r^2}\int_{\Delta(z,r)} |f(w)|^p\,d\sigma\right\}^{1/p} \\
&\le C(r,p)\,|\varphi_\zeta(z)| \left\{\int_{\Delta(z,r)} |f(w)|^p\,d\sigma\right\}^{1/p}
\end{aligned}$$

for $|\varphi_\zeta(z)| = \rho(z,\zeta) < r/4$.

Next apply Cauchy's integral formula to obtain

$$\begin{aligned}
\Phi_2 &= \big|f_\zeta(\varphi_\zeta(z)) - f_\zeta(0)\big| = \left|\int_0^{\varphi_\zeta(z)} f_\zeta'(w)\,dw\right| \\
&\le |\varphi_\zeta(z)| \max_{|w|\le|\varphi_\zeta(z)|} |f_\zeta'(w)| \\
&= |\varphi_\zeta(z)| \max_{|w|\le|\varphi_\zeta(z)|} \frac{1}{2\pi}\left|\int_{|t-w|=\frac{r}{4}} \frac{f_\zeta(t)}{(t-w)^2}\,dt\right| \\
&\le \tfrac{4}{r}\,|\varphi_\zeta(z)| \max_{|t|\le|\varphi_\zeta(z)|+\frac{r}{4}} |f_\zeta(t)|.
\end{aligned}$$

Note that $|\varphi_\zeta(z)| + \frac{r}{4} < \frac{r}{2} < \frac{1}{2}$ by our initial assumption that $\rho(z,\zeta) < \frac{r}{4}$.

Now apply (1) and (2) to write

$$\begin{aligned}(1-|t|^2)^2|f_\zeta(t)|^p &= (1-|\varphi_\zeta(t)|^2)^2|f(\varphi_\zeta(t))|^p \\ &\le \left(\tfrac{4}{r}\right)^2 \int_{\Delta(\varphi_\zeta(t),\frac{r}{4})} |f(w)|^p\, d\sigma \\ &\le \left(\tfrac{4}{r}\right)^2 \int_{\Delta(z,r)} |f(w)|^p\, d\sigma\end{aligned}$$

for $|t| \le |\varphi_\zeta(z)| + \frac{r}{4}$ and $|\varphi_\zeta(z)| < \frac{r}{4}$, because these inequalities imply the inclusion $\Delta(\varphi_\zeta(t),\frac{r}{4}) \subset \Delta(z,r)$. Indeed, it follows that $|t| < \frac{r}{2}$ and so $\rho(\varphi_\zeta(t),\zeta) < \frac{r}{2}$ by the Möbius invariance of the pseudohyperbolic metric. Thus if $w \in \Delta(\varphi_\zeta(t),\frac{r}{4})$, then

$$\rho(w,z) \le \rho(w,\varphi_\zeta(t)) + \rho(\varphi_\zeta(t),\zeta) + \rho(\zeta,z) < \frac{r}{4} + \frac{r}{2} + \frac{r}{4} = r\,.$$

Consequently, for $|\varphi_\zeta(z)| < r/4$ we have

$$\max_{|t|\le|\varphi_\zeta(z)|+\frac{r}{4}} |f_\zeta(t)| \le \left(\tfrac{4}{r}\right)^{\frac{2}{p}} \left(1-\tfrac{r^2}{4}\right)^{-\frac{2}{p}} \left\{\int_{\Delta(z,r)} |f(w)|^p\, d\sigma\right\}^{1/p},$$

so that

$$\Phi_2 \le \left(\tfrac{4}{r}\right)^{1+\frac{2}{p}} \left(1-\tfrac{r^2}{4}\right)^{-\frac{2}{p}} |\varphi_\zeta(z)| \left\{\int_{\Delta(z,r)} |f(w)|^p\, d\sigma\right\}^{1/p}$$

for $\rho(z,\zeta) < r/4$. This proves Lemma 2. □

PROOF OF LEMMA 1. Let $\widetilde{\Gamma} = \{\zeta_j\}$ be a sequence for which $[\Gamma,\widetilde{\Gamma}] < r < 1$. Then $\rho^*(\Gamma,\widetilde{\Gamma}) < r$ and so $\inf_j \rho(z_k,\zeta_j) < r$ for each index k. In other words, each point z_k belongs to some disk $\Delta(\zeta_j,r)$. Similarly, $\rho^*(\widetilde{\Gamma},\Gamma) < r$ says that $\inf_k \rho(\zeta_j,z_k) < r$ for each index j, so that each disk $\Delta(\zeta_j,r)$ contains some point z_k. Now take $r = \delta/2$. Then no disk $\Delta(\zeta_j,r)$ can contain two distinct points z_k, since the sequence Γ is uniformly discrete with separation constant δ and so the disks $\Delta(z_k,\delta/2)$ are disjoint. Therefore, if $[\Gamma,\widetilde{\Gamma}] < \delta/2$, the points of Γ and $\widetilde{\Gamma}$ are in one-to-one correspondence and $\widetilde{\Gamma}$ can be reordered in such a way that $\rho(z_k,\zeta_k) \le [\Gamma,\widetilde{\Gamma}]$ for every k.

Apply Lemma 2 with $r = \delta/2$ to obtain

$$\begin{aligned}&\left|(1-|z_k|^2)^{\frac{2}{p}}|f(z_k)| - (1-|\zeta_k|^2)^{\frac{2}{p}}|f(\zeta_k)|\right| \\ &\le C\,\rho(z_k,\zeta_k) \left\{\int_{\Delta(z_k,\delta/2)} |f(w)|^p\, d\sigma\right\}^{1/p} \\ &\le C\,[\Gamma,\widetilde{\Gamma}] \left\{\int_{\Delta(z_k,\delta/2)} |f(w)|^p\, d\sigma\right\}^{1/p},\end{aligned}$$

provided $[\Gamma, \widetilde{\Gamma}] < \delta/8$. Thus by Minkowski's inequality

$$\begin{aligned}
&\left| \left\{ \sum_{k=1}^{\infty} (1 - |z_k|^2)^2 |f(z_k)|^p \right\}^{1/p} - \left\{ \sum_{k=1}^{\infty} (1 - |\zeta_k|^2)^2 |f(\zeta_k)|^p \right\}^{1/p} \right| \\
&\le \left\{ \sum_{k=1}^{\infty} \left| (1 - |z_k|^2)^{\frac{2}{p}} |f(z_k)| - (1 - |\zeta_k|^2)^{\frac{2}{p}} |f(\zeta_k)| \right|^p \right\}^{1/p} \\
&\le C\,[\Gamma, \widetilde{\Gamma}] \left\{ \sum_{k=1}^{\infty} \int_{\Delta(z_k, \delta/2)} |f(w)|^p \, d\sigma \right\}^{1/p} \\
&\le C\,[\Gamma, \widetilde{\Gamma}] \left\{ \int_{\mathbb{D}} |f(w)|^p \, d\sigma \right\}^{1/p} = C\,[\Gamma, \widetilde{\Gamma}]\, \|f\|_p
\end{aligned} \tag{3}$$

since $[\Gamma, \widetilde{\Gamma}] < \delta/8$.

By the definition of $K_1(\Gamma)$, for each $\varepsilon > 0$ there is a function $f \in A^p$ of norm $\|f\|_p = 1$ such that

$$\left\{ \sum_{k=1}^{\infty} (1 - |z_k|^2)^2 |f(z_k)|^p \right\}^{1/p} \le K_1(\Gamma)^{\frac{1}{p}} + \varepsilon\,.$$

On the other hand,

$$K_1(\widetilde{\Gamma})^{\frac{1}{p}} \le \left\{ \sum_{k=1}^{\infty} (1 - |\zeta_k|^2)^2 |f(\zeta_k)|^p \right\}^{1/p}.$$

Combining these inequalities with (3), we see that

$$K_1(\widetilde{\Gamma})^{\frac{1}{p}} - K_1(\Gamma)^{\frac{1}{p}} \le C\,[\Gamma, \widetilde{\Gamma}] + \varepsilon$$

if $[\Gamma, \widetilde{\Gamma}] < \delta/8$. Letting $\varepsilon \to 0$, and then interchanging the roles of Γ and $\widetilde{\Gamma}$, we obtain the desired result. □

In the next section we will prove the necessity of the condition for sampling as stated in Theorem 2. Before embarking on the proof, however, we can make some preliminary observations. First of all, a sampling sequence must be a finite union of uniformly discrete sequences. This is an immediate consequence of Theorem 15 in Chapter 2, where it is shown that the upper sampling inequality holds for all $f \in A^p$ if and only if Γ is a finite union of uniformly discrete sequences.

Lemma 2 allows us to reduce the proof of necessity to sampling sequences that are uniformly discrete. Specifically, we have the following lemma.

LEMMA 3. *Let $1 \le p < \infty$. Every sampling sequence for A^p has a uniformly discrete subsequence that is also a sampling sequence for A^p.*

PROOF. As just remarked, every sampling set is a finite union of uniformly discrete sets. The strategy of proof is to show that if Γ is a union of n uniformly discrete sets, then some subset is a union of $n-1$ uniformly discrete sets and is again a sampling set. The lemma then follows by induction. Suppose, then, that the given sampling sequence Γ has the form

$$\Gamma = \Gamma_1 \cup \cdots \cup \Gamma_{n-1} \cup \Lambda = \Gamma' \cup \Lambda ,$$

where each of the sets Γ_j and Λ is uniformly discrete. Let $\delta = \delta(\Lambda)$ be the separation constant of Λ and fix $\varepsilon < \delta/8$. Let

$$\Lambda_1 = \{\lambda \in \Lambda : \rho(\lambda, \Gamma') < \varepsilon\} \qquad \text{and} \qquad \Lambda_2 = \{\lambda \in \Lambda : \rho(\lambda, \Gamma') \geq \varepsilon\} .$$

By construction, Λ_2 can be adjoined to one of the sets Γ_j, thus representing Γ as a union of $n-1$ uniformly discrete sets. We will show that $\Gamma' \cup \Lambda_2$ is a sampling set for A^p. That will complete the proof if Λ_1 is empty. If $\Lambda_1 \neq \emptyset$, write $\Lambda_1 = \{\lambda_k\}$. For each index k, there is a point $z_k \in \Gamma'$ such that $\rho(z_k, \lambda_k) < \varepsilon$. Since Λ is uniformly discrete and $\varepsilon < \delta/2$, the points z_k are distinct. Write $\Gamma_1' = \{z_k\}$ and $\Gamma_2' = \Gamma' \setminus \Gamma_1'$.

We claim that

$$\left| \sum_{\lambda_k \in \Lambda_1} (1-|\lambda_k|^2)^2 |f(\lambda_k)|^p - \sum_{z_k \in \Gamma_1'} (1-|z_k|^2)^2 |f(z_k)|^p \right| \leq C\varepsilon \|f\|_p^p \tag{4}$$

for all functions $f \in A^p$, where C is a constant depending only on p and δ. Deferring a proof of (4), write

$$\sum_{z \in \Gamma} (1-|z|^2)^2 |f(z)|^p = \sum_{\Gamma_1'} + \sum_{\Gamma_2'} + \sum_{\Lambda_1} + \sum_{\Lambda_2} .$$

An application of (4) gives $\sum_{\Lambda_1} \leq \sum_{\Gamma_1'} + C\varepsilon \|f\|_p^p$, so that

$$\sum_{z \in \Gamma} (1-|z|^2)^2 |f(z)|^p \leq 2 \sum_{z \in \Gamma' \cup \Lambda_2} (1-|z|^2)^2 |f(z)|^p + C\varepsilon \|f\|_p^p .$$

Thus, since Γ is a sampling set,

$$\begin{aligned} 2 \sum_{z \in \Gamma' \cup \Lambda_2} (1-|z|^2)^2 |f(z)|^p &\geq \sum_{z \in \Gamma} (1-|z|^2)^2 |f(z)|^p - C\varepsilon \|f\|_p^p \\ &\geq (K_1(\Gamma) - C\varepsilon) \|f\|_p^p . \end{aligned}$$

Now choose $\varepsilon > 0$ small enough that $K_1(\Gamma) - C\varepsilon > 0$ to conclude that $\Gamma' \cup \Lambda_2$ is a sampling set.

For a proof of (4), begin with the inequality

$$|b^p - a^p| \leq p \max\{a^{p-1}, b^{p-1}\} \, |b-a| , \qquad a > 0, \ b > 0 , \tag{5}$$

which follows from the mean value theorem. Since both Λ_1 and Γ_1' are uniformly discrete, the upper sampling inequality applies, and we find from (5) that the left-hand side of (4) is bounded above by

$$C\,\|f\|_p^{p-1}\left|\left\{\sum_{\lambda_k\in\Lambda_1}(1-|\lambda_k|^2)^2|f(\lambda_k)|^p\right\}^{\frac{1}{p}}-\left\{\sum_{z_k\in\Gamma_1'}(1-|z_k|^2)^2|f(z_k)|^p\right\}^{\frac{1}{p}}\right|.$$

Since $\rho(z_k,\lambda_k)<\varepsilon$, Lemma 2 can be applied as in the proof of Lemma 1 to see that the last difference of sums is estimated by $C\varepsilon\|f\|_p$, and (4) follows. This completes the proof of Lemma 3. □

In the next section Lemma 1 will be applied to sequences that are shifted radially outward. To that end, consider the function

$$g_\eta(z)=\frac{|z|+\eta}{1+\eta|z|}\frac{z}{|z|}\,,\qquad 0<\eta<1\,,\quad z\neq 0\,,$$

and $g_\eta(0)=\eta$. Then $|z|<|g_\eta(z)|<1$ and $|g_\eta(z)|\le|z|+\eta$, so for any sequence $\Gamma=\{z_k\}$ in $\mathbb{D}$ the sequence $g_\eta(\Gamma)=\{g_\eta(z_k)\}$ lies in $\mathbb{D}$ and is obtained from Γ by displacing each point radially toward the boundary by Euclidean distance at most η. Note also that $|g_\eta(z)|\ge\eta$. A calculation shows that $\rho(z,g_\eta(z))=\eta$, and it follows that $[\Gamma,g_\eta(\Gamma)]\le\eta$. We will need to compare the density of $g_\eta(\Gamma)$ with that of Γ. It will be useful to have an inequality in terms of the "predensity"

$$D(\Gamma,\zeta,r)=\left(\log\frac{1}{1-r}\right)^{-1}\sum_{z_k\in\Omega(\zeta,\frac{1}{2},r)}\log\frac{1}{|\varphi_\zeta(z_k)|}\,,$$

as defined in Section 6.5.

Lemma 4. *If Γ is a uniformly discrete sequence with separation constant $\delta=\delta(\Gamma)$, then*

$$D(g_\eta(\Gamma),0,r)\le(1-\eta)\,D(\Gamma,0,r)+C\left(\log\frac{1}{1-r}\right)^{-1}$$

for $0<\eta\le\frac{1}{4}$ and $\frac{1}{2}<r<1$, where C is a constant depending only on δ.

Proof. First note the elementary inequality

$$\log\frac{1}{|g_\eta(z)|}\le(1-\eta)\log\frac{1}{|z|}\,.$$

It reduces to the inequality

$$f(x)=\log\frac{a+x}{1+ax}-(1-a)\log x\ge 0\,,\qquad 0<x\le 1\,,$$

for $0 < a < 1$, which can be verified by observing that $f(1) = 0$ and $f'(x) < 0$.

According to Lemma 15 of Chapter 2, the counting function satisfies

$$n(\Gamma, \alpha, r) \leq \left(\tfrac{2}{\delta} + 1\right)^2 \frac{1}{1 - r^2} . \tag{6}$$

Therefore,

$$\begin{aligned}
\left(\log \frac{1}{1-r}\right) D(g_\eta(\Gamma), 0, r) &= \sum_{\frac{1}{2} < |g_\eta(z_k)| < r} \log \frac{1}{|g_\eta(z_k)|} \\
&\leq \sum_{\frac{1}{2} < |z_k| < r} \log \frac{1}{|g_\eta(z_k)|} + \sum_{\frac{1}{4} < |z_k| < \frac{1}{2}} \log \frac{1}{|g_\eta(z_k)|} \\
&= \sum_{\frac{1}{2} < |z_k| < r} \log \frac{1}{|g_\eta(z_k)|} + (\log 4)\, n(\Gamma, 0, \tfrac{1}{2}) \\
&\leq (1 - \eta) \sum_{\frac{1}{2} < |z_k| < r} \log \frac{1}{|z_k|} + C = (1 - \eta) \left(\log \frac{1}{1-r}\right) D(\Gamma, 0, r) + C ,
\end{aligned}$$

which gives the desired result. □

§7.2. Necessity of the sampling condition.

The proof of necessity in Theorem 2 follows an argument of Seip [3], who modified the techniques of Beurling [3] to prove a sampling theorem for a certain growth space of analytic functions, to be discussed in the next section. The proof will require another fact about uniformly discrete sequences.

LEMMA 5. *Let $\Gamma = \{z_k\}$ be uniformly discrete with separation constant $\delta = \delta(\Gamma)$. Then*

$$\sum_{|z_k| > r} (1 - |z_k|^2)^2 \leq \frac{12}{\delta^2}(1 - r^2) , \qquad \tfrac{\delta}{2} < r < 1 .$$

PROOF. By the triangle inequality for the pseudohyperbolic metric, the pseudohyperbolic disks $\Delta(z_k, \frac{\delta}{2})$ are disjoint. According to a formula in Section 2.5, the normalized area of $\Delta(z_k, \frac{\delta}{2})$ is

$$\sigma(\Delta(z_k, \tfrac{\delta}{2})) = \frac{(\frac{\delta}{2})^2 (1 - |z_k|^2)^2}{\left(1 - (\frac{\delta}{2})^2 |z_k|^2\right)^2} \geq \frac{\delta^2}{4}(1 - |z_k|^2)^2 .$$

The strong form of the triangle inequality shows that if $r < |z_k| < 1$, then $\Delta(z_k, \frac{\delta}{2})$ lies in the annulus $R < |z| < 1$, where

$$R = \frac{r - \frac{\delta}{2}}{1 - r\frac{\delta}{2}}.$$

Therefore,

$$\begin{aligned}\sum_{|z_k|>r} (1 - |z_k|^2)^2 &\leq \frac{4}{\delta^2} \sum_{|z_k|>r} \sigma(\Delta(z_k, \tfrac{\delta}{2})) \\ &\leq \frac{4}{\delta^2}(1 - R^2) < \frac{4}{\delta^2}\frac{1+\frac{\delta}{2}}{1-\frac{\delta}{2}}(1 - r^2) < \frac{12}{\delta^2}(1 - r^2). \qquad \square\end{aligned}$$

We now turn to the proof of necessity in Theorem 2. As noted in the previous section, every sampling sequence for A^p is a finite union of uniformly discrete sequences. By Lemma 3, every sampling sequence for A^p has a uniformly discrete subsequence Γ' that is a sampling sequence for A^p. We are to show that Γ' has lower density $D^-(\Gamma') > \frac{1}{p}$. This reduces to showing that if Γ is a *uniformly discrete* sampling sequence for A^p, then $D^-(\Gamma) > \frac{1}{p}$.

Recall that

$$D^-(\Gamma) = \liminf_{r\to 1} \inf_{\zeta\in\mathbb{D}} D(\Gamma, \zeta, r).$$

Let $0 < \varepsilon_j < \frac{1}{2}$ and $\varepsilon_j \to 0$. Then for each j there exist a point $\zeta_j \in \mathbb{D}$ and a radius r_j with $1 - \varepsilon_j < r_j < 1$ for which

$$D(\Gamma, \zeta_j, r_j) < D^-(\Gamma) + \varepsilon_j.$$

Since $D(\Gamma, \zeta, r) = D(\varphi_\zeta(\Gamma), 0, r)$, where $\varphi_\zeta(z) = \frac{\zeta - z}{1-\bar{\zeta}z}$, this inequality can be written

$$D(\Gamma_j, 0, r_j) < D^-(\Gamma) + \varepsilon_j, \tag{7}$$

with the notation $\Gamma_j = \varphi_{\zeta_j}(\Gamma)$.

Now suppose that Γ is a uniformly discrete sampling sequence for A^p. Let $\delta = \delta(\Gamma)$ be its separation constant, and let $K_1(\Gamma)$ be its (positive) sampling constant. Note that $\delta(\Gamma_j) = \delta(\Gamma)$ and $K_1(\Gamma_j) = K_1(\Gamma)$ by Möbius invariance. Apply Lemma 1 with $\eta < \delta/8$ and recall that $\rho(z, g_\eta(z)) = \eta$ to obtain

$$K_1(g_\eta(\Gamma_j))^{\frac{1}{p}} \geq K_1(\Gamma_j)^{\frac{1}{p}} - C\,[\Gamma_j, g_\eta(\Gamma_j)] \geq K_1(\Gamma)^{\frac{1}{p}} - C\eta,$$

where C is a constant depending only on p and δ. Therefore,

$$K_1(g_\eta(\Gamma_j)) \geq \left(K_1(\Gamma)^{\frac{1}{p}} - C\eta\right)^p > 0$$

if $\eta > 0$ is chosen sufficiently small, which says that $g_\eta(\Gamma_j)$ satisfies the sampling inequality for A^p. Moreover, the sampling constant $K_1(g_\eta(\Gamma_j))$ has a positive lower bound independent of j.

Next observe that $g_\eta(\Gamma_j)$ is uniformly discrete and its separation constant $\delta(g_\eta(\Gamma_j))$ also has a positive lower bound independent of j. To see this, write $\Gamma_j = \{\zeta_{jk}\}$ and $g_\eta(\Gamma_j) = \{w_{jk}\} = \{g_\eta(\zeta_{jk})\}$. Then for $k \neq \ell$

$$\delta \leq \rho(\zeta_{jk}, \zeta_{j\ell}) \leq \rho(\zeta_{jk}, w_{jk}) + \rho(w_{jk}, w_{j\ell}) + \rho(w_{j\ell}, \zeta_{j\ell}) = 2\eta + \rho(w_{jk}, w_{j\ell}),$$

so $\rho(w_{jk}, w_{j\ell}) \geq \delta - 2\eta \geq 3\delta/4 > 0$, since $\eta < \delta/8$.

We are to prove that $D^-(\Gamma) > \frac{1}{p}$. The plan is to estimate $D(\Gamma_j, 0, r_j)$ and use (7). For this purpose, consider the finite product

$$f_j(w) = \prod_{|w_{j\ell}| < r_j} \frac{1}{|w_{j\ell}|} \frac{w_{j\ell} - w}{1 - \overline{w_{j\ell}} w}.$$

Then

$$\begin{aligned}
\sum_{k=1}^{\infty} (1 - |w_{jk}|^2)^2 |f_j(w_{jk})|^p &= \sum_{|w_{jk}| \geq r_j} (1 - |w_{jk}|^2)^2 |f_j(w_{jk})|^p \\
&\leq \sum_{|w_{jk}| \geq r_j} (1 - |w_{jk}|^2)^2 \prod_{|w_{j\ell}| < r_j} \frac{1}{|w_{j\ell}|^p} \\
&\leq C \sum_{|w_{jk}| \geq r_j} (1 - |w_{jk}|^2)^2 \prod_{\frac{1}{2} < |w_{j\ell}| < r_j} \frac{1}{|w_{j\ell}|^p},
\end{aligned}$$

where C is a constant depending only on p and δ, since $|w_{j\ell}| = |g_\eta(\zeta_{j\ell})| \geq \eta$ and the counting function satisfies

$$n(g_\eta(\Gamma_j), 0, r) \leq \left(\frac{8}{3\delta} + 1\right)^2 \frac{1}{1 - r^2}, \tag{8}$$

by (6) and the estimate $\delta(g_\eta(\Gamma_j)) \geq 3\delta/4$ found earlier. But Lemma 4 shows that

$$\begin{aligned}
\prod_{\frac{1}{2} < |w_{j\ell}| < r_j} \frac{1}{|w_{j\ell}|^p} &= \exp\left\{ p \sum_{\frac{1}{2} < |w_{j\ell}| < r_j} \log \frac{1}{|w_{j\ell}|} \right\} \\
&= \exp\left\{ p \left(\log \frac{1}{1 - r_j} \right) D(g_\eta(\Gamma_j), 0, r_j) \right\} \\
&\leq \exp\left\{ p(1 - \eta) \left(\log \frac{1}{1 - r_j} \right) D(\Gamma_j, 0, r_j) + pC \right\}.
\end{aligned}$$

In view of (7) and Lemma 5, this gives the estimate

$$\sum_{k=1}^{\infty}(1-|w_{jk}|^2)^2|f_j(w_{jk})|^p \leq C(1-r_j)^{-p(1-\eta)(D^-(\Gamma)+\varepsilon_j)} \sum_{|w_{jk}|\geq r_j}(1-|w_{jk}|^2)^2$$
$$\leq C(1-r_j)^{1-p(1-\eta)(D^-(\Gamma)+\varepsilon_j)},$$

where C depends only on p and δ.

On the other hand, since $\|f_j\|_p \geq |f_j(0)| = 1$, the lower sampling inequality gives

$$K_1(g_\eta(\Gamma_j)) \leq \sum_{k=1}^{\infty}(1-|w_{jk}|^2)^2|f_j(w_{jk})|^p .$$

Combining the last two inequalities and letting $\varepsilon_j \to 0$, so that $r_j \to 1$, we conclude that the exponent

$$1-p(1-\eta)D^-(\Gamma) \leq 0,$$

because the sampling constants $K_1(g_\eta(\Gamma_j))$ have a positive lower bound. Thus $D^-(\Gamma) > \frac{1}{p}$. This completes the proof of necessity in Theorem 2.

It may be remarked that the sole purpose of the radial shift from Γ_j to $g_\eta(\Gamma_j)$ is to obtain strict inequality. Without that device the proof would yield only the weaker result $D^-(\Gamma) \geq \frac{1}{p}$.

§7.3. Sampling in the growth space.

The proofs of sufficiency in Theorem 2 and necessity in Theorem 1 will depend on the characterization of sampling sequences for the *growth space* $A^{-\beta}$, which consists of analytic functions f satisfying

$$\|f\|_{-\beta} = \sup_{z\in\mathbb{D}}(1-|z|^2)^\beta|f(z)| < \infty, \qquad \beta > 0.$$

The subspace $A_0^{-\beta}$, consisting of analytic functions with the property

$$\lim_{|z|\to 1}(1-|z|^2)^\beta|f(z)| = 0,$$

will also come into play. Here is some basic information about these spaces.

PROPOSITION 1.

(*a*) *Both* $A^{-\beta}$ *and* $A_0^{-\beta}$ *are Banach spaces.*

(*b*) *The space* $A_0^{-\beta}$ *is the closure of the polynomials in the norm of* $A^{-\beta}$.

The proofs are similar to those of corresponding results for the Bloch space (see Section 2.6) and will be omitted.

The growth spaces are closely related to the Bergman spaces A^p. Theorem 1 of Chapter 3 says that $A^p \subset A_0^{-\frac{2}{p}}$, while $A^{-\beta} \subset A^p$ for every $\beta < \frac{1}{p}$.

A sequence $\Gamma = \{z_k\}$ of distinct points in $\mathbb{D}$ is said to be a *sampling sequence for* $A^{-\beta}$ if there is a positive constant L such that

$$L\|f\|_{-\beta} \le \sup_k (1 - |z_k|^2)^\beta |f(z_k)| \tag{9}$$

for all $f \in A^{-\beta}$. Note that the analogue of the upper sampling inequality in the Bergman space is automatically satisfied here. This implies in particular that if any subsequence of Γ is an $A^{-\beta}$ sampling sequence, then Γ is itself an $A^{-\beta}$ sampling sequence. We define $L(\Gamma)$ to be the largest constant L for which (9) holds, and we set $L(\Gamma) = 0$ if (9) fails. The number $L(\Gamma)$ will be called the *sampling constant* of Γ for the space $A^{-\beta}$. A simple calculation shows that the mapping

$$f(z) \longmapsto f(\varphi_\zeta(z))(\varphi_\zeta'(z))^\beta, \qquad \varphi_\zeta(z) = \frac{\zeta - z}{1 - \overline{\zeta} z},$$

is an isometry in $A^{-\beta}$ for each $\zeta \in \mathbb{D}$. This fact can be used to show that the sampling sequences for $A^{-\beta}$ are Möbius invariant and $L(\varphi_\zeta(\Gamma)) = L(\Gamma)$. The following theorem is due to Seip [3].

THEOREM 3. *A sequence Γ of points in the unit disk is a sampling sequence for $A^{-\beta}$ if and only if it has a uniformly discrete subsequence Γ' with lower density $D^-(\Gamma') > \beta$.*

The theorem says that a sampling set must be in some sense evenly distributed throughout the disk. In the course of the proof we will use the weaker fact that a sampling set cannot be contained in any subdisk $|z| < r < 1$. Indeed, this would imply that

$$\sup_{z\in\mathbb{D}} (1 - |z|^2)^\beta |f(z)| \le K \sup_{|z|<r} (1 - |z|^2)^\beta |f(z)|$$

for some constant K and all $f \in A^{-\beta}$. To see that this is impossible, take for instance $f_n(z) = (1 - \overline{\alpha_n} z)^{-\beta-1}$ with $\alpha_n \in \mathbb{D}$ and $|\alpha_n| \to 1$. Then $f_n \in H^\infty \subset A^{-\beta}$ and

$$\sup_{|z|<r} (1 - |z|^2)^\beta |f_n(z)| \le (1 - r)^{-\beta-1},$$

while

$$\sup_{z\in\mathbb{D}} (1 - |z|^2)^\beta |f_n(z)| \ge (1 - |\alpha_n|^2)^{-1} \to \infty \qquad \text{as } n \to \infty,$$

which contradicts the sampling inequality.

The proof of necessity in Theorem 3 is similar to that in Theorem 2, as given in Section 7.2. We begin with two lemmas analogous to Lemmas 1 and 2, respectively.

LEMMA 6. *Let Γ be a sampling sequence for $A^{-\beta}$, and let $L(\Gamma)$ be its sampling constant. Then there is a constant C, depending only on β, such that*

$$L(\widetilde{\Gamma}) \geq L(\Gamma) - C\,[\Gamma, \widetilde{\Gamma}]$$

for every sequence $\widetilde{\Gamma}$ with $[\Gamma, \widetilde{\Gamma}] < \frac{1}{2}$.

LEMMA 7. *There is a constant C, depending only on β, such that*

$$\big|(1-|z|^2)^\beta |f(z)| - (1-|\zeta|^2)^\beta |f(\zeta)|\big| \leq C\,\rho(z,\zeta)\|f\|_{-\beta}$$

for all $f \in A^{-\beta}$, provided that $\rho(z,\zeta) \leq \frac{1}{2}$.

PROOF OF LEMMA 7. The proof is very similar to that of Lemma 2. Defining

$$S(z) = (1-|z|^2)^\beta f(z) \qquad \text{and} \qquad f_\zeta(z) = f(\varphi_\zeta(z))\varphi_\zeta'(z)^\beta\,,$$

we find in the same way that $\big||S(z)| - |S(\zeta)|\big| \leq \Phi_1 + \Phi_2$, where

$$\Phi_1 = \big(1 - (1-|\varphi_\zeta(z)|^2)^\beta\big)|f_\zeta(\varphi_\zeta(z))|\,, \qquad \Phi_2 = |f_\zeta(\varphi_\zeta(z)) - f_\zeta(0)|\,.$$

An application of the identity (1) yields

$$\Phi_1 \leq \frac{1-(1-|\varphi_\zeta(z)|^2)^\beta}{(1-|\varphi_\zeta(z)|^2)^\beta}\|f\|_{-\beta} \leq C\,|\varphi_\zeta(z)|^2\|f\|_{-\beta}$$

for $|\varphi_\zeta(z)| = \rho(z,\zeta) < \frac{1}{2}$. Applying Cauchy's integral formula as in the proof of Lemma 2, and recalling that $\|f_\zeta\|_{-\beta} = \|f\|_{-\beta}$, we find

$$\begin{aligned}\Phi_2 &\leq 4\,|\varphi_\zeta(z)| \max_{|t|\leq|\varphi_\zeta(z)|+\frac{1}{4}} |f_\zeta(t)| \\ &\leq 4\,|\varphi_\zeta(z)|\,\|f_\zeta\|_{-\beta} \max_{|t|\leq\frac{3}{4}} \frac{1}{(1-|t|^2)^\beta} = 4\big(\tfrac{16}{7}\big)^\beta|\varphi_\zeta(z)|\,\|f\|_{-\beta}\,.\end{aligned}$$

Combining the two estimates, we obtain the desired result. □

PROOF OF LEMMA 6. Let $\Gamma = \{z_k\}$ and choose $\varepsilon > 0$. Then by definition of the Hausdorff distance $[\Gamma, \widetilde{\Gamma}]$, to each point $z_k \in \Gamma$ there corresponds a point $\zeta_k \in \widetilde{\Gamma}$ with $\rho(z_k, \zeta_k) \leq [\Gamma, \widetilde{\Gamma}] + \varepsilon$. Let $\Gamma' = \{\zeta_k\}$ be the resulting subset of $\widetilde{\Gamma}$. Then there exists a function $f \in A^{-\beta}$ of norm $\|f\|_{-\beta} = 1$ such that

$$\sup_k (1-|\zeta_k|^2)^\beta |f(\zeta_k)| < L(\Gamma') + \varepsilon \leq L(\widetilde{\Gamma}) + \varepsilon\,.$$

Also, $L(\Gamma) \leq \sup_k (1-|z_k|^2)^\beta |f(z_k)|$, so

$$L(\Gamma) < (1-|z_n|^2)^\beta |f(z_n)| + \varepsilon$$

for some index n. Combining these inequalities, we find by Lemma 7 that

$$\begin{aligned} L(\Gamma) - L(\widetilde{\Gamma}) &\leq (1-|z_n|^2)^\beta |f(z_n)| - (1-|\zeta_n|^2)^\beta |f(\zeta_n)| + 2\varepsilon \\ &\leq C\,\rho(z_n,\zeta_n) + 2\varepsilon < C([\Gamma,\widetilde{\Gamma}]+\varepsilon) + 2\varepsilon\,, \end{aligned}$$

provided that $[\Gamma,\widetilde{\Gamma}]+\varepsilon < \frac{1}{2}$. Now let $\varepsilon \to 0$ to arrive at the desired inequality. □

As in the case of Bergman spaces, the proof of necessity in Theorem 3 can be reduced to the consideration of a *uniformly discrete* sampling sequence Γ for $A^{-\beta}$, so that one need only prove $D^-(\Gamma) > \beta$. This reduction is a consequence of the following lemma.

LEMMA 8. *If Γ is a sampling sequence for $A^{-\beta}$, then for each $\varepsilon > 0$ it has a uniformly discrete subsequence Γ' with separation constant $\delta(\Gamma') \geq \varepsilon$ and pseudohyperbolic Hausdorff distance $[\Gamma,\Gamma'] \leq \varepsilon$.*

PROOF. The required subsequence $\Gamma' = \{z_{k_j}\}$ will be constructed inductively. Choose $k_1 = 1$. Let k_2 be the smallest integer k for which $\rho(z_k, z_1) \geq \varepsilon$. Having chosen $k_1 < k_2 < \cdots < k_n$, let k_{n+1} be the smallest integer $k > k_n$ such that $\rho(z_k, z_{k_j}) \geq \varepsilon$ for $j = 1,2,\dots,n$. Such an integer k_{n+1} exists because a sampling set cannot be contained in any proper subdisk $|z| < r < 1$. (See the remarks following the statement of Theorem 3.) By construction, $\delta(\Gamma') \geq \varepsilon$. Since $\Gamma' \subset \Gamma$, it is clear that $\rho^*(\Gamma',\Gamma) = 0$. To see that $\rho^*(\Gamma,\Gamma') \leq \varepsilon$, choose any point $z_k \in \Gamma$ with $z_k \notin \Gamma'$, and let $k_n < k < k_{n+1}$. Then by construction $\rho(z_k, z_{k_j}) < \varepsilon$ for some $j = 1,2,\dots,n$. Thus $\rho(z_k,\Gamma') < \varepsilon$ and $\rho^*(\Gamma,\Gamma') \leq \varepsilon$. □

Combining Lemma 8 with Lemma 6, we obtain the desired analogue of Lemma 3.

LEMMA 9. *Every sampling sequence for $A^{-\beta}$ has a uniformly discrete subsequence that is also a sampling sequence for $A^{-\beta}$.*

In view of Lemma 9, the proof of necessity in Theorem 3 reduces to showing that if Γ is a uniformly discrete sampling set for $A^{-\beta}$, then it has lower density $D^-(\Gamma) > \beta$. The argument proceeds along the same lines as for Bergman spaces (*cf.* Section 7.2). Let $0 < \varepsilon_j < \frac{1}{2}$ with $\varepsilon_j \to 0$, and let Γ be any sequence in $\mathbb{D}$. Then for each j we can choose a point $\zeta_j \in \mathbb{D}$ and a radius r_j with $1 - \varepsilon_j < r_j < 1$ such that

$$D(\Gamma_j, 0, r_j) < D^-(\Gamma) + \varepsilon_j\,, \tag{10}$$

where $\Gamma_j = \varphi_{\zeta_j}(\Gamma) = \{\zeta_{jk}\}$. Recall that $L(\Gamma_j) = L(\Gamma)$ by Möbius invariance.

Suppose now that Γ is a uniformly discrete sampling sequence for $A^{-\beta}$ with separation constant $\delta = \delta(\Gamma)$ and sampling constant $L(\Gamma)$. We are to prove that $D^-(\Gamma) > \beta$. For $0 < \eta < \delta/8$, let $g_\eta(\Gamma_j) = \{w_{jk}\}$ be the radial shift of Γ_j as defined in Section 7.1. Then $[\Gamma_j, g_\eta(\Gamma_j)] \le \eta$ and Lemma 6 gives

$$L(g_\eta(\Gamma_j)) \ge L(\Gamma_j) - C\,[\Gamma_j, g_\eta(\Gamma_j)] > L(\Gamma) - C\eta > 0$$

if η is small enough, so that $g_\eta(\Gamma_j)$ is a sampling sequence for $A^{-\beta}$ and the sampling constants have a positive lower bound independent of j. Moreover, $g_\eta(\Gamma_j)$ is uniformly discrete, as shown in Section 7.2, with separation constant $\delta(g_\eta(\Gamma_j)) > 3\delta/4$. By Lemma 4,

$$D(g_\eta(\Gamma_j), 0, r_j) \le (1-\eta) D(\Gamma_j, 0, r_j) + C\left(\log \frac{1}{1-r_j}\right)^{-1}, \tag{11}$$

where the constant C depends only on δ. Now let $f_j(w)$ be the finite product defined in Section 7.2. Then $f_j(w_{jk}) = 0$ for $|w_{jk}| < r_j$. For $|w_{jk}| \ge r_j$ the inequality (11) gives

$$\begin{aligned}
(1-|w_{jk}|^2)^\beta |f_j(w_{jk})| &\le (1-r_j^2)^\beta \prod_{|w_{j\ell}|<r_j} \frac{1}{|w_{j\ell}|} \\
&\le C(1-r_j^2)^\beta \prod_{\frac{1}{2}<|w_{j\ell}|<r_j} \frac{1}{|w_{j\ell}|} \\
&= C(1-r_j^2)^\beta \exp\left\{\log\frac{1}{1-r_j} D(g_\eta(\Gamma_j), 0, r_j)\right\} \\
&\le C(1-r_j^2)^\beta \exp\left\{(1-\eta)\log\frac{1}{1-r_j} D(\Gamma_j, 0, r_j)\right\},
\end{aligned}$$

where the constant C depends only on δ, in view of the uniform bound (8) on the counting function for $g_\eta(\Gamma_j) = \{w_{jk}\}$. Since $1 = |f_j(0)| \le \|f_j\|_{-\beta}$, it now follows from (10) that

$$L(g_\eta(\Gamma_j)) \le \sup_k (1-|w_{jk}|^2)^\beta |f_j(w_{jk})| \le C(1-r_j)^{\beta-(1-\eta)(D^-(\Gamma)+\varepsilon_j)},$$

where the constant C depends only on δ and β. Letting $\varepsilon_j \to 0$, so that $r_j \to 1$, and bearing in mind that the sampling constants $L(g_\eta(\Gamma_j))$ have a uniform positive lower bound, we conclude that $D^-(\Gamma) > \beta$. This completes the proof of necessity in Theorem 3.

Before turning to the converse, which will be used in the proof of sufficiency in Theorem 2, we define a concept of weak convergence of sequences of sequences of points in the unit disk. This concept was used by Beurling [3] in the context of the real line and was modified by Seip [3] to work in the setting of the disk. Weak convergence will be a useful tool in the proofs of Theorems 1 and 2, but a more immediate application is to Theorem 3. The present formulation is adapted from the book of Hedenmalm, Korenblum, and Zhu [2].

A sequence $\Gamma = \{z_k\}$ in the disk is said to be *naturally ordered* if $|z_1| \leq |z_2| \leq \dots$. A sequence $\Lambda = \{\lambda_k\}$ is a *rearrangement* of Γ if there is a bijection $\psi : \mathbb{N} \to \mathbb{N}$ such that $\lambda_k = z_{\psi(k)}$ for $k = 1, 2, \dots$. A sequence $\{\Gamma_j\}$ of sequences Γ_j is called *equidiscrete* if $\inf_j \delta(\Gamma_j) > 0$, where $\delta(\Gamma_j)$ is the separation constant of Γ_j. In particular, each sequence Γ_j is uniformly discrete. A sequence $\{\Gamma_j\}$ is said to be *weakly convergent* to an infinite sequence $\Gamma = \{z_k\}$ of points in $\mathbb{D}$, denoted by $\Gamma_j \rightharpoonup \Gamma$, if there are naturally ordered rearrangements $\{z_{jk}\}$ of the sequences Γ_j such that

$$\lim_{j\to\infty} z_{jk} = z_k\,, \qquad k = 1, 2, \dots . \tag{12}$$

A sequence $\{\Gamma_j\}$ converges weakly to a finite sequence $\Gamma = \{z_1, z_2, \dots, z_N\}$ of points in $\mathbb{D}$ if the sequences Γ_j have naturally ordered rearrangements $\{z_{jk}\}$ such that

$$\begin{aligned} &\lim_{j\to\infty} z_{jk} = z_k\,, && k = 1, 2, \dots, N\,; \qquad \text{and} \\ &\lim_{j\to\infty} |z_{jk}| = 1\,, && k = N+1, N+2, \dots . \end{aligned} \tag{13}$$

The number N is called the *length* of the limit sequence Γ. In the case (12) we set $N = \infty$. If

$$\lim_{j\to\infty} |z_{jk}| = 1\,, \qquad k = 1, 2, \dots\,,$$

then we set $N = 0$ and say that $\{\Gamma_j\}$ converges weakly to the empty sequence, written $\Gamma_j \rightharpoonup \emptyset$.

LEMMA 10. *Every equidiscrete sequence $\{\Gamma_j\}$ has a subsequence that converges weakly to a (possibly empty) uniformly discrete sequence Γ.*

PROOF. Suppose without loss of generality that each sequence $\{\Gamma_j\} = \{z_{jk}\}$ is naturally ordered. If $|z_{j1}| \to 1$, then $|z_{jk}| \to 1$ for all k (since the Γ_j are naturally ordered) and $\Gamma_j \rightharpoonup \emptyset$. If $|z_{j1}| \not\to 1$, then $\{\Gamma_j\}$ has a subsequence $\{\Gamma_{j_n}\}$ such that $z_{j_n 1} \to z_1 \in \mathbb{D}$. Then either $|z_{j_n 2}| \to 1$, which implies that $|z_{j_n k}| \to 1$ for all $k \geq 2$; or $\{\Gamma_{j_n}\}$ has a further subsequence whose second components converge to a point $z_2 \in \mathbb{D}$. In the latter case the process continues. There are now two possibilities. The process may terminate after a finite number N of iterations, producing a subsequence of $\{\Gamma_j\}$ that converges weakly to a finite sequence $\Gamma = \{z_1, z_2, \dots, z_N\}$. Otherwise, the process never terminates. Then $N = \infty$ and the diagonal subsequence converges weakly to an infinite sequence $\Gamma = \{z_1, z_2, \dots\}$. The uniform discreteness of Γ follows from the hypothesis that $\{\Gamma_j\}$ is equidiscrete. □

For a given uniformly discrete sequence Γ, it is useful to consider the set of weak limits under Möbius transformations. We denote by $W(\Gamma)$ the collection of all sequences Λ such that $\varphi_{\zeta_j}(\Gamma) \rightharpoonup \Lambda$ for some points $\zeta_j \in$

$\mathbb{D}$. Since $\delta(\varphi_\zeta(\Gamma)) = \delta(\Gamma)$ for every $\zeta \in \mathbb{D}$, it follows that every sequence $\{\varphi_{\zeta_j}(\Gamma)\}$ is equidiscrete, and so Lemma 10 guarantees the existence of a subsequence converging weakly to some $\Lambda \in W(\Gamma)$. All members of $W(\Gamma)$ are uniformly discrete.

It will be important to know how the upper and lower densities behave under passage to weak limits. The following lemma gives the required information.

LEMMA 11. *Let Γ be a uniformly discrete sequence and suppose that $\Lambda \in W(\Gamma)$. Then*

$$D^-(\Lambda) \geq D^-(\Gamma) \qquad \text{and} \qquad D^+(\Lambda) \leq D^+(\Gamma).$$

PROOF. The hypothesis $\Lambda \in W(\Gamma)$ says that $\varphi_{\zeta_j}(\Gamma) \rightharpoonup \Lambda$ for some sequence of points $\zeta_j \in \mathbb{D}$. Write $\Gamma_j = \varphi_{\zeta_j}(\Gamma)$ and note that by the definition of weak convergence, $\lim_{j\to\infty} n(\Gamma_j, z, s) = n(\Lambda, z, s)$ for each point $z \in \mathbb{D}$ and $0 < s < 1$. But $n(\Gamma_j, z, s) \leq \frac{C}{1-s}$ by Lemma 15 of Chapter 2 and the equidiscreteness of $\{\Gamma_j\}$. Thus by the Lebesgue dominated convergence theorem,

$$\lim_{j\to\infty} \int_0^r n(\Gamma_j, z, s)\, ds = \int_0^r n(\Lambda, z, s)\, ds\,.$$

Consequently, for $z \in \mathbb{D}$ and $0 < r < 1$,

$$\begin{aligned}\int_0^r n(\Lambda, z, s)\, ds &= \lim_{j\to\infty} \int_0^r n(\Gamma_j, z, s)\, ds = \lim_{j\to\infty} \int_0^r n(\Gamma, \varphi_{\zeta_j}(z), s)\, ds \\ &\geq \inf_{w\in\mathbb{D}} \int_0^r n(\Gamma, w, s)\, ds\,.\end{aligned}$$

This implies $D^-(\Lambda) \geq D^-(\Gamma)$. The proof of the other inequality proceeds in a similar way. □

The next fact about weak convergence is elementary but is recorded as a lemma for convenient reference. The proof is straightforward and will be omitted.

LEMMA 12. *Suppose $\Gamma_j \rightharpoonup \Gamma$ and let $\{f_j\}$ be a sequence of analytic functions in $\mathbb{D}$ that converges locally uniformly to a function f. If $\Gamma = \{z_k\}$ and the Γ_j have naturally ordered rearrangements $\{z_{jk}\}$ that satisfy* (13) *if Γ has finite length or* (12) *if $N = \infty$, then $f_j(z_{jk}) \to f(z_k)$ as $j \to \infty$ for $k = 1, 2, \ldots, N$ if N is finite, or for $k = 1, 2, \ldots$ if $N = \infty$.*

Lemma 12 implies in particular that, if in addition to the stated hypotheses, f_j vanishes on Γ_j, then f vanishes on Γ, a principle that will be applied later.

We now turn to the proof of sufficiency in Theorem 3. First note that every $A^{-\beta}$ sampling set Γ is a set of uniqueness. In other words, the only element of $A^{-\beta}$ vanishing on Γ is the zero function. This fact comes directly from the definition of a sampling set. The following lemma is a partial converse.

LEMMA 13. *If* Γ *is uniformly discrete and every element of* $W(\Gamma)$ *is a set of uniqueness for* $A^{-\beta}$, *then* Γ *is a sampling sequence for* $A^{-\beta}$.

PROOF. If Γ is not a sampling sequence, then

$$\lim_{j\to\infty} \sup_k (1-|z_k|^2)^\beta |f_j(z_k)| = 0$$

for some sequence of functions $f_j \in A^{-\beta}$ of unit norm. These two properties of f_j imply that

$$(1-|\zeta_j|^2)^\beta |f_j(\zeta_j)| = \tfrac{1}{2}$$

for some points $\zeta_j \in \mathbb{D}$ when j is sufficiently large. The sequence $\{\Gamma_j\} = \{\varphi_{\zeta_j}(\Gamma)\}$ is equidiscrete, and so by Lemma 10 we may suppose (passing to a subsequence if necessary) that $\Gamma_j \rightharpoonup \Lambda \in W(\Gamma)$. Define

$$g_j(z) = f_j(\varphi_{\zeta_j}(z))\varphi'_{\zeta_j}(z)^\beta, \qquad \text{so that } \|g_j\|_{-\beta} = \|f_j\|_{-\beta} = 1\,.$$

The sequence $\{g_j\}$ is locally bounded, so by a normal family argument (Montel's theorem) it has a subsequence that converges locally uniformly to some function $g \in A^{-\beta}$. Because

$$|g_j(0)| = |f_j(\zeta_j)|(1-|\zeta_j|^2)^\beta = \tfrac{1}{2}$$

for large j, it is clear that $|g(0)| = \frac{1}{2}$ and so $g(z) \not\equiv 0$. To show that g vanishes on Λ, use (1) to see that

$$\lim_{j\to\infty} \sup_{w\in\Gamma_j} (1-|w|^2)^\beta |g_j(w)| = \lim_{j\to\infty} \sup_k (1-|z_k|^2)^\beta |f_j(z_k)| = 0\,.$$

An application of Lemma 12 now shows that g vanishes on Λ. Thus we have found a nontrivial function in $A^{-\beta}$ that vanishes on a sequence in $W(\Gamma)$, contradicting the hypothesis that every element of $W(\Gamma)$ is a set of uniqueness for $A^{-\beta}$. This contradiction shows Γ is a sampling sequence for $A^{-\beta}$. □

We can now complete the proof of sufficiency in Theorem 3. It is enough to show that a uniformly discrete sequence Γ with lower density $D^-(\Gamma) > \beta$ is a sampling set for $A^{-\beta}$. If not, then by Lemma 13 some sequence Λ in $W(\Gamma)$ admits a nontrivial function $f \in A^{-\beta}$ vanishing on Λ. Lemma 11 shows that $D^-(\Lambda) > \beta$, so we can choose a number γ with $\beta < \gamma < D^-(\Lambda)$. Then since

$$\gamma < D^-(\Lambda) = \liminf_{r\to 1} \inf_{\zeta\in\mathbb{D}} D(\Lambda,\zeta,r) \le \liminf_{r\to 1} D(\Lambda,0,r)\,,$$

we infer that $D(\Lambda,0,r) > \gamma$ for all r in some interval $R < r < 1$. We will take $R > \frac{1}{2}$.

If $0 \in \Lambda$, the function f can be modified so that $f(0) \neq 0$ but f vanishes elsewhere on Λ. Thus we can assume without loss of generality that $0 \notin \Lambda$ and $f(0) = 1$. With $\Lambda = \{\lambda_k\}$ and $R < r < 1$, Jensen's formula then gives

$$\begin{aligned}
\frac{1}{2\pi}\int_0^{2\pi} \log|f(re^{i\theta})|\,d\theta &\geq \sum_{|\lambda_k|<r} \log\frac{r}{|\lambda_k|} \\
&\geq \sum_{\frac{1}{2}<|\lambda_k|<r} \log\frac{1}{|\lambda_k|} + n(\Lambda, 0, r)\log r \\
&\geq \sum_{\frac{1}{2}<|\lambda_k|<r} \log\frac{1}{|\lambda_k|} + C\frac{\log r}{1-r} \\
&= \left(\log\frac{1}{1-r}\right) D(\Lambda, 0, r) + C\frac{\log r}{1-r} \\
&> \gamma\log\frac{1}{1-r} + C\frac{\log r}{1-r},
\end{aligned}$$

where the counting function $n(\Lambda, 0, r)$ has been estimated by Lemma 15 of Chapter 2 and the constant C depends only on the separation constant $\delta(\Lambda)$.

On the other hand, the function $f \in A^{-\beta}$ satisfies

$$|f(z)| \leq \|f\|_{-\beta}(1-|z|)^{-\beta},$$

and so

$$\frac{1}{2\pi}\int_0^{2\pi} \log|f(re^{i\theta})|\,d\theta \leq \log\|f\|_{-\beta} + \beta\log\frac{1}{1-r}.$$

This combines with the previous inequality to give

$$\gamma\log\frac{1}{1-r} + C\frac{\log r}{1-r} < \log\|f\|_{-\beta} + \beta\log\frac{1}{1-r}$$

for $r > R$. Now let $r \to 1$ to conclude that $\gamma \leq \beta$, a contradiction. Therefore, $W(\Gamma)$ contains only sets of uniqueness for $A^{-\beta}$ and Lemma 13 says that Γ is a sampling sequence for $A^{-\beta}$. This completes the proof of sufficiency in Theorem 3.

There is also a natural way to discuss interpolation in the setting of $A^{-\beta}$. A sequence $\Gamma = \{z_k\}$ is said to be an *interpolation sequence for* $A^{-\beta}$ if for every sequence $\{w_k\}$ of complex numbers satisfying

$$\sup_k (1-|z_k|^2)^\beta |w_k| < \infty,$$

there is a function $f \in A^{-\beta}$ such that $f(z_k) = w_k$ for $k = 1, 2, \ldots$. The following theorem of Seip [3] is the companion to Theorem 3.

THEOREM 4. *A sequence Γ is an interpolation sequence for $A^{-\beta}$ if and only if Γ is uniformly discrete and $D^+(\Gamma) < \beta$.*

The proof is omitted, since the result plays no role in our discussion of the corresponding theorem for Bergman spaces.

§7.4. Sufficiency of the sampling condition in A^p.

Sampling sequences for the growth space are closely related to sampling sequences for the Bergman space. In this section we will derive a proof of sufficiency in Theorem 2 from the corresponding part of Theorem 3. The argument is due to Seip [3].

Let Γ be a uniformly discrete sequence with lower density $D^-(\Gamma) > 0$. Then Theorem 3 implies that Γ is a sampling sequence for each space $A_0^{-\beta}$ with $\beta < D^-(\Gamma)$, where sampling sequences for $A_0^{-\beta}$ are defined by the same inequality (9), which now need hold only for functions $f \in A_0^{-\beta}$. The following lemma provides a key link to the sampling theorem for Bergman spaces. It reconstructs a function of class $A_0^{-\beta}$ from sampling data.

LEMMA 14. *Let $\Gamma = \{z_k\}$ be a uniformly discrete sampling sequence for $A_0^{-\beta}$. Then there are functions $a_k(\zeta)$ defined in $\mathbb{D}$ such that*

$$f(\zeta) = (1 - |\zeta|^2)^{-\beta} \sum_{k=1}^{\infty} (1 - |z_k|^2)^{\beta} f(z_k)\, a_k(\zeta)\,, \qquad \zeta \in \mathbb{D}\,, \tag{14}$$

for every function $f \in A_0^{-\beta}$, where

$$\sum_{k=1}^{\infty} |a_k(\zeta)| \le C \tag{15}$$

and

$$|a_k(\zeta)| \le C\, \frac{(1 - |\zeta|^2)(1 - |z_k|^2)}{|1 - \overline{z_k}\zeta|^2}\,. \tag{16}$$

for some constant C depending only on the sampling constant of Γ.

COROLLARY. *Under the same hypotheses,*

$$f(\zeta) = (1 - |\zeta|^2)^{s-\beta} \sum_{k=1}^{\infty} \frac{(1 - |z_k|^2)^{\beta} f(z_k) a_k(\zeta)}{(1 - \overline{\zeta} z_k)^s}$$

for every $f \in A_0^{-\beta}$ and every number $s \in \mathbb{R}$.

The corollary is remarkable in that the formula does not actually depend on the choice of s. To deduce it from the lemma, simply replace f by

$$f(z) \left(\frac{1 - |\zeta|^2}{1 - \overline{\zeta} z} \right)^s$$

for fixed $\zeta \in \mathbb{D}$.

Proof of Lemma 14. In terms of the sequence Γ, let $T : A_0^{-\beta} \to c_0$ be the bounded linear operator defined by

$$T(f) = \{(1-|z_k|^2)^\beta f(z_k)\}, \qquad f \in A_0^{-\beta},$$

where c_0 is the subspace of ℓ^∞ consisting of sequences that converge to 0. Then T is bounded below because Γ is a sampling sequence. Thus T has closed range, and T^{-1} maps $T(A_0^{-\beta})$ boundedly onto $A_0^{-\beta}$. Note that $\|T^{-1}\| = L_0(\Gamma)$, the sampling constant of Γ with respect to the space $A_0^{-\beta}$.

For fixed $\zeta \in \mathbb{D}$, let the bounded linear functional ψ_ζ be defined by

$$\psi_\zeta(f) = (1-|\zeta|^2)^\beta f(\zeta), \qquad f \in A_0^{-\beta}.$$

It induces the linear functional

$$\widetilde{\psi}_\zeta(\{w_k\}) = \psi_\zeta(T^{-1}(\{w_k\})), \qquad \{w_k\} \in T(A_0^{-\beta}),$$

with norm $\|\widetilde{\psi}_\zeta\| \le \|T^{-1}\|$, since $\|\psi_\zeta\| \le 1$.

Appealing to the Hahn–Banach theorem, extend $\widetilde{\psi}_\zeta$ to a linear functional $\widetilde{\Psi}_\zeta$ on c_0 of norm $\|\widetilde{\Psi}_\zeta\| = \|\widetilde{\psi}_\zeta\| \le \|T^{-1}\|$. But the dual space $c_0{}^*$ can be identified with ℓ^1, and $\widetilde{\Psi}_\zeta$ has a representation of the form

$$\widetilde{\Psi}_\zeta(\{w_k\}) = \sum_{k=1}^\infty w_k\, b_k(\zeta), \qquad \{w_k\} \in c_0,$$

where

$$\sum_{k=1}^\infty |b_k(\zeta)| = \|\widetilde{\Psi}_\zeta\| \le \|T^{-1}\|. \tag{17}$$

Restrict the representation to $T(A_0^{-\beta}) \subset c_0$ to see that

$$(1-|\zeta|^2)^\beta f(\zeta) = \sum_{k=1}^\infty (1-|z_k|^2)^\beta f(z_k)\, b_k(\zeta), \qquad f \in A_0^{-\beta}. \tag{18}$$

Now consider the subset

$$\Lambda_\zeta = \left\{ z_k : |\varphi_\zeta(z_k)| > \tfrac{1}{2}, \;\; |b_k(\zeta)| > 1 - |\varphi_\zeta(z_k)|^2 \right\}.$$

In view of the standard identity

$$1 - |\varphi_\zeta(z)|^2 = \frac{(1-|\zeta|^2)(1-|z|^2)}{|1-\overline{\zeta}z|^2},$$

the estimate (17) shows that Λ_ζ forms a Blaschke sequence, with associated Blaschke product $B_\zeta(z)$. Note that $B_\zeta(\zeta) \ne 0$ since $\zeta \notin \Lambda_\zeta$. Now apply the formula to the function $B_\zeta f$ to obtain

$$(1-|\zeta|^2)^\beta f(\zeta) = \sum_{k=1}^\infty (1-|z_k|^2)^\beta f(z_k)\, a_k(\zeta), \qquad f \in A_0^{-\beta},$$

where $a_k(\zeta) = B_\zeta(z_k) b_k(\zeta) B_\zeta(\zeta)^{-1}$. Applying the inequality $x \geq e^{-2(1-x)}$ for $\frac{1}{4} \leq x \leq 1$, we find

$$\begin{aligned} |B_\zeta(\zeta)|^2 &= \prod_{z_k \in \Lambda_\zeta} |\varphi_\zeta(z_k)|^2 \geq \prod_{z_k \in \Lambda_\zeta} \exp\{-2(1 - |\varphi_\zeta(z_k)|^2)\} \\ &\geq \prod_{z_k \in \Lambda_\zeta} \exp\{-2|b_k(\zeta)|\} = \exp\Big\{-2 \sum_{z_k \in \Lambda_\zeta} |b_k(\zeta)|\Big\} \\ &\geq \exp\{-2\|T^{-1}\|\}, \end{aligned}$$

so that $|a_k(\zeta)| \leq e^{\|T^{-1}\|} |b_k(\zeta)|$. Hence (15) follows from (17).

It remains to show that the coefficients $a_k(\zeta)$ satisfy (16). But $a_k(\zeta) = 0$ if $z_k \in \Lambda_\zeta$, since $B_\zeta(z_k) = 0$. If $z_k \notin \Lambda_\zeta$, then either $|b_k(\zeta)| \leq 1 - |\varphi_\zeta(z_k)|^2$ or $|\varphi_\zeta(z_k)| \leq \frac{1}{2}$. The first inequality is

$$|b_k(\zeta)| \leq 1 - |\varphi_\zeta(z_k)|^2 = \frac{(1-|\zeta|^2)(1-|z_k|^2)}{|1-\overline{z_k}\zeta|^2}.$$

If $|\varphi_\zeta(z_k)| \leq \frac{1}{2}$, then $(1-|\zeta|^2)(1-|z_k|^2)|1-\overline{z_k}\zeta|^{-2} \geq \frac{3}{4}$, so

$$|b_k(\zeta)| \leq \sum_{n=1}^{\infty} |b_n(\zeta)| \leq \|T^{-1}\| \leq \tfrac{4}{3}\, \|T^{-1}\| \,\frac{(1-|\zeta|^2)(1-|z_k|^2)}{|1-\overline{z_k}\zeta|^2}.$$

In either case, (16) follows from the inequality $|a_k(\zeta)| \leq e^{\|T^{-1}\|} |b_k(\zeta)|$. □

We are now in position to give a proof of sufficiency in Theorem 2, for $1 \leq p < \infty$. Let $\Gamma = \{z_k\}$ be a uniformly discrete set with lower density $D^-(\Gamma) > \frac{1}{p}$. We are to show that Γ is a sampling sequence for A^p. The upper sampling inequality holds only because Γ is uniformly discrete, so the problem is to establish the lower sampling inequality

$$K_1 \|f\|_p^p \leq \sum_{k=1}^{\infty} (1-|z_k|^2)^2 |f(z_k)|^p$$

for all functions $f \in A^p$. It is sufficient to prove this for all polynomials f, since the polynomials are dense in A^p and the upper sampling inequality holds. Choose β in the interval $\frac{1}{p} < \beta < D^-(\Gamma)$ with $\beta < \frac{2}{p}$. Then Γ is a sampling sequence for $A_0^{-\beta}$, by Theorem 3, so that Lemma 14 applies to every function $f \in A_0^{-\beta}$, hence to every polynomial.

Consider first the case $p = 1$. Here we require that $1 < \beta < 2$. By (14) and (16),

$$|f(\zeta)| \leq C(1-|\zeta|^2)^{1-\beta} \sum_{k=1}^{\infty} \frac{(1-|z_k|^2)^{1+\beta}}{|1-\overline{z_k}\zeta|^2} |f(z_k)|$$

for every polynomial f. Therefore, by Lemma 1 of Chapter 6,

$$\begin{aligned}\int_{\mathbb{D}} |f(\zeta)|\, d\sigma &\le C \sum_{k=1}^{\infty} (1-|z_k|^2)^{1+\beta} |f(z_k)| \int_{\mathbb{D}} \frac{(1-|\zeta|^2)^{1-\beta}}{|1-\overline{z_k}\zeta|^2}\, d\sigma \\ &\le C \sum_{k=1}^{\infty} (1-|z_k|^2)^2 |f(z_k)|\,,\end{aligned}$$

which shows that Γ is a sampling sequence for A^1.

Next consider the case $1 < p < \infty$. Then we choose $\beta < D^-(\Gamma)$ with $\frac{1}{p} < \beta < \frac{2}{p}$, and again let f be a polynomial. An application of Hölder's inequality to (14) gives

$$\begin{aligned}|f(\zeta)|^p &\le (1-|\zeta|^2)^{-\beta p} \left\{ \sum_{k=1}^{\infty} (1-|z_k|^2)^{\beta p} |f(z_k)|^p |a_k(\zeta)| \right\} \left\{ \sum_{k=1}^{\infty} |a_k(\zeta)| \right\}^{p/q} \\ &\le C(1-|\zeta|^2)^{1-\beta p} \sum_{k=1}^{\infty} \frac{(1-|z_k|^2)^{1+\beta p}}{|1-\overline{z_k}\zeta|^2} |f(z_k)|^p\,,\end{aligned}$$

where (15) and (16) have been used. Since $\frac{1}{p} < \beta < \frac{2}{p}$, Lemma 1 of Chapter 6 can again be applied to give

$$\int_{\mathbb{D}} |f(\zeta)|^p\, d\sigma \le C \sum_{k=1}^{\infty} (1-|z_k|^2)^2 |f(z_k)|^p\,.$$

This shows that Γ is a sampling sequence for A^p, and the proof of Theorem 2 is complete.

§7.5. Sufficiency of the interpolation condition.

In Section 6.3 we considered a special family of sequences in the disk and classified them as interpolating or sampling for the Bergman space A^p. Specifically, we saw that if a uniformly discrete set Γ admits an analytic function g with the property

$$|g(z)| \simeq \rho(z,\Gamma)(1-|z|^2)^{-\alpha}\,,$$

then Γ is an interpolation set for A^p if and only if $\alpha < \frac{1}{p}$, and a sampling set if and only if $\alpha > \frac{1}{p}$. (Recall that the notation $F(z) \simeq G(z)$ means $C_1 F(z) \le G(z) \le C_2 F(z)$ for some positive constants C_1 and C_2 and all $z \in \mathbb{D}$.) Our main goal is now to establish a connection between arbitrary uniformly discrete sets and the special sets examined in Section 6.3. This connection will be exploited to extract a proof of Theorem 1 from the results of Chapter 6. Our aim is to prove the following theorem.

THEOREM 5. *Let $\alpha > 0$. If Γ is a uniformly discrete sequence in $\mathbb{D}$ with $D^+(\Gamma) < \alpha$, then there exist a sequence Λ and an analytic function f such that $\Gamma \cup \Lambda$ is uniformly discrete and*

$$|f(z)| \simeq \rho(z, \Gamma \cup \Lambda)(1 - |z|^2)^{-\alpha}.$$

The sufficiency direction of Theorem 1 is an immediate corollary of Theorem 5 and the results of Section 6.3. If Γ is uniformly discrete and $D^+(\Gamma) < \frac{1}{p}$, one can choose α so that $D^+(\Gamma) < \alpha < \frac{1}{p}$. Then Theorem 5 combines with the earlier result (Theorem 1 of Chapter 6) to show that $\Gamma \cup \Lambda$ is an interpolation sequence for A^p, which implies the same for Γ.

The full meaning of Theorem 5 becomes apparent only in light of a result to be obtained later in this chapter. Lemma 19 (in Section 7.6) says that the property of $\Gamma \cup \Lambda$ given by Theorem 5 actually implies $D^+(\Gamma \cup \Lambda) = D^-(\Gamma \cup \Lambda) = \alpha$. Thus Theorem 5 says that it is always possible to add points to any given uniformly discrete set, at the cost of arbitrarily small increase in upper uniform density, to create a set of the type studied in Section 6.3. Furthermore, the set so formed is regularly distributed in the sense that its upper and lower densities are equal.

Theorem 5 is analogous to a result of Schuster and Seip [2], who considered it in the context of the Bargmann–Fock space. The proof in the present setting was communicated to us by Seip and relies heavily on Lemma 16 below, a variant of which was obtained in $\mathbb{C}$ by Lyubarskii and Sodin [1] and in $\mathbb{D}$ by Seip [5]. Also playing an important role in the proof of Theorem 5 are the techniques of Berndtsson and Ortega-Cerdà [1], who examined sampling and interpolation sequences for a class of spaces of analytic functions that includes the Bergman space. The idea of applying Theorem 5 to give a proof of sufficiency in Theorem 1 is due to Seip [6]. His original proof (for $p = 2$) in [3] appealed to techniques of Korenblum [1].

We begin by presenting some of the tools to be used in the proof of Theorem 5. The *invariant convolution* of a measure μ and a measurable function g is defined by the formula

$$(\mu * g)(\zeta) = \int_{\mathbb{D}} g(\varphi_\zeta(z)) \frac{d\mu(z)}{(1 - |z|^2)^2}$$

whenever the integral exists. Absolutely continuous measures have the commutative property

$$(f\, d\sigma) * g = (g\, d\sigma) * f$$

(written with abuse of notation), which can be verified by a change of variables and the identity (1). The associative relation

$$\mu * \{(f\, d\sigma) * g\} = \{(\mu * f)d\sigma\} * g$$

holds whenever f is a radial function, meaning that $f(z) = f(|z|)$. This is a consequence of Fubini's Theorem, a change of variables, and the fact that

$$\left|\varphi_{\varphi_\zeta(z)}(w))\right| = \left|\varphi_{\varphi_\zeta(w)}(z))\right|.$$

The *invariant Laplacian* $\widetilde{\Delta}$ is defined for functions $f \in C^2(\mathbb{D})$ by

$$\widetilde{\Delta} f(z) = (1-|z|^2)^2 f_{z\bar{z}}(z) = \tfrac{1}{4}(1-|z|^2)^2 \Delta f(z),$$

where Δ is the standard Laplacian. It has the Möbius invariance property

$$\widetilde{\Delta}(f \circ \varphi_\zeta)(z) = \widetilde{\Delta}(f)(\varphi_\zeta(z)), \qquad \zeta \in \mathbb{D}.$$

To see this, apply the chain rule to obtain

$$(f \circ \varphi_\zeta)_z = (f_z \circ \varphi_\zeta)(\varphi_\zeta)_z + (f_{\bar{z}} \circ \varphi_\zeta)(\overline{\varphi_\zeta})_z = (f_z \circ \varphi_\zeta)\varphi_\zeta{}',$$

so that

$$\begin{aligned}(f \circ \varphi_\zeta)_{z\bar{z}} &= \{(f_{zz} \circ \varphi_\zeta)(\varphi_\zeta)_{\bar{z}} + (f_{z\bar{z}} \circ \varphi_\zeta)(\overline{\varphi_\zeta})_{\bar{z}}\}\, \varphi_\zeta{}' + (f_z \circ \varphi_\zeta)(\varphi_\zeta{}')_{\bar{z}} \\ &= (f_{z\bar{z}} \circ \varphi_\zeta)|\varphi_\zeta{}'|^2 .\end{aligned}$$

The result then follows from the identity (1).

The definition of the invariant Laplacian can be extended using the theory of distributions. For a locally integrable function h, define the distribution T_h by

$$T_h(\psi) = \int_{\mathbb{D}} \psi(z) h(z) \frac{d\sigma(z)}{(1-|z|^2)^2},$$

where $\psi \in C_0^\infty(\mathbb{D})$, the set of real-valued infinitely differentiable functions with compact support in the disk. Then the invariant Laplacian of T_h, denoted by $\widetilde{\Delta} T_h$, is the distribution defined by

$$\widetilde{\Delta} T_h(\psi) = \int_{\mathbb{D}} \widetilde{\Delta}\psi(z) h(z) \frac{d\sigma(z)}{(1-|z|^2)^2}.$$

This definition of the invariant Laplacian can be viewed as an extension of the earlier definition. More precisely, if $h \in C^2(\mathbb{D})$, then $\widetilde{\Delta} T_h = T_{\widetilde{\Delta} h}$. To see this we apply Green's formula (*cf.* Section 1.5) with $h \in C^2(\mathbb{D})$, $\psi \in C_0^\infty(\mathbb{D})$, and $r < 1$ so close to 1 that the support of ψ is contained in $r\mathbb{D}$. Then

$$\begin{aligned}\widetilde{\Delta} T_h(\psi) &= \int_{r\mathbb{D}} \widetilde{\Delta}\psi(z) h(z) \frac{d\sigma(z)}{(1-|z|^2)^2} = \int_{r\mathbb{D}} \widetilde{\Delta} h(z) \psi(z) \frac{d\sigma(z)}{(1-|z|^2)^2} \\ &= \int_{\mathbb{D}} \widetilde{\Delta} h(z) \psi(z) \frac{d\sigma(z)}{(1-|z|^2)^2} = T_{\widetilde{\Delta} h}(\psi).\end{aligned}$$

In particular, if h is harmonic in $\mathbb{D}$, then

$$\int_{\mathbb{D}} \widetilde{\Delta}\psi(z) h(z) \frac{d\sigma(z)}{(1-|z|^2)^2} = 0 \tag{19}$$

for all $\psi \in C_0^\infty(\mathbb{D})$. The converse is given by Weyl's lemma (*cf.* Section 5.3), which states that every locally integrable function h satisfying (19) for all $\psi \in C_0^\infty(\mathbb{D})$ is harmonic.

A real-valued function h defined on $\mathbb{D}$ is *subharmonic* if it is upper semicontinuous and has the local submean value property on $\mathbb{D}$; that is, for each $z \in \mathbb{D}$, there exists $r_0 > 0$ such that

$$h(z) \leq \frac{1}{2\pi} \int_0^{2\pi} h(z + re^{i\theta})\, d\theta$$

whenever $0 \leq r < r_0$. Since subharmonic functions are locally integrable, both T_h and $\widetilde{\Delta} T_h$ are well-defined when h is subharmonic. In fact, a characterization of subharmonicity can be given in terms of the invariant Laplacian $\widetilde{\Delta} T_h$. In particular, it can be shown that if h is a locally integrable upper semicontinuous function, then h is subharmonic if and only if $\widetilde{\Delta} T_h$ is a positive distribution, which means that $\widetilde{\Delta} T_h(\psi) \geq 0$ for all nonnegative functions $\psi \in C_0^\infty(\mathbb{D})$. (See Lieb and Loss [1], Theorem 9.3.)

More can be said about one direction of this result. Recall that a *Radon measure* on $\mathbb{D}$ is a positive measure μ with the property that $\mu(K) < \infty$ for every compact subset K of $\mathbb{D}$. It can be shown by a simple approximation lemma (see Ransford [1], pp. 71–74) that if h is subharmonic, then $\widetilde{\Delta} T_h$ can be extended to a positive linear functional on all of $C_0(\mathbb{D})$, the set of continuous functions in $\mathbb{D}$ with compact support. Then, by the Riesz representation theorem, there is a unique Radon measure μ such that

$$\widetilde{\Delta} T_h(\psi) = \int_{\mathbb{D}} \psi(z) \frac{d\mu(z)}{(1 - |z|^2)^2}$$

for all $\psi \in C_0(\mathbb{D})$. In particular,

$$\int_{\mathbb{D}} \widetilde{\Delta}\psi(z) h(z) \frac{d\sigma(z)}{(1 - |z|^2)^2} = \int_{\mathbb{D}} \psi(z) \frac{d\mu(z)}{(1 - |z|^2)^2}$$

for all $\psi \in C_0^\infty(\mathbb{D})$. It is customary to express this relationship by the equation $\widetilde{\Delta} h = \mu$. When $h \in C^2(\mathbb{D})$, that notation conflicts with the ordinary interpretation of the invariant Laplacian as a function $f = \widetilde{\Delta} h$. The conflict is resolved by identifying f with the absolutely continuous measure μ for which $d\mu = f\, d\sigma$. Then we will write, with abuse of notation, $\widetilde{\Delta} h = f\, d\sigma$.

Letting $E(z) = 2 \log |z|$, we can express the reproducing property of Green's function (see Section 1.5 for the definition and basic properties) by the equation

$$\int_{\mathbb{D}} E(\varphi_\zeta(z)) \widetilde{\Delta}\psi(z) \frac{d\sigma(z)}{(1 - |z|^2)^2} = \psi(\zeta)\,, \tag{20}$$

for all functions $\psi \in C^2(\overline{\mathbb{D}})$ that vanish on $\partial\mathbb{D}$.

Therefore, if μ is a measure for which $\mu * E$ is well-defined, Fubini's theorem implies that

$$\begin{aligned}\int_{\mathbb{D}} \widetilde{\Delta}\psi(\zeta)(\mu * E)(\zeta)\frac{d\sigma(\zeta)}{(1-|\zeta|^2)^2} &= \int_{\mathbb{D}} \widetilde{\Delta}\psi(\zeta)\int_{\mathbb{D}} E(\varphi_\zeta(z))\frac{d\mu(z)}{(1-|z|^2)^2}\frac{d\sigma(\zeta)}{(1-|\zeta|^2)^2} \\ &= \int_{\mathbb{D}}\int_{\mathbb{D}} E(\varphi_z(\zeta))\widetilde{\Delta}\psi(\zeta)\frac{d\sigma(\zeta)}{(1-|\zeta|^2)^2}\frac{d\mu(z)}{(1-|z|^2)^2} \\ &= \int_{\mathbb{D}} \psi(z)\frac{d\mu(z)}{(1-|z|^2)^2}.\end{aligned}$$

In other words,

$$\widetilde{\Delta}(\mu * E) = \mu\,. \tag{21}$$

An important example of a subharmonic function that is in general not twice continuously differentiable is $\log|g(z)|$, where g is an analytic function. In fact, if $\Gamma = \{z_k\}$ is the zero-set of g, then

$$\widetilde{\Delta}\log|g(z)| = \tfrac{1}{2}\nu_\Gamma\,,$$

where

$$\nu_\Gamma = \sum_k (1-|z_k|^2)^2\delta_{z_k} \tag{22}$$

is a weighted sum of Dirac-delta measures concentrated at the points of Γ. To prove this, we need to show that

$$\int_{\mathbb{D}} \widetilde{\Delta}\psi(z)\log|g(z)|\frac{d\sigma(z)}{(1-|z|^2)^2} = \tfrac{1}{2}\sum_k \psi(z_k) \tag{23}$$

for all $\psi \in C_0^\infty(\mathbb{D})$. Since the support of any such ψ is a compact subset of $\mathbb{D}$, it will be enough to consider the case where $\Gamma = \{z_1, z_2, \dots, z_n\}$ is finite. Then

$$g(z) = \left(\frac{z_1 - z}{1-\overline{z_1}z}\right)\cdots\left(\frac{z_n - z}{1-\overline{z_n}z}\right)h(z)\,,$$

where h is a nonvanishing analytic function, which implies that

$$\log|g(z)| = \log|h(z)| + \tfrac{1}{2}\sum_{k=1}^n E(\varphi_{z_k}(z))\,.$$

Then by (20) and the fact that $\log|h(z)|$ is harmonic, the left-hand side of (23) becomes

$$\begin{aligned}\int_{\mathbb{D}} \widetilde{\Delta}\psi(z)\log|h(z)|\frac{d\sigma(z)}{(1-|z|^2)^2} + \tfrac{1}{2}\sum_{k=1}^n\int_{\mathbb{D}}\widetilde{\Delta}\psi(z)E(\varphi_{z_k}(z))\frac{d\sigma(z)}{(1-|z|^2)^2} \\ = \tfrac{1}{2}\sum_{k=1}^n \psi(z_k)\,.\end{aligned}$$

The converse is also true. Namely, if Γ is a sequence of points in $\mathbb{D}$ and Φ is a subharmonic function with the property $\widetilde{\Delta}\Phi = \frac{1}{2}\nu_\Gamma$, then $\Phi(z) = \log|g(z)|$ for some analytic function g with zero-set Γ. To see this, let G be any analytic function that vanishes precisely on Γ. Then $\widetilde{\Delta}\log|G| = \frac{1}{2}\nu_\Gamma$, which implies that $\widetilde{\Delta}(\Phi - \log|G|) = 0$, and so by Weyl's lemma the function $\Phi - \log|G|$ is harmonic. Thus $\Phi - \log|G| = \mathrm{Re}\{f\}$ for some analytic function f, and $\Phi = \log|e^f G|$.

We now turn to the proof of Theorem 5. Part of the argument will be developed in a series of three lemmas. Before stating them, we need to introduce some further notation.

For $\frac{1}{2} < r < 1$, define the function

$$\chi_r(\zeta) = \begin{cases} c_r \log \frac{1}{|\zeta|^2} & \text{if } \frac{1}{2} < |\zeta| < r, \\ 0 & \text{otherwise,} \end{cases}$$

where the constant c_r is chosen so that

$$\int_{\mathbb{D}} \chi_r(\zeta) \frac{d\sigma(\zeta)}{(1-|\zeta|^2)^2} = 1,$$

or, after passing to polar coordinates,

$$4c_r \int_{\frac{1}{2}}^{r} \left(\log \frac{1}{s}\right) \frac{s}{(1-s^2)^2}\, ds = 1.$$

An integration by parts shows that

$$\int_{\frac{1}{2}}^{r} \left(\log \frac{1}{s}\right) \frac{s}{(1-s^2)^2}\, ds = \tfrac{1}{4} \log \frac{1}{1-r} + O(1), \qquad r \to 1,$$

so that $c_r \log \frac{1}{1-r} \to 1$ as $r \to 1$.

LEMMA 15. *Let $\Gamma = \{z_k\}$ be a uniformly discrete sequence with separation constant $\delta = \delta(\Gamma)$. Let the measure ν_Γ be defined by (22), and let $E(z) = 2\log|z|$. Then for $0 < r < 1$ the function*

$$v_r = (\nu_\Gamma - (\nu_\Gamma * \chi_r)d\sigma) * E$$

is well defined and has the properties $v_r(z) \le 0$ in $\mathbb{D}$,

$$v_r(z) \ge -K_r \qquad \text{for } \rho(z,\Gamma) \ge \tfrac{\delta}{2}, \tag{24}$$

and

$$|v_r(z) - 2\log\rho(z,\Gamma)| \le C_r \qquad \text{for } \rho(z,\Gamma) < \tfrac{\delta}{2}, \tag{25}$$

where K_r and C_r are positive constants depending only on r and δ. Moreover,

$$\widetilde{\Delta} v_r = \nu_\Gamma - (\nu_\Gamma * \chi_r)d\sigma.$$

LEMMA 16. *Suppose Φ is a subharmonic function in $\mathbb{D}$ with the property that $\widetilde{\Delta}\Phi = h\,d\sigma$, where h satisfies $h(z) \simeq 1$. Then for any uniformly discrete sequence Γ, there exist an analytic function g and a sequence Λ such that $\Gamma \cup \Lambda$ is uniformly discrete and*

$$|g(z)| \simeq \rho(z, \Lambda) e^{\Phi(z)} . \tag{26}$$

LEMMA 17. *Any two uniformly discrete sequences Λ_1 and Λ_2 satisfy*

$$\tfrac{1}{2}\,\rho(\Lambda_1, \Lambda_2)\,\rho(z, \Lambda_1 \cup \Lambda_2) \le \rho(z, \Lambda_1)\,\rho(z, \Lambda_2) \le \rho(z, \Lambda_1 \cup \Lambda_2)$$

for all $z \in \mathbb{D}$, where $\rho(A, B) = \inf\{\rho(a, b) : a \in A, b \in B\}$.

Deferring the proofs of the lemmas to the end of this section, we show first how they lead to a proof of Theorem 5.

DEDUCTION OF THEOREM 5. Suppose $\Gamma = \{z_k\}$ is a uniformly discrete sequence with upper density $D^+(\Gamma) < \alpha$, and choose β with $D^+(\Gamma) < \beta < \alpha$. Note that

$$\begin{aligned}(\nu_\Gamma * \chi_r)(\zeta) &= 2\,c_r \sum_{\frac{1}{2} < |\varphi_\zeta(z_k)| < r} \log \frac{1}{|\varphi_\zeta(z_k)|} \\ &= 2\,c_r \left(\log \frac{1}{1-r}\right) D(\Gamma, \zeta, r) .\end{aligned}$$

Since $c_r \log \frac{1}{1-r} \to 1$ as $r \to 1$, and

$$D^+(\Gamma) = \limsup_{r \to 1} \sup_{\zeta \in \mathbb{D}} D(\Gamma, \zeta, r) ,$$

we infer that $(\nu_\Gamma * \chi_r)(\zeta) < 2\beta$ for all $\zeta \in \mathbb{D}$ and all r in some interval $R < r < 1$, where $R > \frac{1}{2}$.

Next define

$$\Phi(z) = \alpha \log \left(\frac{1}{1 - |z|^2}\right) - \log|H(z)| + \tfrac{1}{2} v_r(z) ,$$

where H is any analytic function that vanishes precisely on Γ. For example, we can choose H to be the Horowitz product (*cf.* Section 4.4)

$$H(z) = \prod_{k=1}^{\infty} b_{z_k}(z)(2 - b_{z_k}(z)) ,$$

where $b_\zeta(z) = \frac{|\zeta|}{\zeta} \varphi_\zeta(z)$ if $\zeta \ne 0$ and $b_0(z) = z$. Because $\sum (1 - |z_k|^2)^2 < \infty$ for any uniformly discrete sequence (*cf.* Section 2.11), the Horowitz product converges locally uniformly in $\mathbb{D}$, as demonstrated in Chapter 4. Then, since $\widetilde{\Delta} \log\left(\frac{1}{1-|z|^2}\right) = d\sigma$ and $\widetilde{\Delta} v_r = \nu_\Gamma - (\nu_\Gamma * \chi_r)d\sigma$, by (21), we find that

$$\widetilde{\Delta}\Phi = \alpha\, d\sigma - \tfrac{1}{2}\nu_\Gamma + \tfrac{1}{2}(\nu_\Gamma - (\nu_\Gamma * \chi_r)d\sigma) = (\alpha - \tfrac{1}{2}\,\nu_\Gamma * \chi_r)d\sigma\,.$$

But for $R < r < 1$, we have seen that

$$0 < \alpha - \beta \le \alpha - \tfrac{1}{2}(\nu_\Gamma * \chi_r) \le \alpha\,.$$

Thus Φ is subharmonic and its invariant Laplacian has the form $\widetilde{\Delta}\Phi = h\, d\sigma$ for some function h bounded above and below by positive constants. By Lemma 16, there exist an analytic function g and a sequence Λ such that $\Gamma \cup \Lambda$ is uniformly discrete and (26) holds. Note that the properties of v_r, as given in Lemma 15, imply that

$$\exp\{\tfrac{1}{2}v_r(z)\} \simeq \rho(z,\Gamma)\,.$$

Note also that $\rho(\Gamma,\Lambda) > 0$ because $\Gamma\cup\Lambda$ is uniformly discrete. Consequently, with the definition $f = Hg$, we conclude from Lemmas 16, 15, and 17 that

$$\begin{aligned} |f(z)| = |H(z)g(z)| &\simeq |H(z)|\,\rho(z,\Lambda)\,e^{\Phi(z)} \\ &= \rho(z,\Lambda)\,(1-|z|^2)^{-\alpha}\,\exp\left\{\tfrac{1}{2}v_r(z)\right\} \\ &\simeq \rho(z,\Gamma)\,\rho(z,\Lambda)\,(1-|z|^2)^{-\alpha} \\ &\simeq \rho(z,\Gamma\cup\Lambda)\,(1-|z|^2)^{-\alpha}\,. \end{aligned}$$

This completes the deduction of Theorem 5 and therefore the proof of sufficiency in Theorem 1. □

We turn now to the proofs of Lemmas 15, 16, and 17.

PROOF OF LEMMA 15. Let $\Gamma_n = \{z_1, z_2, \dots, z_n\}$ and write ν_n for ν_{Γ_n}. For $\frac{1}{2} < r < 1$, define the function

$$v_{nr} = (\nu_n - (\nu_n * \chi_r)d\sigma) * E\,.$$

Unravelling this definition, we find that

$$v_{nr}(z) = 2\sum_{j=1}^{n}\left\{\log|\varphi_{z_j}(z)| - I(z_j,r,z)\right\}\,, \tag{27}$$

where

$$I(z_j,r,z) = \int_{\mathbb{D}} \chi_r(\varphi_{z_j}(\zeta))\log|\varphi_z(\zeta)|\frac{d\sigma(\zeta)}{(1-|\zeta|^2)^2}\,.$$

We claim that for each r and each compact set $K \subset \mathbb{D}$, there exists an a integer m such that $v_{nr}(z) = v_{mr}(z)$ for all $z \in K$ whenever $n \ge m$. To see this, fix $z \in \mathbb{D}$ and suppose that $|\varphi_{z_j}(z)| \geq r$, or equivalently $\rho(z_j, z) \geq r$, for some point $z_j \in \Gamma$. If w is any point in the disk $|w| < r$, then

$$|\varphi_z(\varphi_{z_j}(w))| = \rho(z, \varphi_{z_j}(w)) = \rho(\varphi_{z_j}(z), w) > 0\,,$$

and so the function $\log|\varphi_z(\varphi_{z_j}(w))|$ is harmonic in the annulus $\frac{1}{2} < |w| < r$. Thus

$$\begin{aligned} I(z_j, r, z) &= \int_{\frac{1}{2}<|\varphi_{z_j}(\zeta)|<r} \chi_r(\varphi_{z_j}(\zeta)) \log|\varphi_z(\zeta)| \frac{d\sigma(\zeta)}{(1-|\zeta|^2)^2} \\ &= \int_{\frac{1}{2}<|w|<r} \chi_r(w) \log|\varphi_z(\varphi_{z_j}(w))| \frac{d\sigma(w)}{(1-|w|^2)^2} \\ &= \log|\varphi_z(z_j)| = \log|\varphi_{z_j}(z)|\,, \end{aligned}$$

where the evaluation of the last integral uses the mean value property of harmonic functions and the fact that $\chi_r(w)(1-|w|^2)^{-2}d\sigma(w)$ is a unit radial measure. This shows that when $|\varphi_{z_j}(z)| \geq r$, the corresponding term in the sum (27) is equal to 0, and so

$$v_{nr}(z) = 2 \sum_{z_j \in \Gamma_n \cap \Delta(z,r)} \left\{\log|\varphi_{z_j}(z)| - I(z_j, r, z)\right\}.$$

But since Γ is uniformly discrete, there are at most a finite number of points $z_j \in \Gamma$ in the disk $\Delta(z, r)$, and the estimate (6) shows that this number has an upper bound depending only on r and $\delta = \delta(\Gamma)$. Finally, since K is compact, it is covered by finitely many disks $\Delta(z, r)$ with $z \in K$. The conclusion is that

$$v_r(z) = \lim_{n\to\infty} v_{nr}(z)\,,$$

or equivalently

$$v_r(z) = 2 \sum_{z_j \in \Delta(z,r)} \left\{\log|\varphi_{z_j}(z)| - I(z_j, r, z)\right\}, \tag{28}$$

where the sum extends over a finite number of terms depending only on r, δ, and K, not on the particular point $z \in K$. Therefore, $v_{nr}(z) \to v_r(z)$ as $n \to \infty$, uniformly on each compact subset of $\mathbb{D}$. It follows that

$$v_r = (\nu_\Gamma - (\nu_\Gamma * \chi_r)d\sigma) * E$$

is a well-defined function, and so by (21) that $\widetilde{\Delta} v_r = \nu_\Gamma - (\nu_\Gamma * \chi_r)d\sigma$.

Since $\widetilde{\Delta}(\nu_n * E) = \nu_n$ is a positive measure, $\nu_n * E$ and therefore $(\nu_n * E) \circ \varphi_z$ are subharmonic. Because $\chi_r(\zeta)(1-|\zeta|^2)^{-2}d\sigma(\zeta)$ is a unit radial measure,

$$((\nu_n * E) \circ \varphi_z)(0) \leq \int_{\mathbb{D}} ((\nu_n * E) \circ \varphi_z)(\zeta)\chi_r(\zeta)\frac{d\sigma(\zeta)}{(1-|\zeta|^2)^2}.$$

This means that

$$\nu_n * E \leq \chi_r d\sigma * (\nu_n * E) . \tag{29}$$

By the associative and commutative properties of invariant convolution,

$$\begin{aligned}(\nu_n * \chi_r) d\sigma * E &= \nu_n * (\chi_r d\sigma * E) = \nu_n * (E d\sigma * \chi_r) \\ &= (\nu_n * E) d\sigma * \chi_r = \chi_r d\sigma * (\nu_n * E) ,\end{aligned}$$

which implies that

$$v_{nr} = \nu_n * E - (\nu_n * \chi_r) d\sigma * E = \nu_n * E - \chi_r d\sigma * (\nu_n * E) .$$

Thus $v_{nr}(z) \leq 0$ by (29), and so $v_r(z) \leq 0$ in $\mathbb{D}$.

To verify (25), we must show that there is a constant C_r such that

$$\left| v_r(z) - 2 \log |\varphi_{z_k}(z)| \right| \leq C_r \tag{30}$$

for every point $z_k \in \Gamma$ and all $z \in \Delta(z_k, \frac{\delta}{2})$. But $v_r(z)$ is represented by the sum (28), which extends over all points $z_j \in \Delta(z, r)$. Thus $\rho(z, z_k) < \frac{\delta}{2}$ and $\rho(z_j, z) < r$, so the strong form of the triangle inequality shows that $\rho(z_j, z_k) < \varepsilon$, where

$$\varepsilon = \frac{r + \frac{\delta}{2}}{1 + r\frac{\delta}{2}} > \frac{\delta}{2} .$$

Thus for $z \in \Delta(z_k, \frac{\delta}{2})$, the sum (28) extends only over the points $z_j \in \Delta(z_k, \varepsilon)$. The estimate (6) shows that there are at most a finite number $N = N(r, \delta)$ of such points. Consequently, we can express v_r in the equivalent form

$$v_r(z) = 2 \sum_{z_j \in \Delta(z_k, \varepsilon)} \left\{ \log |\varphi_{z_j}(z)| - I(z_j, r, z) \right\} , \qquad z \in \Delta(z_k, \tfrac{\delta}{2}) . \tag{31}$$

Observe now that

$$\begin{aligned} -I(z_j, r, z) &= 2c_r \int_{\frac{1}{2} < |\varphi_{z_j}(\zeta)| < r} \log \frac{1}{|\varphi_{z_j}(\zeta)|} \log \frac{1}{|\varphi_z(\zeta)|} \frac{d\sigma(\zeta)}{(1 - |\zeta|^2)^2} \\ &\leq 2c_r \log 2 \int_{|\varphi_z(\zeta)| < \frac{1}{2}} \log \frac{1}{|\varphi_z(\zeta)|} \frac{d\sigma(\zeta)}{(1 - |\zeta|^2)^2} \\ &\qquad + \log 2 \int_{\mathbb{D}} \chi_r(\varphi_{z_j}(\zeta)) \frac{d\sigma(\zeta)}{(1 - |\zeta|^2)^2} \\ &= 4c_r \log 2 \int_0^{\frac{1}{2}} \left(\log \frac{1}{s} \right) \frac{s}{(1 - s^2)^2} \, ds + \log 2 = B_r , \end{aligned}$$

by the definition of χ_r. Also, since $\rho(z, z_k) < \frac{\delta}{2}$ and Γ is uniformly discrete with separation constant δ, the triangle inequality shows that $\rho(z, z_j) > \frac{\delta}{2}$

for all $j \neq k$. In other words, $|\varphi_{z_j}(z)| > \frac{\delta}{2}$ for all $j \neq k$. Consequently, it follows from (31) that for $z \in \Delta(z_k, \frac{\delta}{2})$,

$$\begin{aligned}
&\left|v_r(z) - 2\log|\varphi_{z_k}(z)|\right| \\
&\quad \leq -I(z_k, r, z) + 2 \sum_{z_j \in \Delta(z_k,\varepsilon),\, j\neq k} \left\{\log\frac{1}{|\varphi_{z_j}(z)|} - I(z_j, r, z)\right\} \\
&\quad \leq 2N\left\{\log\tfrac{2}{\delta} + B_r\right\}.
\end{aligned}$$

This completes the proof of (30), which is equivalent to (25).

It is now a short step to the proof of (24). If $\rho(z, z_j) \geq \frac{\delta}{2}$ for all $z_j \in \Gamma$, then by (28) and (6),

$$|v_{nr}(z)| \leq 2\,n(\Gamma, z, r)\left\{\log\tfrac{2}{\delta} + B_r\right\} \leq 2\left(\tfrac{2}{\delta}+1\right)^2 \frac{1}{1-r^2}\left\{\log\tfrac{2}{\delta} + B_r\right\}.$$

This proves (24) and completes the proof of Lemma 15. □

Proof of Lemma 16. We begin with the construction of Λ and note that it is similar to the example in Section 6.6. We define the measure

$$d\mu(z) = \frac{2h(z)d\sigma(z)}{(1-|z|^2)^2}$$

and once again partition the disk into disjoint annuli

$$R_n = \{z : t_{n-1} \leq |z| < t_n\}, \qquad n = 1, 2, 3, \ldots,$$

where the radii t_n are defined inductively by $t_0 = 0$ and $\mu(R_n) = 2^{n-1}$. A straightforward calculation, using the hypothesis that $h(z) \simeq 1$, shows there exist positive constants C_1 and C_2 such that

$$C_1 \leq \frac{1-t_n}{2^{-n}} \leq C_2 \qquad \text{and} \qquad C_1 \leq \frac{t_n - t_{n-1}}{2^{-n}} \leq C_2 \tag{32}$$

for $n = 1, 2, \ldots$. In the construction of the example in Section 6.6, the measure μ under consideration was a constant multiple of hyperbolic area, while here one knows only that μ lies *between* constant multiples of hyperbolic area. For that reason, here one achieves merely the estimates (32) on t_n instead of the exact formula obtained previously, but these estimates turn out to be sufficient for our purposes. Another difference is that subsequently the annuli are partitioned into cells that are twice as "big" as in the example of Chapter 6. In fact, one divides each annulus R_n into 2^{n-2} cells Q_{nj} of equal area $\mu(Q_{nj}) = 2$, but the method of partitioning is different. In this case we cannot use equally spaced radial segments since h may not be a radial function. Instead, we choose numbers $0 = \alpha_{n,0} < \alpha_{n,1} < \ldots \alpha_{n,2^{n-2}} = 2\pi$ such that

$$\frac{2}{\pi}\int_{\alpha_{n,j-1}}^{\alpha_{n,j}}\int_{t_{n-1}}^{t_n} h(re^{i\theta})\frac{r}{(1-r^2)^2}\,dr d\theta = 2$$

for $j = 1, 2, \ldots, 2^{n-2}$. Defining

$$Q_{nj} = \left\{z \in \mathbb{D} : t_{n-1} \le |z| < t_n,\ \alpha_{n,j-1} \le \arg\{z\} < \alpha_{n,j}\right\},$$

we then have $\mu(Q_{nj}) = 2$. We next define the center of mass of Q_{nj} by

$$\zeta_{nj} = \frac{1}{2}\int_{Q_{nj}} z\, d\mu(z)\,.$$

Using (32), one can show that there are radii r_1 and r_2 such that

$$\Delta(\zeta_{nj}, r_1) \subset Q_{nj} \subset \Delta(\zeta_{nj}, r_2)\,,$$

at least for all n sufficiently large. If we define $\Sigma = \{s_k\}$ to be an enumeration of $\{\zeta_{nj}\}$, and $\{Q_k\}$ to be the corresponding enumeration of $\{Q_{nj}\}$, we obtain the following analogue of Proposition 1 in Chapter 6.

PROPOSITION 2.
(a) *The sequence Σ is uniformly discrete.*
(b) *There is a constant $K < 1$ such that $d(Q_k) \le K$ for all k, where $d(E)$ denotes the pseudohyperbolic diameter of a set E, as defined in Chapter 6.*
(c) *There is a constant C such that if a and b belong to the same cell Q_k, then*

$$\left|\frac{1-\bar{a}z}{1-\bar{b}z}\right| \le C$$

for all $z \in \mathbb{D}$.

With the notation $\eta = \min\{\frac{1}{2}\delta(\Gamma), r_1\}$, our sequence Λ is constructed as follows. For each k, choose any pair of points w_k and y_k in the disk such that $s_k = \frac{1}{2}(w_k + y_k)$ is the midpoint of the line segment connecting w_k and y_k, with the extra stipulation that

$$\rho(w_k, y_k) = \frac{\eta}{20} \qquad \text{if} \quad \rho(s_k, \Gamma) \ge \frac{\eta}{10}\,,$$
$$\rho(w_k, y_k) = \frac{\eta}{2} \qquad \text{if} \quad \rho(s_k, \Gamma) < \frac{\eta}{10}\,.$$

Then define $\Lambda = \{w_1, y_1, w_2, y_2, \ldots\} = \{\lambda_1, \lambda_2, \lambda_3, \lambda_4, \ldots\}$.

We first show that $\Gamma \cup \Lambda$ is uniformly discrete. Since for any pair of points $a, b \in \mathbb{D}$,

$$\tfrac{1}{3}\rho(a,b) \le \rho\left(a, \tfrac{1}{2}(a+b)\right) \le \rho(a,b)\,,$$

it follows from the triangle inequality that

$$\Delta\left(w_k, \frac{r_1}{2}\right) \subset \Delta(s_k, r_1) \qquad \text{and} \qquad \Delta\left(y_k, \frac{r_1}{2}\right) \subset \Delta(s_k, r_1)$$

for each k, which implies that Λ is uniformly discrete and moreover that w_k and y_k belong to the same cell Q_k as s_k.

If $\rho(s_k, \Gamma) \geq \frac{\eta}{10}$ and $z_n \in \Gamma$, then

$$\rho(w_k, z_n) \geq \rho(z_n, s_k) - \rho(s_k, w_k) \geq \rho(z_n, s_k) - \rho(y_k, w_k) \geq \frac{\eta}{10} - \frac{\eta}{20} = \frac{\eta}{20}.$$

On the other hand, suppose there is a point $z_n \in \Gamma$ satisfying $\rho(s_k, z_n) < \frac{\eta}{10}$. Then

$$\rho(w_k, z_n) \geq \rho(s_k, w_k) - \rho(z_n, s_k) \geq \frac{\eta}{6} - \frac{\eta}{10} = \frac{\eta}{15},$$

and, if $m \neq n$,

$$\rho(w_k, z_m) \geq \rho(z_n, z_m) - \rho(s_k, z_n) - \rho(s_k, w_k) \geq 2\eta - \frac{\eta}{10} - \frac{\eta}{2} = \frac{7\eta}{5}.$$

One obtains the same estimates when w_k is replaced by y_k. It follows that $\Gamma \cup \Lambda$ is uniformly discrete.

Now define

$$\nu = \sum_{k=1}^{\infty} (\delta_{w_k} + \delta_{y_k}) = \sum_{j=1}^{\infty} \delta_{\lambda_j},$$

where δ_z is the Dirac-delta measure at the point z. For $\alpha \in \mathbb{D}$, consider the function

$$w(\alpha) = \int_{\mathbb{D}} \log|\varphi_\alpha(z)|\, d(\mu(z) - \nu(z)) = \sum_{k=1}^{\infty} \int_{Q_k} \log|\varphi_\alpha(z)|\, d(\mu(z) - \nu(z)).$$

We will show that there is a constant C such that

$$|w(\alpha) + \log \rho(\alpha, \Lambda)| \leq C \tag{33}$$

for all $\alpha \in \mathbb{D}$. To this end, split the integral and write

$$w(\alpha) = \int_{\Delta(\alpha, \frac{1}{2})} \log|\varphi_\alpha(z)|\, d(\mu(z) - \nu(z)) + \int_{\rho(z,\alpha) \geq \frac{1}{2}} \log|\varphi_\alpha(z)|\, d(\mu(z) - \nu(z)).$$

To show that the second term is bounded, we prove that

$$\sup_{\alpha \in \mathbb{D},\, r<1} \left| \int_{\Omega(\alpha, \frac{1}{2}, r)} \log|\varphi_\alpha(z)|\, d(\mu(z) - \nu(z)) \right| < \infty, \tag{34}$$

where $\Omega(\alpha, \frac{1}{2}, r)$ is the pseudohyperbolic annulus $\{z : \frac{1}{2} < \rho(z,\alpha) < r\}$. Denote by Ω_1 the union of all the cells Q_k entirely contained in $\Omega(\alpha, \frac{1}{2}, r)$, and write $\Omega_2 = \Omega(\alpha, \frac{1}{2}, r) \setminus \Omega_1$. The proof that the part of the integral in (34) over Ω_2 is bounded is exactly the same as the corresponding proof in Section 6.6 and will not be repeated here. The proof for Ω_1 requires slight modifications and will now be given.

With $L(z) = \log\left(\frac{z-\alpha}{1-\overline{\alpha}z}\right)$, we can write

$$L(z) = L(s_k) + L'(s_k)(z - s_k) + \int_{s_k}^{z} L''(w)(z-w)\,dw\,.$$

Then

$$\int_{Q_k} \log|\varphi_\alpha(z)|\,d(\mu(z)-\nu(z)) = \mathrm{Re}\Big\{\int_{Q_k}\int_{s_k}^{z} L''(w)(z-w)\,dw\,d\mu(z) - \int_{s_k}^{w_k} L''(w)(w_k-w)\,dw - \int_{s_k}^{y_k} L''(w)(y_k-w)\,dw\Big\}.$$

Here we have used the relations

$$\int_{Q_k} d(\mu(z)-\nu(z)) = 2-2 = 0\,, \qquad \text{and}$$

$$\int_{Q_k} z\,d(\mu(z)-\nu(z)) = 2s_k - (w_k+y_k) = 0\,.$$

As in Chapter 6, we show that

$$\left|\int_{s_k}^{z} L''(w)(z-w)\,dw\right| \le C\,\frac{(1-|\alpha|^2)(1-|z|^2)^2}{|1-\overline{\alpha}z|^3}$$

for $z \in Q_k$. Since $w_k, y_k \in Q_k$, we obtain the same inequality with z replaced by w_k or y_k. Therefore, since $h(z) \simeq 1$,

$$\left|\sum_{Q_k\subset\Omega_1}\int_{Q_k} \log|\varphi_\alpha(z)|\,d(\mu(z)-\nu(z))\right| \le C(1-|\alpha|^2)\sum_{k=1}^{\infty}\int_{Q_k}\frac{d\sigma(z)}{|1-\overline{\alpha}z|^3} + C(1-|\alpha|^2)\sum_{k=1}^{\infty}\frac{(1-|w_k|^2)^2}{|1-\overline{\alpha}w_k|^3} + C(1-|\alpha|^2)\sum_{k=1}^{\infty}\frac{(1-|y_k|^2)^2}{|1-\overline{\alpha}y_k|^3}.$$

The last two sums in this inequality are bounded, by Lemma 3 of Chapter 6. On the other hand, Lemma 1 of Chapter 6 implies that

$$(1-|\alpha|^2)\sum_{k=1}^{\infty}\int_{Q_k}\frac{d\sigma(z)}{|1-\overline{\alpha}z|^3} = (1-|\alpha|^2)\int_{\mathbb{D}}\frac{d\sigma(z)}{|1-\overline{\alpha}z|^3} \le C\,.$$

This shows that the part of the integral in (34) taken over Ω_1 is bounded for $\alpha \in \mathbb{D}$ and $\frac{1}{2} < r < 1$, and so the proof of (34) is complete.

To show that (33) holds, we note first that changing variables yields

$$\int_{\Delta(\alpha,\frac{1}{2})} \log \frac{1}{|\varphi_\alpha(z)|} \frac{d\sigma(z)}{(1-|z|^2)^2} = 2\int_0^{\frac{1}{2}} \frac{s \log s^{-1}}{(1-s^2)^2}\, ds\,.$$

Suppose now that $\rho(\alpha, \Lambda) \geq \frac{1}{2}\delta(\Lambda)$. Then by (34), the estimate (6) on the counting function, and the above,

$$\begin{aligned} \big|w(\alpha)+\log\rho(\alpha,\Lambda)\big| \leq C \int_{\Delta(\alpha,\frac{1}{2})} & \log \frac{1}{|\varphi_\alpha(z)|} \frac{d\sigma(z)}{(1-|z|^2)^2} \\ & + \left| \int_{\rho(\alpha,z)\geq\frac{1}{2}} \log|\varphi_\alpha(z)|\, d(\mu(z)-\nu(z)) \right| \\ & + \sum_{\lambda_j\in\Delta(\alpha,\frac{1}{2})} \log\frac{1}{|\varphi_\alpha(\lambda_j)|} + \log\frac{1}{\rho(\alpha,\Lambda)} \\ \leq C \int_0^{\frac{1}{2}} \frac{s\log s^{-1}}{(1-s^2)^2}\, ds + C + & \big(n(\Lambda,\alpha,\tfrac{1}{2})+1\big)\log\frac{2}{\delta(\Lambda)} \leq C\,. \end{aligned}$$

If $\rho(\alpha, \Lambda) < \frac{1}{2}\delta(\Lambda)$, then $\rho(\alpha,\Lambda) = \rho(\alpha, \lambda_k)$ for some unique k. Consequently,

$$\begin{aligned} w(\alpha) + \log\rho(\alpha,\Lambda) = & \int_{\Delta(\alpha,\frac{1}{2})} \log|\varphi_\alpha(z)|\, d\mu(z) \\ & - \sum_{\lambda_j\in\Delta(\alpha,\frac{1}{2}),\, j\neq k} \log|\varphi_\alpha(\lambda_j)| \\ & + \int_{\rho(\alpha,z)\geq\frac{1}{2}} \log|\varphi_\alpha(z)|\, d(\mu(z)-\nu(z))\,, \end{aligned}$$

and the resulting terms can be estimated as above. This proves (33).

Recall now that $\nu_\Lambda = \sum_j (1-|\lambda_j|^2)^2 \delta_{\lambda_j}$. Writing

$$w = \big(\widetilde{\Delta}\Phi - \tfrac{1}{2}\nu_\Lambda\big) * E\,,$$

we have

$$\widetilde{\Delta} w = \widetilde{\Delta}\Phi - \tfrac{1}{2}\nu_\Lambda \qquad \text{and so} \qquad \widetilde{\Delta}(\Phi - w) = \tfrac{1}{2}\nu_\Lambda\,.$$

By the remarks made earlier in this section, this implies the existence of an analytic function g, with zero-set Λ, such that

$$(\Phi - w)(z) = \log|g(z)|\,.$$

By the estimate (33) on w, we have

$$\big|\log|g(z)| - \Phi(z) - \log\rho(z,\Lambda)\big| \leq C\,.$$

Exponentiating this inequality, we obtain the desired result. $\square$

PROOF OF LEMMA 17. The second inequality is obvious, so we consider the first, which is also clear if $\Lambda_1 \cup \Lambda_2$ is not uniformly discrete. Assume then that $\Lambda_1 \cup \Lambda_2$ is uniformly discrete and note that $\rho(z, \Lambda_1) \geq \frac{1}{2}\rho(\Lambda_1, \Lambda_2)$ or $\rho(z, \Lambda_2) \geq \frac{1}{2}\rho(\Lambda_1, \Lambda_2)$. Without loss of generality, assume the former, so that

$$\frac{\rho(z, \Lambda_1 \cup \Lambda_2)}{\rho(z, \Lambda_1)\rho(z, \Lambda_2)} \leq \frac{1}{\rho(z, \Lambda_1)} \leq \frac{2}{\rho(\Lambda_1, \Lambda_2)},$$

since $\rho(z, \Lambda_1 \cup \Lambda_2) \leq \rho(z, \Lambda_2)$. □

§7.6. Necessity of the interpolation condition.

The following proof is based on work of Schuster and Seip [2] in the setting of the Bargmann–Fock space. Recall that if $\Gamma = \{z_k\}$ is an interpolation sequence for A^p, then there is a constant M such that the interpolation problem $f(z_k) = w_k$ can be solved by a function satisfying

$$\|f\|_p^p \leq M \sum_{k=1}^{\infty} (1 - |z_k|^2)^2 |w_k|^p,$$

and $M(\Gamma)$ is the smallest such constant M. The number $M(\Gamma)$ is called the interpolation constant of Γ for the space A^p.

LEMMA 18. *For $0 < p < \infty$, let Γ be an interpolation sequence for A^p with interpolation constant $M(\Gamma)$. Then Γ is uniformly discrete, with separation constant $\delta(\Gamma) \geq c$, where c is a positive constant depending only on p and $M(\Gamma)$.*

PROOF. Let $\Gamma = \{z_k\}$. For each positive integer n, let $\{w_{nk}\}$ be the sequence defined by

$$w_{nk} = \begin{cases} (1 - |z_n|^2)^{-\frac{2}{p}} & \text{if } k = n, \\ 0 & \text{if } k \neq n. \end{cases}$$

Then there is a function f_n satisfying $f_n(z_k) = w_{nk}$ and

$$\|f_n\|_p^p \leq M(\Gamma) \sum_{k=1}^{\infty} (1 - |z_k|^2)^2 |w_{nk}|^p = M(\Gamma).$$

In view of Lemma 2, it follows that for $m \neq n$,

$$\begin{aligned} 1 &= \left| (1 - |z_m|^2)^{\frac{2}{p}} |f_n(z_m)| - (1 - |z_n|^2)^{\frac{2}{p}} |f_n(z_n)| \right| \\ &\leq C\,\rho(z_m, z_n)\,\|f_n\|_p \leq C\,\rho(z_m, z_n)\,M(\Gamma)^{\frac{1}{p}} \end{aligned}$$

if $\rho(z_m, z_n) < \frac{1}{4}$, where C depends only on p. Thus

$$\rho(z_m, z_n) \geq \min\left\{\frac{1}{4}, \frac{1}{C\,M(\Gamma)^{\frac{1}{p}}}\right\}, \qquad m \neq n,$$

which shows that Γ is uniformly discrete with $\delta(\Gamma) \geq c$. □

In Section 6.3 we showed that if a uniformly discrete set Γ admits an analytic function g for which

$$|g(z)| \simeq \rho(z, \Gamma)(1 - |z|^2)^{-\alpha},$$

then Γ is an interpolation set for A^p if $\alpha < 1/p$ and a sampling set if $\alpha > 1/p$. From this result we deduced, by appeal to Theorems 1 and 2 (the main interpolation and sampling theorems, stated as Theorems 4 and 5 in Chapter 6 but not proved there), that $D^-(\Gamma) = D^+(\Gamma) = \alpha$. We now give a direct proof.

LEMMA 19. *Let $\alpha > 0$. If Γ is uniformly discrete and there is an analytic function g such that*

$$|g(z)| \simeq \rho(z, \Gamma)(1 - |z|^2)^{-\alpha}, \tag{35}$$

then $D^-(\Gamma) = D^+(\Gamma) = \alpha$.

It should be observed that while Lemma 19 is one of the tools that, together with Theorem 1 of Chapter 6, will lead to a proof of Theorem 1 of the present chapter, it also allows for the process to be reversed. Specifically, an application of Lemma 19 followed by an appeal to the main sampling and interpolation theorems gives a proof of Theorems 1 and 2 of Chapter 6 that is valid for $0 < p < \infty$. These latter results then imply Theorem 3 of Chapter 6, for $0 < p < \infty$.

PROOF OF LEMMA. First note that the given property of Γ is Möbius invariant. For fixed $\zeta \in \mathbb{D}$, define the function

$$g_\zeta(z) = g(\varphi_\zeta(z))(\varphi_\zeta'(z))^\alpha,$$

where g is the function in (35). In view of the identity (1) and the Möbius invariance of the pseudohyperbolic metric, the relation (35) implies that

$$C_1\,\rho(z, \varphi_\zeta(\Gamma))(1 - |z|^2)^{-\alpha} \leq |g_\zeta(z)| \leq C_2\,\rho(z, \varphi_\zeta(\Gamma))(1 - |z|^2)^{-\alpha} \tag{36}$$

for some positive constants C_1 and C_2 and all $z, \zeta \in \mathbb{D}$. These inequalities show that the zero-set of g_ζ is precisely $\varphi_\zeta(\Gamma)$. Let B_ζ be the finite Blaschke product with zero-set $\varphi_\zeta(\Gamma) \cap \Delta(0, \frac{1}{2})$, and let $\widetilde{g_\zeta} = g_\zeta / B_\zeta$. Then for $\frac{1}{2} < r < 1$, Jensen's formula gives

$$\frac{1}{2\pi}\int_0^{2\pi} \log|\widetilde{g_\zeta}(re^{i\theta})|\,d\theta = \sum_{\frac{1}{2} < |\varphi_\zeta(z_k)| < r} \log\frac{r}{|\varphi_\zeta(z_k)|} + \log|\widetilde{g_\zeta}(0)|. \tag{37}$$

The inequality (6) provides the bound

$$n\big(\varphi_\zeta(\Gamma), 0, \tfrac{1}{2}\big) \le 2\left(\tfrac{2}{\delta} + 1\right)^2$$

on the counting function, where δ is the separation constant of Γ. In view of Lemma 17, this implies that

$$C_3\, \rho\big(z, \varphi_\zeta(\Gamma) \cap \Delta(0, \tfrac{1}{2})\big) \le |B_\zeta(z)| \le \rho\big(z, \varphi_\zeta(\Gamma) \cap \Delta(0, \tfrac{1}{2})\big)$$

for some constant C_3. Combining these estimates with (36), we arrive at

$$\frac{C_1\, \rho\big(z, \varphi_\zeta(\Gamma)\big)(1-|z|^2)^{-\alpha}}{\rho\big(z, \varphi_\zeta(\Gamma) \cap \Delta(0, \frac{1}{2})\big)} \le |\widetilde{g_\zeta}(z)| \le \frac{C_2\, \rho\big(z, \varphi_\zeta(\Gamma)\big)(1-|z|^2)^{-\alpha}}{C_3\, \rho\big(z, \varphi_\zeta(\Gamma) \cap \Delta(0, \frac{1}{2})\big)}\,.$$

On the other hand,

$$\begin{aligned}\rho\big(z, \varphi_\zeta(\Gamma) \cap \Delta(0, \tfrac{1}{2})\big)\, \rho\big(z, \varphi_\zeta(\Gamma) \setminus \Delta(0, \tfrac{1}{2})\big) &\le \rho\big(z, \varphi_\zeta(\Gamma)\big) \\ &\le \rho\big(z, \varphi_\zeta(\Gamma) \cap \Delta(0, \tfrac{1}{2})\big)\,,\end{aligned}$$

so the above inequalities become

$$C_1\, \rho\big(z, \varphi_\zeta(\Gamma) \setminus \Delta(0, \tfrac{1}{2})\big)(1-|z|^2)^{-\alpha} \le |\widetilde{g_\zeta}(z)| \le \frac{C_2}{C_3}(1-|z|^2)^{-\alpha}\,.$$

In particular,

$$\log \frac{C_1}{2} \le \log |\widetilde{g_\zeta}(0)| \le \log \frac{C_2}{C_3}\,, \tag{38}$$

and so

$$\begin{aligned}&\log C_1 - \alpha \log(1-r^2) + \frac{1}{2\pi}\int_0^{2\pi} \log \rho\big(re^{i\theta}, \varphi_\zeta(\Gamma) \setminus \Delta(0, \tfrac{1}{2})\big)\, d\theta \\ \le &\frac{1}{2\pi}\int_0^{2\pi} \log |\widetilde{g_\zeta}(re^{i\theta})|\, d\theta \le \log \frac{C_2}{C_3} - \alpha \log(1-r^2)\,.\end{aligned}$$

Suppose for the moment that

$$\int_0^{2\pi} \log \rho\big(re^{i\theta}, \varphi_\zeta(\Gamma) \setminus \Delta(0, \tfrac{1}{2})\big)\, d\theta \ge C \tag{39}$$

for some constant C independent of ζ and r. Then

$$\begin{aligned}\log C_1 - \alpha \log(1-r^2) - C &\le \frac{1}{2\pi}\int_0^{2\pi} \log |\widetilde{g_\zeta}(re^{i\theta})|\, d\theta \\ &\le \log \frac{C_2}{C_3} - \alpha \log(1-r^2)\,.\end{aligned} \tag{40}$$

Apply (38) and (40) to (37) to see that

$$\begin{aligned}\log \frac{C_1C_3}{C_2} - C - \alpha \log(1-r^2) &\le \sum_{\frac{1}{2}<|\varphi_\zeta(z_k)|<r} \log \frac{r}{|\varphi_\zeta(z_k)|} \\ &\le \log \frac{2C_2}{C_1C_3} - \alpha \log(1-r^2)\,.\end{aligned}$$

Since Γ is uniformly discrete, it follows from (6) that

$$\begin{aligned}\sum_{\frac{1}{2}<|\varphi_\zeta(z_k)|<r} \log \frac{1}{|\varphi_\zeta(z_k)|} &\le \sum_{\frac{1}{2}<|\varphi_\zeta(z_k)|<r} \log \frac{r}{|\varphi_\zeta(z_k)|} + n(\Gamma,\zeta,r)\log\frac{1}{r} \\ &\le \sum_{\frac{1}{2}<|\varphi_\zeta(z_k)|<r} \log \frac{r}{|\varphi_\zeta(z_k)|} + O(1)\end{aligned}$$

as $r \to 1$. Therefore,

$$B_1 - \alpha\log(1-r^2) \le \sum_{\frac{1}{2}<|\varphi_\zeta(z_k)|<r} \log \frac{1}{|\varphi_\zeta(z_k)|} \le B_2 - \alpha \log(1-r^2)\,.$$

for some constants B_1 and B_2. Now divide by $\log \frac{1}{1-r}$ and let $r \to 1$ to conclude that $D^-(\Gamma) = D^+(\Gamma) = \alpha$.

To prove (39), note first that for any $\lambda \in \mathbb{D}$,

$$\frac{1}{2\pi}\int_0^{2\pi} \log \rho(re^{i\theta},\lambda)\,d\theta = \log \max\{r,|\lambda|\}\,.$$

Write $\varphi_\zeta(\Gamma) \setminus \Delta(0,\frac{1}{2}) = \{\lambda_k\} = \Lambda_1 \cup \Lambda_2$, where

$$\Lambda_1 = \left\{\lambda_k : \tfrac{1}{2} < |\lambda_k| < \frac{r+\frac{1}{2}}{1+\frac{r}{2}}\right\}, \qquad \Lambda_2 = \left\{\lambda_k : \frac{r+\frac{1}{2}}{1+\frac{r}{2}} \le |\lambda_k|\right\}.$$

In view of Lemma 17, it suffices to show that $\int_0^{2\pi} \log\rho(re^{i\theta},\Lambda_j)d\theta$ is bounded below for $j=1,2$. This is obvious for $j=2$, since $\rho(re^{i\theta},\Lambda_2) \ge \frac{1}{2}$. Furthermore,

$$\begin{aligned}\int_0^{2\pi} \log\rho(re^{i\theta},\Lambda_1)\,d\theta &\ge \sum_{\frac{1}{2}<|\lambda_k|<\frac{r+\frac{1}{2}}{1+\frac{r}{2}}} \int_0^{2\pi} \log\rho(re^{i\theta},\lambda_k)\,d\theta \\ &= 2\pi \sum_{\frac{1}{2}<|\lambda_k|<\frac{r+\frac{1}{2}}{1+\frac{r}{2}}} \log\max\{r,|\lambda_k|\} \\ &\ge 2\pi n\left(\varphi_\zeta(\Gamma)\setminus\Delta(0,\tfrac{1}{2}), 0, \frac{r+\frac{1}{2}}{1+\frac{r}{2}}\right)\log r\,,\end{aligned}$$

which is bounded below by virtue of the estimate (6) on the counting function. This proves (39), which completes the proof of the lemma. □

It was shown in Chapter 6 that interpolation sequences Γ and their interpolation constants $M(\Gamma)$ are Möbius invariant. It will be important to know that this invariance is preserved under passage to weak limits.

LEMMA 20. *Let $0 < p < \infty$ and let $M(\Gamma)$ denote the interpolation constant of Γ for the space A^p. If Γ_j are sequences with $\Gamma_j \rightharpoonup \Lambda$, then $M(\Lambda) \le \liminf_{j\to\infty} M(\Gamma_j)$.*

PROOF. We may assume that $\Lambda = \{\lambda_k\}$ is an infinite sequence and that

$$\lim_{j\to\infty} M(\Gamma_j) = L < \infty .$$

Since $\Gamma_j \rightharpoonup \Lambda$, there is a naturally ordered rearrangement $\{z_{jk}\}$ of Γ_j such that $\lim_{j\to\infty} z_{jk} = \lambda_k$ for $k = 1, 2, \ldots$, where $\Lambda = \{\lambda_k\}$. Given a sequence $\{w_k\}$ for which

$$\sum_{k=1}^{\infty}(1-|\lambda_k|^2)^2|w_k|^p < \infty ,$$

define

$$w_{jk} = \frac{(1-|\lambda_k|^2)^{\frac{2}{p}}}{(1-|z_{jk}|^2)^{\frac{2}{p}}} w_k .$$

Then

$$\sum_{k=1}^{\infty}(1-|z_{jk}|^2)^2|w_{jk}|^p = \sum_{k=1}^{\infty}(1-|\lambda_k|^2)^2|w_k|^p < \infty ,$$

and so $f_j(z_{jk}) = w_{jk}$ for some function $f_j \in A^p$ with

$$\|f_j\|_p^p \le M(\Gamma_j)\sum_{k=1}^{\infty}(1-|\lambda_k|^2)^2|w_k|^p < (L+\varepsilon)\sum_{k=1}^{\infty}(1-|\lambda_k|^2)^2|w_k|^p$$

for each $\varepsilon > 0$ and all j sufficiently large. By a normal family argument, a subsequence of $\{f_j\}$ converges uniformly on compact subsets of $\mathbb{D}$ to a function $f \in A^p$. Lemma 12 implies that $f(\lambda_k) = w_k$, and Fatou's lemma gives

$$\|f\|_p^p \le (L+\varepsilon)\sum_{k=1}^{\infty}(1-|\lambda_k|^2)^2|w_k|^p .$$

Thus $M(\Lambda) \le L + \varepsilon$ for each $\varepsilon > 0$, and so $M(\Lambda) \le L$, as claimed. □

Since $M(\varphi_\zeta(\Gamma)) = M(\Gamma)$ for every $\zeta \in \mathbb{D}$, it follows from Lemma 20 that if Γ is an interpolation sequence for A^p, then so is every member of $W(\Gamma)$. In particular, $W(\Gamma)$ can contain only A^p zero-sets. Thus the proof of necessity in Theorem 1 will be completed by applying the following lemma.

LEMMA 21. *Let $0 < p < \infty$. If Γ is uniformly discrete and every element of $W(\Gamma)$ is an A^p zero-set, then $D^+(\Gamma) < \frac{1}{p}$.*

PROOF. Since Γ is uniformly discrete, we know that $D^+(\Gamma) < \infty$. Choose $\alpha > \max\{D^+(\Gamma), \frac{1}{p}\}$. Then by Theorem 5 there exist a sequence Λ and an analytic function f such that $\Gamma \cup \Lambda$ is uniformly discrete and

$$|f(z)| \simeq \rho(z, \Gamma \cup \Lambda)(1-|z|^2)^{-\alpha}.$$

According to Lemma 19, this implies that $D^+(\Gamma \cup \Lambda) = \alpha$. Therefore, by Lemma 7 of Chapter 6,

$$D^+(\Gamma) + D^-(\Lambda) \le D^+(\Gamma \cup \Lambda) = \alpha,$$

so that $D^+(\Gamma) \le \alpha - D^-(\Lambda)$.

We claim now that every element of $W(\Lambda)$ is a set of uniqueness for the growth space $A^{-(\alpha - \frac{1}{p})}$. If so, then by Lemma 13 we can infer that Λ is a sampling sequence for $A^{-(\alpha-\frac{1}{p})}$. Theorem 3 then says that $D^-(\Lambda) > \alpha - \frac{1}{p}$, and so

$$D^+(\Gamma) \le \alpha - D^-(\Lambda) < \alpha - \left(\alpha - \tfrac{1}{p}\right) = \tfrac{1}{p},$$

which is the desired inequality.

Consequently, the proof reduces to showing that every element of $W(\Lambda)$ is a set of uniqueness for $A^{-(\alpha-\frac{1}{p})}$. If not, then for some $\widetilde{\Lambda} \in W(\Lambda)$ there is a nonzero function $h \in A^{-(\alpha-\frac{1}{p})}$ that vanishes on $\widetilde{\Lambda}$. By definition, $\varphi_{\zeta_j}(\Lambda) \rightharpoonup \widetilde{\Lambda}$ for some sequence of points ζ_j in $\mathbb{D}$. In view of Lemma 10, we may assume, by passing to a subsequence if necessary, that $\varphi_{\zeta_j}(\Gamma) \rightharpoonup \widetilde{\Gamma}$ for some sequence $\widetilde{\Gamma}$. Then $\varphi_{\zeta_j}(\Gamma \cup \Lambda) \rightharpoonup \widetilde{\Gamma} \cup \widetilde{\Lambda}$. Now define the functions

$$f_j(z) = f(\varphi_{\zeta_j}(z))\varphi'_{\zeta_j}(z)^\alpha$$

and use the identity (1) to show that

$$C_1\, \rho\big(z, \varphi_{\zeta_j}(\Gamma \cup \Lambda)\big)\big(1-|z|^2\big)^{-\alpha} \le |f_j(z)| \le C_2\, \rho\big(z, \varphi_{\zeta_j}(\Gamma \cup \Lambda)\big)\big(1-|z|^2\big)^{-\alpha}$$

for some constants C_1 and C_2 and for all $z \in \mathbb{D}$ and $j = 1, 2, \dots$. In particular, $\|f_j\|_{-\alpha} \le C_2$. By a normal family argument, some subsequence of $\{f_j\}$ converges locally uniformly to an analytic function F. Since

$$\rho\big(z, \varphi_{\zeta_j}(\Gamma \cup \Lambda)\big) \to \rho\big(z, \widetilde{\Gamma} \cup \widetilde{\Lambda}\big) \qquad \text{for each } z \in \mathbb{D},$$

we conclude that

$$C_1\, \rho\big(z, \widetilde{\Gamma} \cup \widetilde{\Lambda}\big)\big(1-|z|^2\big)^{-\alpha} \le |F(z)| \le C_2\, \rho\big(z, \widetilde{\Gamma} \cup \widetilde{\Lambda}\big)\big(1-|z|^2\big)^{-\alpha}.$$

It was shown earlier (see the remarks following the statements of Theorems 1 and 2 of Chapter 6), that this pair of inequalities implies that $\widetilde{\Gamma} \cup \widetilde{\Lambda}$ is a set of uniqueness for $A^{\frac{1}{\alpha}}$. Strictly speaking, this was proved only for $0 < \alpha < 1$, but the proof in Chapter 6 goes through without change for $\alpha \geq 1$.

By hypothesis, $\widetilde{\Gamma}$ is a zero-set for A^p, so some nonzero function $g \in A^p$ vanishes on $\widetilde{\Gamma}$. The inequality $|g(z)| \leq \|g\|_p(1-|z|^2)^{-\frac{2}{p}}$ shows that

$$|g(z)|^{\frac{1}{\alpha}-p} \leq C\big(1-|z|^2\big)^{2-\frac{2}{p\alpha}} \qquad \text{for all } \ z \in \mathbb{D}\,.$$

Then, since $h \in A^{-(\alpha-\frac{1}{p})}$ and $\alpha > \frac{1}{p}$, it follows that

$$\begin{aligned}
\int_{\mathbb{D}} |g(z)h(z)|^{\frac{1}{\alpha}}\, d\sigma &\leq C \int_{\mathbb{D}} (1-|z|^2)^{\frac{1}{p\alpha}-1} |g(z)|^{\frac{1}{\alpha}-p} |g(z)|^p\, d\sigma \\
&\leq C \int_{\mathbb{D}} (1-|z|^2)^{1-\frac{1}{p\alpha}} |g(z)|^p\, d\sigma \\
&\leq C \int_{\mathbb{D}} |g(z)|^p\, d\sigma < \infty\,.
\end{aligned}$$

Thus gh is a nonzero function in $A^{\frac{1}{\alpha}}$ that vanishes on $\widetilde{\Gamma} \cup \widetilde{\Lambda}$, contradicting the fact that $\widetilde{\Gamma} \cup \widetilde{\Lambda}$ is a set of uniqueness for $A^{\frac{1}{\alpha}}$. Therefore, $\widetilde{\Lambda}$ is a set of uniqueness for $A^{-(\alpha-\frac{1}{p})}$, and the proof of the lemma is complete. □

By combining Theorem 1 with Lemmas 20 and 21, we arrive at the following alternate characterization of interpolation sequences for the Bergman space.

THEOREM 6. *Let $0 < p < \infty$. A sequence Γ is an interpolation sequence for A^p if and only if Γ is uniformly discrete and every element of $W(\Gamma)$ is an A^p zero-set.*

§7.7. Weak interpolation.

Interpolation sequences for the Hardy space were discussed in Section 6.2. Recall that a Blaschke sequence $\Gamma = \{z_k\}$ is an interpolation sequence for H^p if and only if Γ is uniformly separated:

$$\prod_{j \neq k} \left| \frac{z_j - z_k}{1 - \overline{z_j} z_k} \right| \geq \delta\,, \qquad k = 1, 2, \dots\,, \tag{41}$$

for some constant $\delta > 0$. In particular, the H^p interpolation sequences are independent of p.

Interpolation sequences for the Bergman space A^p are defined in a similar way but are characterized by a rather different separation condition that exhibits their dependence on p. According to Theorem 1, a sequence Γ is interpolating for A^p if and only if it is uniformly discrete and $D^+(\Gamma) < \frac{1}{p}$.

Our purpose is now to develop an equivalent condition for interpolation in Bergman spaces that, while less explicit and geometric than the density condition of Theorem 1, provides a direct analogue of the theorem for Hardy spaces. To demonstrate the analogy, we first rewrite the uniform separation condition (41) in slightly different form. For each index n, let $B_n(z)$ denote the Blaschke product associated with the sequence $\varphi_{z_n}(\Gamma \setminus z_n)$, a Möbius transform of the sequence Γ with the point z_n omitted. Then one can say that Γ is uniformly separated if and only if $B_n(0) \geq \delta$ for $n = 1, 2, \dots$. The idea is to formulate a similar separation condition for an A^p zero-set, with canonical divisors playing the role of Blaschke products.

Let $\Gamma = \{z_k\}$ be an A^p zero-set consisting of distinct points z_k. For each index n, let $G_n(z)$ denote the canonical divisor in A^p of the zero-set $\varphi_{z_n}(\Gamma \setminus z_n)$. This function G_n is the unique solution to the extremal problem of maximizing $\operatorname{Re}\{f(0)\}$ among all functions $f \in A^p$ that vanish on the set $\varphi_{z_n}(\Gamma \setminus z_n)$.

THEOREM 7. *Let $0 < p < \infty$. Suppose $\Gamma = \{z_k\}$ is an A^p zero-set with distinct points z_k in $\mathbb{D}$. Let G_n be the canonical divisor in A^p of the zero-set $\varphi_{z_n}(\Gamma \setminus z_n)$. Then Γ is an interpolation sequence for A^p if and only if $G_n(0) \geq \delta$ for some $\delta > 0$ and all $n = 1, 2, \dots$.*

This theorem was first proved by Schuster and Seip [1] by a completely different argument from the one given below, which is taken essentially from their paper [2]. The proof depends on a lemma that says, roughly speaking, that the hypothesis of the theorem is preserved under passage to weak limits.

LEMMA 22. *Let $0 < p < \infty$. If Γ has the property $G_n(0) \geq \delta$ of Theorem 7, then so does every element of $W(\Gamma)$. In particular, every element of $W(\Gamma)$ is an A^p zero-set.*

PROOF OF LEMMA. Suppose first that $\Lambda = \{\lambda_k\}$ is a sequence in $W(\Gamma)$ of the special form $\Lambda = \varphi_\zeta(\Gamma)$ for some $\zeta \in \mathbb{D}$. For $n = 1, 2, \dots$, let H_n be the canonical divisor in A^p of the zero-set $\varphi_{\lambda_n}(\Lambda \setminus \lambda_n)$. Define

$$\widetilde{G_n}(z) = G_n(\gamma_n z), \qquad \text{where } \gamma_n = -\frac{1 - \overline{\zeta} z_n}{1 - \overline{z_n}\zeta},$$

a unimodular constant. Straightforward calculation gives

$$\begin{aligned}\varphi_{\lambda_n}(\Lambda \setminus \lambda_n) &= \{\varphi_{\lambda_n}(\lambda_k) : k \neq n\} = \{\varphi_{\varphi_\zeta(z_n)}(\varphi_\zeta(z_k)) : k \neq n\} \\ &= \{\gamma_n^{-1}\varphi_{z_n}(z_k) : k \neq n\} = \gamma_n^{-1}\varphi_{z_n}(\Gamma \setminus z_n).\end{aligned}$$

This shows that $\widetilde{G_n}$ vanishes on $\varphi_{\lambda_n}(\Lambda \setminus \lambda_n)$, since G_n vanishes on $\varphi_{z_n}(\Gamma \setminus z_n)$. Moreover, $\|\widetilde{G_n}\|_p = \|G_n\|_p = 1$, so $\widetilde{G_n}$ is an admissible candidate for the canonical divisor of $\varphi_{\lambda_n}(\Lambda \setminus \lambda_n)$, which implies that

$$H_n(0) \geq \widetilde{G_n}(0) = G_n(0) \geq \delta.$$

Next suppose that $\Lambda = \{\lambda_k\}$ is an arbitrary element of $W(\Gamma)$. Then there is a sequence $\{\zeta_j\}$ of points in $\mathbb{D}$ such that $\Gamma_j = \varphi_{\zeta_j}(\Gamma) \rightharpoonup \Lambda$. Let $\Gamma_j = \{z_{jk}\}$. By the previous paragraph, $G_n^j(0) \geq \delta$ for all j and all n, where G_n^j denotes the canonical divisor of the A^p zero-set $\varphi_{z_{jn}}(\Gamma_j \setminus z_{jn})$. For each fixed n, a normal family argument shows that a subsequence of $\{G_n^j\}$ converges locally uniformly to an analytic function h_n of norm $\|h_n\|_p \leq 1$. By Lemma 12, this function h_n vanishes on the set $\varphi_{\lambda_n}(\Lambda \setminus \lambda_n)$. Also, since $G_n^j(0) \geq \delta$, it follows that $h_n(0) \geq \delta$. In particular, h_n is not the zero function, so $\varphi_{\lambda_n}(\Lambda \setminus \lambda_n)$ is an A^p zero-set, and it has a canonical divisor H_n. Finally, $H_n(0) \geq h_n(0) \geq \delta$, and the proof of the lemma is complete. □

PROOF OF THEOREM 7. It is not difficult to see that the condition $G_n(0) \geq \delta$ is necessary. For each n, consider the sequence

$$w_{nk} = \begin{cases} (1 - |z_k|^2)^{-\frac{2}{p}} & \text{if } k = n, \\ 0 & \text{if } k \neq n. \end{cases}$$

Define the function

$$f_n(z) = \frac{G_n(\varphi_{z_n}(z))(-\varphi'_{z_n}(z))^{\frac{2}{p}}}{G_n(0)}.$$

Then f_n is an A^p function with $f_n(z_k) = w_{nk}$ for all k. We claim moreover that f_n is the solution of the interpolation problem with smallest norm. To see this, suppose that h_n is any A^p function satisfying $h_n(z_k) = w_{nk}$ for $k = 1, 2, \ldots$. Set

$$g_n(z) = \frac{h_n(\varphi_{z_n}(z))(-\varphi'_{z_n})^{\frac{2}{p}}}{\|h_n\|_p}.$$

Then $\|g_n\|_p = 1$ and g_n vanishes on the zero-set $\varphi_{z_n}(\Gamma \setminus z_n)$, so $g_n(0) \leq G_n(0)$ by the extremal property of the canonical divisor G_n. But $g_n(0) = \|h_n\|_p^{-1}$ and $G_n(0) = \|f_n\|_p^{-1}$, so this inequality implies that $\|h_n\|_p \geq \|f_n\|_p$, as claimed. It now follows from the definition of the interpolation constant $M(\Gamma)$ that $|G_n(0)|^{-p} = \|f_n\|_p^p \leq M(\Gamma)$, and so $G_n(0) \geq M(\Gamma)^{-\frac{1}{p}}$. This proves the necessity of the condition $G_n(0) \geq \delta$.

To prove the sufficiency, we will show that the hypotheses of Theorem 6 are satisfied if $G_n(0) \geq \delta$. By Lemma 2,

$$\begin{aligned} 1 &= \left| (1 - |z_n|^2)^{\frac{2}{p}} |f_n(z_n)| - (1 - |z_k|^2)^{\frac{2}{p}} |f_n(z_k)| \right| \\ &\leq C\,\rho(z_n, z_k)\,\|f_n\|_p = C\,\rho(z_n, z_k)\,G_n(0)^{-1} \\ &\leq C\,\rho(z_n, z_k)\,\delta^{-1} \end{aligned}$$

if $\rho(z_n, z_k) < \frac{1}{4}$, which shows that Γ is uniformly discrete. According to Lemma 22, the condition $G_n(0) \geq \delta$ also implies that every element of $W(\Gamma)$

is an A^p zero-set. Thus it follows from Theorem 6 that Γ is an interpolation sequence for A^p. $\square$

Theorem 7 can be interpreted in terms of uniformly minimal families, as defined by Nikolskii [1] for the Hardy space. For details concerning this interpretation in the setting of the Bergman space, see Krosky and Schuster [1].

For the case $p = 2$, Theorem 7 may be viewed in yet another way. Suppose $\mathcal{H}$ is a Hilbert space of analytic functions on a domain $\Omega \subset \mathbb{C}$, with the additional property that the point evaluation functionals are bounded. Then a reproducing kernel k_ζ exists for each $\zeta \in \Omega$, so that $\langle f, k_\zeta \rangle = f(\zeta)$ for every $f \in \mathcal{H}$. In this setting, a sequence $\Gamma = \{z_n\}$ in Ω is said to be an *interpolation sequence for* $\mathcal{H}$ if for every $\{w_n\}$ satisfying

$$\sum_{n=1}^{\infty} |w_n|^2 \, \|k_{z_n}\|^{-2} < \infty ,$$

there is a function $f \in \mathcal{H}$ with $f(z_n) = w_n$. It may be verified that this definition coincides with our standard definitions of interpolation sequences for both the Bergman space A^2 and the Hardy space H^2.

A sequence Γ is called a *weak interpolation sequence for* $\mathcal{H}$ if for some constant $C > 0$ there is a sequence of functions $f_m \in \mathcal{H}$ satisfying

$$f_m(z_n) = \delta_{mn} \qquad \text{and} \qquad \|f_m\| \le \frac{C}{\|k_{z_m}\|} ,$$

where δ_{mn} is the Kronecker delta. The proof of existence of the interpolation constant $M(\Gamma)$ for the Bergman space extends to this more general setting. In other words, if Γ is interpolating, we can solve the interpolation problem with norm control of the solution, so that an interpolation sequence is in particular a weak interpolation sequence.

It is an interesting problem to determine the class of Hilbert spaces $\mathcal{H}$ with the property that every weak interpolation sequence is interpolating. This property will be called the *weak interpolation property.* An easy argument shows that Γ is a weak interpolation sequence for A^2 if and only if $G_m(0) \ge \delta$ for some $\delta > 0$, where G_m are the canonical divisors defined in Theorem 7. Indeed, if $G_m(0) \ge \delta$, the sequence of functions

$$f_m(z) = \frac{G_m(\varphi_{z_m}(z))(-\varphi'_{z_m}(z))(1 - |z_m|^2)}{G_m(0)}$$

performs the weak interpolation. Conversely, if f_m are solutions to the weak interpolation problem, then

$$g_m(z) = \frac{f_m(\varphi_{z_m}(z))(-\varphi'_{z_m}(z))}{\|f_m\|_2}$$

is a function of unit norm that vanishes on $\varphi_{z_m}(\Gamma \setminus z_m)$ and satisfies $g_m(0) \geq \frac{1}{C}$, which implies that $G_m(0) \geq \delta$.

Essentially the same argument shows that H^2 has the weak interpolation property, in view of the theorem of Shapiro and Shields [1] that describes the interpolation sequences (*cf.* Section 6.2). In regard to Hilbert spaces of entire functions, Schuster and Seip [2] showed that the Bargmann–Fock space has the weak interpolation property while the Paley–Wiener space does not. The book of Seip [7] addresses this problem and contains a wealth of related information about interpolation and sampling in Hilbert spaces of analytic functions.

CHAPTER 8

Invariant Subspaces

In any Banach space B of functions analytic in the unit disk, a (closed) subspace M is said to be *invariant* if $zM \subset M$, where zM denotes the space of functions $zf(z)$ for $f \in M$. An equivalent requirement is that M be preserved under multiplication by polynomials: $Qf \in M$ for every $f \in M$ and every polynomial Q. A basic problem is to describe the invariant subspaces of B and their lattice structure under intersections and "unions". The problem is important both from the standpoint of polynomial approximation and more broadly for its relevance to the spectral theory of operators.

The invariant subspace structure of the Hardy spaces H^p is well understood as a result of pioneering work by Arne Beurling [2] in 1949. In contrast, the invariant subspaces of the Bergman spaces have never been completely described and are known to be extremely complicated. In the last few years, however, some light has begun to emerge. This chapter begins with a brief review of Beurling's theory, which serves as a model. Then we show two ways in which things go wrong for Bergman spaces. First comes the revelation that singular inner functions may generate the whole space. We characterize the cyclic inner functions, which leads us to a general discussion of cyclic elements in Bergman spaces. Then comes the revelation that the invariant subspaces of A^p need not be singly generated, as is shown by an explicit example in A^2. Generalizations to A^p are then considered, and invariant subspaces of arbitrary finite index are constructed.

All of this demonstrates that invariant subspaces of Bergman spaces have a structure quite different from those of Hardy spaces. Nevertheless, we shall see in Chapter 9 that a convincing analogue of Beurling's theorem holds for the invariant subspaces of A^2.

§8.1. Beurling's theory for Hardy spaces.

Beurling [2] found that the invariant subspaces of the Hardy space H^2 have an especially elegant description. His theory extends with little change to H^p for $1 \le p < \infty$. In order to state Beurling's theorem, we need to introduce some additional terminology. In any Banach space B of analytic functions on $\mathbb{D}$, let $[f]$ denote the smallest invariant subspace containing a given function $f \in B$. Equivalently $[f]$ is the closure of the set of polynomial multiples of f; it is called the invariant subspace *generated* by f. More generally, for any subset $E \subset B$, let $[E]$ be the smallest invariant subspace containing E, or the invariant subspace generated by E. An invariant subspace M generated by a single function f is said to be *singly generated* or *cyclic*, and f is called a *generator* of M. A function is said to be *cyclic* if it generates the whole space B.

The intersection $M \cap N$ of two invariant subspaces M and N is again an invariant subspace, as is $M \vee N$, the smallest subspace containing both M and N. Note that $M \vee N = [M \cup N]$. If point-evaluation at each point $z \in \mathbb{D}$ is a bounded linear functional on B, then each function in a cyclic subspace $[f]$ has all of the zeros of the generator f, with the same or higher multiplicity. As a consequence, a cyclic function can have no zeros in $\mathbb{D}$.

Recall now the definitions of inner and outer functions, as given in Section 3.1. For present purposes it is convenient to require that an inner function have the form $\varphi = BS$, where B is a Blaschke product and S is a singular inner function, thus imposing the normalization that the constant factor $e^{i\gamma} = 1$. An inner function ψ is said to *divide* an inner function φ, written $\psi|\varphi$, if the quotient φ/ψ is again an inner function. The *greatest common divisor* of two inner functions φ_1 and φ_2 is the inner function $\varphi = \varphi_1 \wedge \varphi_2$ that divides both φ_1 and φ_2 and is divisible by every other inner function with that property. Thus $\varphi|\varphi_1$ and $\varphi|\varphi_2$; and $\psi|\varphi$ whenever $\psi|\varphi_1$ and $\psi|\varphi_2$. If $\varphi_1 = B_1S_1$ and $\varphi_2 = B_2S_2$, then their greatest common divisor has the form $\varphi = BS$, where B is the Blaschke product formed from the common zeros (with the smaller of the two multiplicities) of φ_1 and φ_2, and S is the singular inner function whose singular measure μ is the greatest common minorant of the singular measures μ_1 and μ_2 of S_1 and S_2. Similarly, the *least common multiple* of φ_1 and φ_2 is the inner function $\varphi = \varphi_1 \vee \varphi_2$ which is divisible by both φ_1 and φ_2 and divides any other inner function with that property. Thus $\varphi_1|\varphi$ and $\varphi_2|\varphi$; and $\varphi|\psi$ whenever $\varphi_1|\psi$ and $\varphi_2|\psi$. The least common multiple of $\varphi_1 = B_1S_1$ and $\varphi_2 = B_2S_2$ is $\varphi = BS$, where B is formed from the union of the zeros (with the larger of the two multiplicities) of φ_1 and φ_2, while S is constructed from the least common majorant of μ_1 and μ_2.

BEURLING'S THEOREM. *For $1 \le p < \infty$, every invariant subspace $M \neq \{0\}$ of H^p is cyclic and has the form $M = [\varphi] = \varphi H^p$ for some inner function φ, uniquely determined up to normalization. If $M = [\varphi]$ and $N = [\psi]$, then $M \vee N = [\varphi \wedge \psi]$ and $M \cap N = [\varphi \vee \psi]$. The cyclic elements of H^p are precisely the outer functions.*

Various consequences of Beurling's theorem can be noted. For a given Blaschke sequence $\{z_k\}$, let M denote the set of functions in H^p that vanish on $\{z_k\}$ with at least the indicated multiplicity. Then M is an invariant subspace, and it is generated by the Blaschke product formed from the zeros $\{z_k\}$. Although singular inner functions do not vanish in $\mathbb{D}$, they can never be cyclic elements of H^p. If φ and ψ are two inner functions, then $[\varphi\psi] \subset [\varphi] \cap [\psi]$.

§8.2. Cyclic inner functions in Bergman spaces.

We shall again use the notation $[f]$ for the subspace generated by a function $f \in A^p$, the closure in A^p of the polynomial multiples of f. Thus $g \in [f]$ if and only if there is a sequence of polynomials Q_n for which $Q_n f \to g$ in A^p norm as $n \to \infty$. A function $f \in A^p$ is said to be *cyclic* for A^p if it generates the whole space: $[f] = A^p$. A cyclic function f cannot have any zeros in $\mathbb{D}$ because every element of $[f]$ vanishes at the zeros of f. The function 1 is cyclic for every A^p space, since the set of polynomials is dense. If $1 \in [f]$, then f is cyclic. In other words, f is cyclic for A^p if and only if $Q_n f \to 1$ for some sequence of polynomials Q_n. More generally, if $g \in [f]$ then $[g] \subset [f]$. Thus if $[f]$ contains any cyclic element, then f is itself cyclic for A^p.

The subspace $[f]$ generated by a function $f \in A^p$ can be described equivalently as the closure of the set of multiples φf by bounded analytic functions φ. In particular, f is cyclic in A^p if and only if its multiples φf by functions $\varphi \in H^\infty$ are dense in A^p. These statements are supported by the following simple fact.

LEMMA 1. *For $f \in A^p$ $(0 < p < \infty)$ and $\varphi \in H^\infty$, there are polynomials Q_n such that $\|Q_n f - \varphi f\|_p \to 0$ as $n \to \infty$. In fact, the polynomials can be chosen so that $\|Q_n\|_\infty \le \|\varphi\|_\infty$.*

PROOF. Let Q_n be the n-th arithmetic mean of the partial sums of the Taylor series of φ. Then $\|Q_n\|_\infty \le \|\varphi\|_\infty$, and $\{Q_n\}$ converges to φ pointwise on $\mathbb{D}$. By the Lebesgue dominated convergence theorem, $\|Q_n f - \varphi f\|_p \to 0$.□

COROLLARY. *If f and g are A^p functions with $f/g \in H^\infty$, then $[f] \subset [g]$. In particular, if φ and ψ are inner functions and $\varphi | \psi$, then $[\psi] \subset [\varphi]$.*

PROOF. By the lemma, the function $f = (f/g)g \in [g]$, so $[f] \subset [g]$. □

Although no inner function can generate a Hardy space H^p, the situation is different for the Bergman spaces A^p. It turns out that a singular inner function is cyclic in A^p if and only if its associated singular measure places no mass on any "thin" subset of the unit circle. Before giving a precise statement, we record a lemma for future reference.

LEMMA 2. *Let $0 < p < \infty$. If S is a singular inner function and $a > 0$, then S is cyclic in A^p if and only if S^a is cyclic.*

PROOF. We need only consider $0 < a < 1$. If S is cyclic, then so is S^a, by the corollary to Lemma 1, since $S^a | S$. Conversely, if S^a is cyclic, then $Q_n S^a \to 1$ in A^p norm for some sequence of polynomials Q_n. Hence $Q_n S^{2a} \to S^a$, so that $S^a \in [S^{2a}]$ and S^{2a} is cyclic. Proceeding inductively, we therefore see that S^{ma} is cyclic for $m = 1, 2, \dots$. But $S | S^{ma}$ when $ma > 1$, so it follows that S is cyclic. □

We are now ready to define the class of "thin" sets relevant to cyclic singular inner functions. Let $K \subset \mathbb{T}$ be a closed set of measure $|K| = 0$, and let $\mathbb{T} \setminus K = \bigcup I_n$ be the canonical representation of its complement as a countable union of disjoint open arcs. Then K is said to be a *Carleson set* if

$$\sum |I_n| \log |I_n| > -\infty .$$

An equivalent formulation is that

$$\int_{\mathbb{T}} \log \rho_K(\zeta)\, |d\zeta| > -\infty , \tag{1}$$

where $\rho_K(\zeta)$ is the distance from ζ to K. Note that every finite subset of $\mathbb{T}$ is a Carleson set. The Cantor "middle-thirds" set (adapted to $\mathbb{T}$) can be shown to be a Carleson set. However, it is easy to construct countable closed subsets of $\mathbb{T}$ that are not Carleson sets. Take for instance the sequence of points $e^{\frac{i}{\log n}}$, for which $|I_n| \sim \frac{1}{n(\log n)^2}$. The criterion (1) shows directly that every closed subset of a Carleson set is itself a Carleson set.

Carleson sets arise as boundary zero-sets of analytic functions that satisfy a Lipschitz condition in $\mathbb{D}$. If a function $f \neq 0$ belongs to the disk algebra $\mathcal{A}$ of functions analytic in $\mathbb{D}$ and continuous in $\overline{\mathbb{D}}$, then its boundary zero-set

$$K = \{\zeta \in \mathbb{D} \;:\; f(\zeta) = 0\}$$

is closed and has measure zero, since (see Section 2.1)

$$\int_{\mathbb{T}} \log |f(\zeta)|\, |d\zeta| > -\infty .$$

Conversely, a classical theorem of Fatou says that every closed set $K \subset \mathbb{T}$ of measure zero is the boundary zero-set of some function $f \in \mathcal{A}$. In fact, f can be chosen to be an outer function. However, if f satisfies a Lipschitz condition

$$|f(z) - f(\zeta)| \leq C\, |z - \zeta|^{\alpha}, \qquad z, \zeta \in \overline{\mathbb{D}},$$

then its boundary zero-set K is a Carleson set. To see this, fix an arbitrary point $\zeta \in \mathbb{T}$ and choose $\zeta^* \in K$ such that $|\zeta - \zeta^*| = \rho_K(\zeta)$, the distance from ζ to K. Then

$$|f(\zeta)| = |f(\zeta) - f(\zeta^*)| \leq C\, |\zeta - \zeta^*|^{\alpha} = C\, \rho_K(\zeta)^{\alpha} ,$$

so ρ_K has the property (1) and K is a Carleson set. This observation is due to Beurling [1]. In the converse direction, Carleson [1] obtained the following theorem.

CARLESON'S THEOREM. *Let $K \subset \mathbb{T}$ be a Carleson set, and let n be any positive integer. Then there is an outer function $F \in C^n(\overline{\mathbb{D}})$ with boundary zero-set $K = \{\zeta \in \mathbb{T} \;:\; F(\zeta) = 0\}$.*

Carleson's theorem was later sharpened and extended to C^{∞} by Taylor and Williams [1] and Novinger [1]. A proof will not be given here.

One consequence of Carleson's theorem is that the union of any pair of Carleson sets is a Carleson set. Indeed, if K and J are Carleson sets and F and G are functions in $C^1(\overline{\mathbb{D}})$ with respective boundary zero-sets K and J, then the product FG is a Lipschitz function with boundary zero-set $K \cup J$, so Beurling's observation (described above) shows that $K \cup J$ is a Carleson set.

It will be convenient to use the notation

$$S_\mu(z) = \exp\left\{ - \int_{\mathbb{T}} \frac{\zeta + z}{\zeta - z} \, d\mu(\zeta) \right\}$$

to indicate the dependence of a singular inner function on its singular measure μ. Our main objective is to prove the following theorem.

THEOREM 1. *Suppose $1 \le p < \infty$. Then a singular inner function S_μ is cyclic in A^p if and only if $\mu(K) = 0$ for every Carleson set $K \subset \mathbb{T}$.*

H. S. Shapiro [3] established the necessity of this condition for cyclicity in 1967, and he showed that some singular inner functions are actually cyclic. Roberts [1] proved the sufficiency of the condition in 1979, and his proof was described in notes by J. H. Shapiro [5], although Roberts' proof did not appear in print until 1985. Meanwhile, Korenblum [3] had given another proof of the sufficiency as an outgrowth of his 1977 paper [2]. Our discussion is based on J. H. Shapiro's notes.

PROOF OF NECESSITY. Under the assumption that $\mu(K) > 0$ for some Carleson set $K \subset \mathbb{T}$, we will show that S_μ is not cyclic in A^p. This will be accomplished by producing a functional $\phi \in (A^p)^*$ that annihilates $[S_\mu]$ but is not the zero functional. For this purpose, recall (*cf.* Section 3.3) that the Taylor coefficients of $f \in A^p$ have the bounds

$$|a_n| \le C \, \|f\|_p \, n^{\frac{1}{p}}, \qquad n = 1, 2, \dots$$

for some constant $C > 0$. Thus for any complex sequence $\{b_n\}$ with the growth property $b_n = O(n^{-\lambda})$ for some $\lambda > \frac{1}{p} + 1$, the sum

$$\phi(f) = \sum_{n=0}^{\infty} a_n \overline{b_n}$$

will define a bounded linear functional on A^p. This will be true, in particular, if the numbers b_n arise from the Fourier expansion

$$g(\zeta) = \sum_{n=-\infty}^{\infty} b_n \zeta^n$$

of a function $g \in C^3(\mathbb{T})$, because then $b_n = O(|n|^{-3})$.

The next step is to produce a suitable function g. Suppose first that the support of the singular measure μ is actually contained in the Carleson set K. In view of Carleson's theorem (take a power of the outer function it offers), there is an outer function $F \in C^6(\overline{\mathbb{D}})$ such that

$$K \subset \{\zeta \in \mathbb{T} \,:\, F^{(j)}(\zeta) = 0\} \qquad \text{for } 0 \le j \le 6\,.$$

Then by Taylor's theorem,

$$F^{(j)}(\zeta) = O(\rho_K(\zeta)^{6-j}) \qquad \text{as } \zeta \to K\,,\ 0 \le j \le 6\,.$$

This shows that the function

$$g(\zeta) = \begin{cases} \overline{\zeta F(\zeta)}\,, & \zeta \in \mathbb{T} \setminus K \\ 0\,, & \zeta \in K \end{cases}$$

has the required property $g \in C^3(\mathbb{T})$ and hence defines a bounded linear functional ϕ on A^p. The assumption that the support of μ lies in K guarantees that S_μ has an analytic continuation across every arc of $\mathbb{T} \setminus K$.

With this choice of g, we can now show that $\phi([S_\mu]) = 0$. Indeed, for any polynomial Q, we have

$$\phi(QS_\mu) = \frac{1}{2\pi}\int_{\mathbb{T}} Q(\zeta)\, S_\mu(\zeta)\, \overline{g(\zeta)}\, |d\zeta| = \frac{1}{2\pi}\int_{\mathbb{T}} \zeta\, Q(\zeta)\, F(\zeta)\, |d\zeta| = 0\,,$$

since F is analytic in $\mathbb{D}$, and $|S_\mu(\zeta)| = 1$ a.e. on $\mathbb{T}$. Thus $\phi([S_\mu]) = 0$. To show that $\phi \neq 0$, suppose on the contrary that it annihilates every polynomial. Then in particular,

$$0 = \phi(z^n) = \frac{1}{2\pi}\int_{\mathbb{T}} \zeta^{n+1} F(\zeta)\, \overline{S_\mu(\zeta)}\, |d\zeta|\,, \qquad n = 0, 1, 2, \dots\,.$$

It follows that $F(\zeta)\overline{S_\mu(\zeta)} = h(\zeta)$ a.e. on $\mathbb{T}$, for some $h \in H^\infty$. But $|S_\mu(\zeta)| = 1$ a.e. on $\mathbb{T}$, so this implies that $F = S_\mu h$, violating the construction of F as an outer function. Thus $\phi(f) \neq 0$ for some $f \in A^p$ and we have proved that $[S_\mu] \neq A^p$ under the assumption that μ is supported on a Carleson set K.

Finally, if $\mu(K) > 0$, we can write $\mu = \nu + \sigma$, where ν and σ are positive singular measures and ν is supported on K. Then $[S_\mu] \subset [S_\nu] \neq A^p$. This completes the proof of the necessity part of Theorem 1. □

The proof of sufficiency is more difficult. Here it is supposed that μ places no mass on any Carleson set, and the aim is to show that S_μ is cyclic. Roberts' proof accomplishes this by applying a quantitative form of the corona theorem to estimate the distance from 1 to $[S_\mu]$. The estimates show in particular that $1 \in [S_\mu]$, and hence that S_μ is cyclic, whenever μ has modulus of continuity $\omega(t) = O(t \log \frac{1}{t})$, a result previously found by H. S. Shapiro [3]. But it is not difficult to construct singular nondecreasing functions with this degree of smoothness. For instance, it can be done with Riesz products or with the Lebesgue function over a suitable Cantor set with variable ratio of dissection (*cf.* Duren [1]). In this way, explicit examples of cyclic singular inner functions can be produced.

The proof is carried out in two main steps. First it is shown that S_μ is cyclic if μ is sufficiently smooth, or more generally if it is "smoothly decomposable" in a sense to be made precise. The proof is then completed by showing that μ is smoothly decomposable if it puts no mass on any Carleson set.

We begin with a lemma due to H. S. Shapiro [1] that gives a lower bound on $|S_\mu(z)|$ in terms of the modulus of continuity of μ. The *modulus of continuity* of a measure μ on $\mathbb{T}$ is defined by

$$\omega(\delta) = \sup\{\mu(I) \,:\, I \subset \mathbb{T},\ |I| < \delta\}, \qquad 0 < \delta < 2\pi,$$

where I is a subarc of the unit circle and $|I|$ is its length. Here it is not actually required that μ be a singular measure.

LEMMA 3. *If the measure μ has modulus of continuity ω, then for each r in the interval $\frac{1}{4} \le r < 1$, the function S_μ satisfies*

$$|S_\mu(z)| \ge \exp\left\{-\frac{50\,\omega(1-r)}{1-r}\right\}, \qquad |z| \le r.$$

In particular, if $\omega(\delta) \le C\,\delta \log\frac{1}{\delta}$ for some δ with $0 < \delta \le \frac{3}{4}$, then

$$|S_\mu(z)| \ge \delta^{50\,C}, \qquad |z| \le 1-\delta.$$

PROOF. Since $|S_\mu(z)| = \exp\{-P(z)\}$, where

$$P(re^{i\theta}) = \int_{-\pi}^{\pi} \frac{1-r^2}{1-2r\cos(\theta-t)+r^2}\,d\mu(t)$$

is the Poisson-Stieltjes integral of μ, it is enough to show that

$$P(r) \le \frac{50\,\omega(1-r)}{1-r}, \qquad \tfrac{1}{4} \le r < 1.$$

For this purpose we choose an integer n such that $\frac{\pi}{n+1} \le 1-r < \frac{\pi}{n}$ and divide the interval $(-\pi,\pi)$ into sets

$$I_k = \left\{t \,:\, \frac{k\pi}{n+1} \le |t| < \frac{(k+1)\pi}{n+1}\right\}, \qquad k = 0, 1, \dots, n.$$

The Poisson kernel is bounded by

$$0 < \frac{1-r^2}{1-2r\cos t + r^2} \le \frac{2(1-r)}{(1-r)^2 + (t/\pi)^2}, \qquad |t| \le \pi,$$

since $\frac{1}{4} \le r < 1$ and

$$1 - 2r\cos t + r^2 = (1-r)^2 + 4r\sin^2\tfrac{t}{2} \ge (1-r)^2 + \frac{4rt^2}{\pi^2}\,.$$

Thus for $t \in I_k$ the Poisson kernel is bounded by

$$\frac{2(1-r)}{(1-r)^2 + \left(\frac{k}{n+1}\right)^2} < \frac{2}{1-r}\,\frac{1}{1+\frac{k^2}{4\pi^2}}\,,$$

and it follows that

$$\begin{aligned} P(r) &= \sum_{k=0}^{n} \int_{I_k} \frac{1-r^2}{1-2r\cos t + r^2}\,d\mu(t) \le \frac{2}{1-r}\sum_{k=0}^{n}\frac{\mu(I_k)}{1+\frac{k^2}{4\pi^2}} \\ &\le \frac{4\,\omega(\frac{\pi}{n+1})}{1-r}\sum_{k=0}^{n}\frac{1}{1+\frac{k^2}{4\pi^2}} < \frac{4\,\omega(1-r)}{1-r}\left\{1 + \int_0^\infty \frac{dx}{1+\frac{x^2}{4\pi^2}}\right\} \\ &= \frac{4(1+\pi^2)\omega(1-r)}{1-r} < \frac{50\,\omega(1-r)}{1-r}\,. \end{aligned}$$

The first estimate in Lemma 3 now follows for $|z| = r$. But since $S_\mu(z) \ne 0$ in $\mathbb{D}$, the bound extends to $|z| \le r$ by the maximum modulus theorem. □

We will require a quantitative form of the corona theorem, due to Carleson [3], which may be found in Duren [5], p. 205. Versions of Thomas Wolff's relatively simple (unpublished) proof of the corona theorem appear in Garnett [1] and Koosis [1], also in quantitative form.

Corona Theorem. *Let $f_1, f_2, \dots, f_n$ be functions in H^∞ with norms $\|f_k\|_\infty \le 1$ for $k = 1, 2, \dots, n$, and*

$$|f_1(z)| + |f_2(z)| + \cdots + |f_n(z)| \ge \delta\,, \qquad z \in \mathbb{D}\,,$$

where $0 < \delta < \frac{1}{2}$. Then there exist functions $g_1, g_2, \dots, g_n$ in H^∞ such that

$$f_1(z)g_1(z) + f_2(z)g_2(z) + \cdots + f_n(z)g_n(z) \equiv 1\,, \qquad z \in \mathbb{D}\,,$$

and $\|g_k\|_\infty \le \delta^{-\alpha}$, where $\alpha > 0$ is a constant depending only on n.

A simple estimate for the A^p norms of the functions z^n will also be useful. Straightforward calculations give

$$\|z^n\|_p = \left\{\frac{2}{np+2}\right\}^{1/p} \le n^{-\beta}\,, \qquad n = 2, 3, \dots, \tag{2}$$

for some constant $\beta > 0$ depending only on p. In terms of β and the number α appearing in the corona theorem, we now define the two constants

$$C_0 = \frac{\beta}{150\,\alpha} \qquad \text{and} \qquad N_0 = \max\{2, 4^{1/(50\,C_0)}\}\,.$$

We are now prepared to show that S_μ is cyclic in A^p if its associated measure μ is sufficiently smooth.

LEMMA 4. *Let ω be the modulus of continuity of a singular measure μ. For some integer $n > N_0$, suppose that*

$$\omega\left(\tfrac{1}{n}\right) \leq C_0 \frac{\log n}{n}.$$

Then there is a function $g \in H^\infty$ with norm $\|g\|_\infty \leq n^{\beta/3}$ such that

$$\|1 - gS_\mu\|_p \leq n^{-2\beta/3}.$$

COROLLARY. *If $\omega(\delta) = O\left(\delta \log \frac{1}{\delta}\right)$ as $\delta \to 0$, then S_μ is cyclic in A^p.*

DEDUCTION OF COROLLARY. In view of Lemma 2, we may suppose that $\omega(\delta) \leq C_0\, \delta \log \frac{1}{\delta}$ for all $\delta > 0$ sufficiently small. Then we infer from Lemma 4 that $g_n S_\mu \to 1$ in A^p norm for some sequence of functions $g_n \in H^\infty$. But $g_n S_\mu \in [S_\mu]$ by Lemma 1, so this shows that $1 \in [S_\mu]$, and S_μ is cyclic. $\square$

PROOF OF LEMMA. The strategy is to find a lower bound for the sum $|S_\mu(z)| + |z^n|$, then to apply the corona theorem to these two functions. Lemma 3 gives

$$|S_\mu(z)| \geq n^{-50\, C_0}, \qquad |z| \leq 1 - \tfrac{1}{n}.$$

On the other hand, for $1 - \frac{1}{n} \leq |z| < 1$ we have

$$|z^n| \geq \left(1 - \tfrac{1}{n}\right)^n \geq \left(1 - \tfrac{1}{2}\right)^2 \geq n^{-50\, C_0},$$

since $n \geq N_0$. Consequently,

$$|S_\mu(z)| + |z^n| \geq n^{-50\, C_0}, \qquad z \in \mathbb{D}.$$

We now apply the corona theorem with $\delta = n^{-50\, C_0}$. Note that $\delta < \frac{1}{2}$ because $n > N_0$. Thus

$$S_\mu(z) g_1(z) + z^n g_2(z) \equiv 1, \qquad z \in \mathbb{D},$$

for some functions $g_1, g_2 \in H^\infty$ with norms $\|g_k\|_\infty \leq n^{50\,\alpha C_0} = n^{\beta/3}$. It follows that

$$\|1 - g_1 S_\mu\|_p = \|z^n g_2\|_p \leq \|z^n\|_p \|g_2\|_\infty \leq n^{-2\beta/3},$$

by (2). This proves Lemma 4. $\square$

The next step is to estimate the distance

$$d(1, [S_\mu]) = \inf_{g \in [S_\mu]} \|1 - g\|_p$$

from 1 to $[S_\mu]$ when μ is a sum of singular measures with the property of Lemma 4. For this purpose we will need to know that the distance function is continuous in the following sense.

LEMMA 5. *Let $\{\mu_j\}$ be a sequence of singular measures that increases to a singular measure μ as $j \to \infty$. Then for each $f \in A^p$, the distance $d_j = d(f, [S_{\mu_j}])$ increases to $d = d(f, [S_\mu])$.*

PROOF. By the corollary to Lemma 1, we see that $[S_{\mu_j}] \supset [S_{\mu_{j+1}}] \supset [S_\mu]$, since $S_{\mu_j} | S_{\mu_{j+1}}$ and $S_{\mu_{j+1}} | S_\mu$. Consequently, $d_j \le d_{j+1} \le d$, and the sequence $\{d_j\}$ converges to a limit no larger than d. To show that $d_j \to d$, let $\{\varepsilon_j\}$ be an arbitrary sequence of positive numbers with $\varepsilon_j \to 0$, and choose polynomials Q_j such that

$$\|f - Q_j S_{\mu_j}\|_p < d_j + \varepsilon_j \,.$$

Write $\mu = \mu_j + \nu_j$, so that ν_j is a singular measure and

$$\|f S_{\nu_j} - Q_j S_\mu\|_p = \|(f - Q_j S_{\mu_j}) S_{\nu_j}\|_p \le \|f - Q_j S_{\mu_j}\|_p < d_j + \varepsilon_j \,.$$

Then

$$\begin{aligned} d \le \|f - Q_j S_\mu\|_p &\le \|f - f S_{\nu_j}\|_p + \|f S_{\nu_j} - Q_j S_\mu\|_p \\ &\le \|(1 - S_{\nu_j}) f\|_p + d_j + \varepsilon_j \,. \end{aligned}$$

But the measures ν_j decrease to 0, so $S_{\nu_j}(z) \to 1$ for each point $z \in \mathbb{D}$, and so $\|(1 - S_{\nu_j}) f\|_p \to 0$ by the Lebesgue dominated convergence theorem. Thus $d \le \lim_{j\to\infty} d_j$, and the lemma is proved. □

We can now generalize the distance estimate of Lemma 4. If $\{n_j\}$ is any finite or infinite sequence of positive integers, we define

$$D = D[\{n_j\}] = {n_1}^{-2\beta/3} + \sum_{j \ge 2} \left(\frac{n_1 n_2 \cdots n_{j-1}}{n_j^2} \right)^{\beta/3} ,$$

where β is the constant in (2). For the moment there is no assumption about convergence of the infinite series, but later we will choose the sequence $\{n_j\}$ to make D arbitrarily small. If $D = \infty$, the following lemma is true by default.

LEMMA 6. *Suppose a singular measure μ can be represented as a finite or infinite sum $\mu = \Sigma\, \mu_j$ of singular measures μ_j with moduli of continuity ω_j satisfying*

$$\omega_j\left(\tfrac{1}{n_j}\right) \le C_0 \, \frac{\log n_j}{n_j} \tag{3}$$

for some sequence of integers $n_j > N_0$. Then

$$d(1, [S_\mu]) \le D[\{n_j\}] \,.$$

PROOF. Suppose first that the sum is finite, so that $\mu = \mu_1 + ... + \mu_m$ for some m and some measures μ_j with the property (3). We need to show that

$$\|1 - gS_\mu\|_p \le D[\{n_1, \dots, n_m\}] \tag{4}$$

for some $g \in H^\infty$. If $m = 1$, the result follows from Lemma 4. The proof for $m > 1$ will proceed by induction. Suppose that (4) is true for some m and for all measures representable as a sum of m measures satisfying (3), and let μ have an admissible decomposition of the form $\mu = \mu_1 + \mu_2 + \cdots + \mu_{m+1}$. By Lemma 4,

$$\|1 - g_1 S_{\mu_1}\|_p \le {n_1}^{-2\beta/3}$$

for some $g_1 \in H^\infty$ with $\|g_1\|_\infty \le {n_1}^{\beta/3}$. By the inductive hypothesis,

$$\|1 - gS_{\mu_2+\cdots+\mu_{m+1}}\|_p \le D[\{n_2, \dots, n_{m+1}\}]$$

for some $g \in H^\infty$. It follows that

$$\begin{aligned}\|1 - g_1 g S_\mu\|_p &= \|(g_1 S_{\mu_1})(1 - gS_{\mu_2+\cdots+\mu_{m+1}}) + (1 - g_1 S_{\mu_1})\|_p \\ &\le \|g_1 S_{\mu_1}\|_\infty \|1 - gS_{\mu_2+\cdots+\mu_{m+1}}\|_p + \|1 - g_1 S_{\mu_1}\|_p \\ &\le {n_1}^{\beta/3} D[\{n_2, \dots, n_{m+1}\}] + {n_1}^{-2\beta/3} = D[\{n_1, n_2, \dots, n_{m+1}\}] .\end{aligned}$$

This completes the inductive step and gives the desired result for finite sums.

Next suppose that the sum is infinite, so that $\mu = \sum_{j=1}^\infty \mu_j$ for some measures μ_j satisfying (3). Let $\nu_m = \mu_1 + \cdots + \mu_m$ be the partial sums. Then ν_m increases to μ as $m \to \infty$. By Lemma 5 and the inequality just proved for finite sums,

$$d(1, [S_\mu]) = \lim_{m\to\infty} d(1, [S_{\nu_m}]) \le \lim_{m\to\infty} D[\{n_1, \dots, n_m\}] = D[\{n_1, n_2, \dots\}] .$$

This completes the proof of Lemma 6. □

Motivated by Lemma 6, we say that a singular measure μ is *smoothly decomposable* if for each $\varepsilon > 0$ it can be represented as a finite or infinite sum $\mu = \sum \mu_j$, where each μ_j is a singular measure satisfying (3) for some integer $n_j > N_0$, and $D[\{n_j\}] < \varepsilon$. It is clear from Lemma 6 that if μ is smoothly decomposable, then $1 \in [S_\mu]$ and therefore S_μ is cyclic. Thus the proof of the sufficiency in Theorem 1 reduces to showing that a singular measure is smoothly decomposable if it places no mass on any Carleson set.

PROOF OF SUFFICIENCY. In view of Lemma 6, it is enough to prove that if a singular measure μ is *not* smoothly decomposable, then $\mu(K) > 0$ for some Carleson set K.

If μ is not smoothly decomposable, then there exists an $\varepsilon > 0$ such that for every decomposition $\mu = \sum \mu_j$ of μ into a finite or infinite sum of singular measures μ_j and for every choice of integers $n_j > N_0$ with $D[\{n_j\}] < \varepsilon$, some measure μ_j fails to satisfy (3). For a sufficiently large positive integer a, the increasing sequence defined by

$$n_j = 2^{2^{a+j}}, \qquad j = 1, 2, \dots,$$

will satisfy the requirements $n_j > N_0$ and $D[\{n_j\}] < \varepsilon$, since

$$n_1 n_2 \cdots n_{j-1} < n_j .$$

Note also that n_j divides n_{j+1}.

With respect to this sequence $\{n_j\}$, let $\mathcal{P}_j$ be a partition of the unit circle $\mathbb{T}$ into n_j *closed* arcs of equal length $2\pi/n_j$, for $j = 1, 2, \dots$. Proceeding inductively, construct $\mathcal{P}_{j+1}$ to be a refinement of $\mathcal{P}_j$. This is possible because n_j divides n_{j+1}. An arc I of the partition $\mathcal{P}_1$ is said to be *light* if

$$\mu(I) \le \tfrac{1}{2}\, C_0 \frac{\log n_1}{n_1},$$

and *heavy* otherwise. If E is any Borel subset of I, define

$$\mu_1(E) = \begin{cases} \mu(E) & \text{if } I \text{ is light,} \\ \frac{\mu(E)}{\mu(I)} \frac{1}{2}\, C_0 \frac{\log n_1}{n_1} & \text{if } I \text{ is heavy.} \end{cases}$$

Then μ_1 is a singular measure dominated by μ, the measure $\mu - \mu_1$ is supported on the union H_1 of the heavy arcs of $\mathcal{P}_1$, and

$$\mu_1(I) = \tfrac{1}{2}\, C_0 \frac{\log n_1}{n_1}$$

for each such heavy arc I. Furthermore, the modulus of continuity ω_1 of μ_1 satisfies

$$\omega_1\big(\tfrac{1}{n_1}\big) \le C_0 \frac{\log n_1}{n_1}.$$

To see this, observe that any arc J of length $|J| \le 1/n_1$ intersects at most 2 (adjacent) arcs I_1 and I_2 of $\mathcal{P}_1$, since these arcs have common length $2\pi/n_1 > 1/n_1$. Thus

$$\mu_1(J) \le \mu_1(I_1) + \mu_1(I_2) \le 2\Big(\tfrac{1}{2}\, C_0 \frac{\log n_1}{n_1}\Big) = C_0 \frac{\log n_1}{n_1},$$

and the bound for $\omega_1(1/n_1)$ follows. If $\mu_1 = \mu$, then μ will have been "decomposed" as a sum of one term with $n_1 > N_0$ and $D < \varepsilon$, contrary to assumption. Thus $\mu - \mu_1$ is not the zero measure. In particular, H_1 is not empty.

The process is now iterated. Having defined the singular measures μ_j satisfying (3) for $j = 1, 2, \dots, m-1$, define μ_m in the same way that μ_1 was defined, but with respect to $\mathcal{P}_m$ and n_m instead of $\mathcal{P}_1$ and n_1, and with respect to the singular measure $\mu - (\mu_1 + \cdots + \mu_{m-1})$ instead of μ. Let H_m be the union of the heavy arcs of the partition $\mathcal{P}_m$. Then we see as above that $\mu - (\mu_1 + \cdots + \mu_m)$ is a singular measure supported on H_m. This shows in particular that $H_1 \supset H_2 \supset \cdots$. The construction also gives

$$\mu_m(I) = \tfrac{1}{2}\, C_0 \frac{\log n_m}{n_m}$$

for each heavy arc I of $\mathcal{P}_m$, and the modulus of continuity ω_m of μ_m satisfies (3) for $j = m$. Thus it follows as before that $\mu - (\mu_1 + \cdots + \mu_m) \neq 0$, since μ is not smoothly decomposable with respect to our choice of ε. In particular, the set H_m is not empty.

Now define the set

$$K = \bigcap_{m=1}^{\infty} H_m \,.$$

Then K is a nonempty closed subset of $\mathbb{T}$, since $H_1 \supset H_2 \supset \cdots$ and each set H_m is a nonempty finite union of closed arcs. Let $\nu = \sum_{j=1}^{\infty} \mu_j$, and let $\nu_m = \sum_{j=1}^{m} \mu_j$. Because $\mu - \nu_m$ is supported on H_m, we see that $\mu - \nu$ is supported on K. Since $\mu - \nu \neq 0$, this implies that $\mu(K) > \nu(K) \geq 0$.

Next we show that K has Lebesgue measure $|K| = 0$. By construction, $|K| = \lim_{j\to\infty} |H_j|$. For each heavy arc I in the partition $\mathcal{P}_j$, we have

$$\mu_j(I) = \tfrac{1}{2}\, C_0 \frac{\log n_j}{n_j} = \frac{1}{4\pi}\, C_0\, |I| \log n_j \,,$$

hence

$$\frac{1}{4\pi}\, C_0\, |H_j| \log n_j = \mu_j(H_j) \leq \mu_j(\mathbb{T}) \,.$$

Consequently,

$$\frac{1}{4\pi}\, C_0 \sum_{j=1}^{\infty} |H_j| \log n_j = \sum_{j=1}^{\infty} \mu_j(H_j) \leq \sum_{j=1}^{\infty} \mu_j(\mathbb{T}) = \nu(\mathbb{T}) < \infty \,,$$

which implies that $|H_j| \to 0$ and so $|K| = 0$.

Finally, we will show that K is a Carleson set. Let L_j be the union of the *interiors* of the light intervals of $\mathcal{P}_j$ that lie in H_{j-1}. The sets L_j are disjoint, by construction of the measures μ_j, and each is a finite disjoint union of open arcs. Therefore,

$$K' = \mathbb{T} \setminus \left\{ \bigcup_{j=1}^{\infty} L_j \right\}$$

is a closed set containing K. But a point $\zeta \in \mathbb{T}$ belongs to $K' \setminus K$ if and only if it is a common endpoint of adjacent light intervals of some partition

$\mathcal{P}_j$, so $K' \setminus K$ is countable. Thus $|K'| = 0$, since $|K| = 0$. To show that K' is a Carleson set, we need to verify that

$$\sum_{j=1}^{\infty} |L_j| \log\left(\frac{2\pi}{n_j}\right) > -\infty ,$$

or equivalently that $\sum |L_j| \log n_j < \infty$. But $L_j \subset H_{j-1}$, so it will suffice to show that $\sum |H_{j-1}| \log n_j < \infty$. But we showed above that $\sum |H_j| \log n_j < \infty$, and it is easily seen from the definition of n_j that $\log n_j = 2 \log n_{j-1}$. This proves that K' is a Carleson set. To infer that K is a Carleson set, we need only recall that every closed subset of a Carleson set is a Carleson set, as observed following the alternate definition (1).

In summary, our assumption that μ is not smoothly decomposable has allowed us to construct a Carleson set K for which $\mu(K) > 0$. This concludes the proof of Theorem 1. □

§8.3. Cyclic elements of Bergman spaces.

We turn now to a more general discussion of cyclic elements in the Bergman space A^p, for $0 < p < \infty$. If $0 < p < q < \infty$, any cyclic element of A^q is also cyclic in A^p. This is because $A^q \subset A^p$, and in fact $\|f\|_p \le \|f\|_q$ for all $f \in A^q$, so that $Q_n f \to 1$ in A^p norm whenever $Q_n f \to 1$ in A^q norm. Similarly, because $H^p \subset A^{2p}$ with $\|f\|_{2p} \le \|f\|_{H^p}$ (see Section 3.2), any cyclic element of H^p is also cyclic in A^{2p}, for $0 < p < \infty$. But the cyclic elements of H^p are known to be precisely the outer functions for H^p, the functions of the form

$$F(z) = e^{i\gamma} \exp\left\{ \frac{1}{2\pi} \int_0^{2\pi} \frac{e^{it} + z}{e^{it} - z} \log \psi(t)\, dt \right\},$$

where $\gamma \in \mathbb{R}$, $\psi(t) \ge 0$, $\log \psi(t) \in L^1$, and $\psi(t) \in L^p$. Thus every such function is also cyclic for A^{2p}. More generally, we can say that a function $f \in H^p$ is cyclic in A^{2p} if and only if it is nonvanishing and its singular inner factor is cyclic. This is a consequence of the following theorem (*cf.* Brown and Shields [1] and Shields [2]).

THEOREM 2. *For $0 < p < \infty$, suppose $f \in A^p$ and $\varphi \in H^\infty$. Then*

(*i*) $[\varphi f] \subset [f] \cap [\varphi]$.

(*ii*) *If f is cyclic, then $[\varphi f] = [\varphi]$.*

(*iii*) *φf is cyclic if and only if both f and φ are cyclic.*

PROOF. (*i*). By Lemma 1 of Section 2, $\varphi f \in [f]$, so $[\varphi f] \subset [f]$. Also $\varphi f \in [\varphi]$, since $\|Q_n - f\|_p \to 0$ for some polynomials Q_n, and so $\|Q_n \varphi - \varphi f\|_p \to 0$. Thus $[\varphi f] \subset [\varphi]$, and so $[\varphi f] \subset [f] \cap [\varphi]$.

(ii). If f is cyclic, then $\|Q_n f - 1\|_p \to 0$ for some polynomials Q_n. Hence $\|Q_n \varphi f - \varphi\|_p \to 0$, which shows that $\varphi \in [\varphi f]$, so that $[\varphi] \subset [\varphi f]$. The reverse inclusion is part of *(i)*.

(iii). If f and φ are cyclic, then $[\varphi f] = [\varphi] = A^p$, by *(ii)*. Thus φf is cyclic. Conversely, if φf is cyclic, then $[\varphi f] = A^p$, so it follows from *(i)* that $[f] = [\varphi] = A^p$. □

COROLLARY 1. *For $0 < p < \infty$, a function $f \in H^{p/2}$ is cyclic in A^p if and only if $f = SF$, where F is an outer function for $H^{p/2}$ and S is a singular inner function whose associated singular measure μ places no mass on any Carleson set.*

PROOF. If $f \in H^{p/2}$ is cyclic in A^p, it cannot vanish anywhere in $\mathbb{D}$, so it has a canonical factorization $f = SF$, where S is a singular inner function and F is an outer function. As noted above, F is cyclic in $H^{p/2}$ and so is cyclic in A^p. Theorem 1 says that S is cyclic in A^p if and only if $\mu(K) = 0$ for every Carleson set $K \subset \mathbb{T}$. But by Theorem 2, SF is cyclic if and only if both S and F are cyclic. □

COROLLARY 2. *If f and g belong to some space A^p $(0 < p < \infty)$, if g is cyclic, and if $|f(z)| \geq |g(z)|$ for all $z \in \mathbb{D}$, then f is cyclic. In particular, if $f \in A^p$ and $|f(z)| \geq c > 0$ in $\mathbb{D}$, then f is cyclic.*

PROOF. The function $\varphi = g/f$ belongs to H^∞, so

$$A^p = [g] = [\varphi f] \subset [f],$$

and therefore f is cyclic. The second statement results from the fact that any nonzero constant function is cyclic. □

Allen Shields conjectured in 1974 that if f and $1/f$ lie in A^p, then f is cyclic in A^p. This is true under the extra assumption that $f \in H^\infty$, for then we may choose polynomials Q_n tending in A^p norm to $1/f$ and conclude that $Q_n f \to (1/f)f = 1$.

Another question, first posed by Harold Shapiro in 1967, is whether a function $f \in A^p$ must be cyclic if $|f(z)| \geq c(1 - |z|^2)^a$ for some positive constants c and a, and for all $z \in \mathbb{D}$. The two questions are closely related, because

$$M_\infty(r, f) = O\big((1-r)^{-\frac{1}{p}+\varepsilon}\big) \implies f \in A^p$$
$$\implies M_\infty(r, f) = o\big((1-r)^{-\frac{2}{p}}\big)$$

(see Section 3.2). Shapiro came to the problem from the standpoint of polynomial approximation, and he called a function $f \in A^p$ *weakly invertible* if $Q_n f \to 1$ in A^p norm for some sequence of polynomials Q_n. It is evident that the terms "weakly invertible" and "cyclic" are synonymous. Shapiro [2,3] was able to establish a slightly more restrictive criterion for cyclicity, as follows.

THEOREM 3. *If $f \in A^q$ for some $q > p$, and if there are positive constants c and a such that $|f(z)| \geq c(1-|z|)^a$ for all $z \in \mathbb{D}$, then f is cyclic in A^p.*

PROOF. Under the given hypotheses, we will show that if $f^t \in [f]$ for some number t in the interval $0 < t \leq 1$, then $f^{t-\varepsilon} \in [f]$ for every ε with $0 < \varepsilon \leq t$ and $\varepsilon < (q-p)/(apq)$. To see that this will suffice, simply choose $t = 1$ to get $f^{1-\varepsilon} \in [f]$. Then iterate the process to obtain $f^{1-2\varepsilon} \in [f]$. If we choose $\varepsilon = 1/n$ for some large integer n, then n iterations will bring us to the conclusion that $1 \in [f]$, which implies that f is cyclic.

Let Q_n be the Fejér means of the Taylor series for $f^{-\varepsilon}$, the arithmetic means of the partial sums. Then $Q_n(z) \to f(z)^{-\varepsilon}$ pointwise in $\mathbb{D}$, and

$$|Q_n(z)| \leq \max\{|f(\zeta)|^{-\varepsilon} \ : \ |\zeta| \leq |z|\} \leq \tfrac{1}{c}(1-|z|)^{-a\varepsilon},$$

by the positivity of the Fejér kernel and the maximum modulus theorem. Therefore,

$$|Q_n(z) - f^{-\varepsilon}(z)| \leq \tfrac{2}{c}(1-|z|)^{-a\varepsilon}, \qquad z \in \mathbb{D},$$

so that

$$\begin{aligned}\big|Q_n(z)f(z)^t - f^{t-\varepsilon}(z)\big|^p &\leq \left(\tfrac{2}{c}\right)^p |f(z)|^{tp}(1-|z|)^{-ap\varepsilon} \\ &\leq \left(\tfrac{2}{c}\right)^p (1+|f(z)|^p)(1-|z|)^{-ap\varepsilon}.\end{aligned}$$

The function $(1-|z|)^{-ap\varepsilon}$ belongs to $L^1(\mathbb{D})$, since $ap\varepsilon < 1$. An application of Hölder's inequality shows that $|f(z)|^p(1-|z|)^{-ap\varepsilon}$ also belongs to $L^1(\mathbb{D})$, because $f \in A^q$ and $ap\varepsilon q/(q-p) < 1$. Consequently,

$$\lim_{n\to\infty} \int_{\mathbb{D}} \big|Q_n(z)f(z)^t - f^{t-\varepsilon}(z)\big|^p \, d\sigma(z) = 0,$$

by the Lebesgue dominated convergence theorem. This shows that $f^{t-\varepsilon} \in [f]$ if $f^t \in [f]$, which is enough to prove the theorem. □

An immediate corollary is a weak form of Shields' conjecture.

COROLLARY. *If $1/f \in A^p$ and $f \in A^q$ for some $q > p$, then f is cyclic in A^p.*

PROOF. If $1/f \in A^p$, then $M_\infty(r, 1/f) = O\big((1-r)^{-2/p}\big)$, which is equivalent to the inequality

$$|f(z)| \geq C(1-|z|)^{2/p}, \qquad z \in \mathbb{D},$$

for some constant C. □

On the other hand, Borichev and Hedenmalm [1,2] have constructed counterexamples showing that a function $f \in A^p$ may have the property $|f(z)| \geq c(1-|z|)^a$ for any prescribed $a > 0$ and yet fail to be cyclic in A^p. This answers the two questions of Shapiro and Shields in the negative.

It remains an open problem to give an explicit description of the cyclic functions in A^p, although we shall see in Chapter 9 that they can be characterized as "A^p outer functions" in a sense to be defined.

§8.4. The index of an invariant subspace.

Let $\mathcal{H}$ be a Hilbert space of functions analytic in $\mathbb{D}$, and assume that point-evaluation at each point $z_0 \in \mathbb{D}$ is a bounded linear functional. Suppose that $\mathcal{H}$ contains all bounded analytic functions, and that the polynomials are dense in $\mathcal{H}$. Suppose also that $\mathcal{H}$ is invariant under multiplication by bounded functions: if $f \in \mathcal{H}$ and $\varphi \in H^\infty$, then $\varphi f \in \mathcal{H}$. The Hardy space H^2 and the Bergman space A^2 will be our main examples.

For any subspaces M and N of $\mathcal{H}$ with $N \subset M$, the orthocomplement of N within M is denoted by $M \ominus N = M \cap N^\perp$, where

$$N^\perp = \{g \in \mathcal{H} \,:\, \langle f, g\rangle = 0 \qquad \text{for all } f \in N\}$$

is the annihilator of N. The *index* of an invariant subspace M is

$$\operatorname{ind}(M) = \dim(M \ominus zM),$$

the codimension of zM within M. Here it is tacitly assumed that zM, the set of functions $zf(z)$ for $f \in M$, is closed. This is trivially true for H^2, and we proved it for A^2 in Section 2.9 by showing that multiplication by z is bounded below. In fact, we showed that multiplication by z is bounded below on A^p for $0 < p < \infty$, so we can assert more generally that zM is closed for every invariant subspace M of A^p.

Certain kinds of invariant subspaces of A^2 can be seen to have index 1. Recall that a subspace M of A^p is said to be *cyclic* (or *singly generated*) if it contains a cyclic element. In other words, there is a function $f \in M$ whose polynomial multiples are dense in M, so that $[f] = M$. It will be tacitly assumed that $M \neq \{0\}$. The proof of the following theorem is adapted from Bourdon [2] (see also Janas [1]).

THEOREM 4.

(*a*) *Every cyclic subspace of A^2 has index 1. More generally, if M is a cyclic subspace of A^2, then* $\dim(M \ominus (z-\lambda)M) = 1$ *for each* $\lambda \in \mathbb{D}$.

(*b*) *Let M be an invariant subspace of A^2 such that $zg \in M$ implies $g \in M$. Then M has index 1. More generally, if some point $\lambda \in \mathbb{D}$ has the property that $(z-\lambda)g \in M$ implies $g \in M$, then* $\dim(M \ominus (z-\lambda)M) = 1$.

(*c*) *If M is an invariant subspace of A^2 and* $\dim(M \ominus (z-\lambda)M) = 1$ *for a point $\lambda \in \mathbb{D}$ at which some function in M does not vanish, then $(z-\lambda)g \in M$ implies $g \in M$.*

COROLLARY. *Let $\{z_k\}$ be an A^2 zero-set with $z_k \neq 0$, and let N^2 be the subspace of functions in A^2 that vanish (to prescribed multiplicity or higher) on $\{z_k\}$. Then N^2 has index 1.*

The corollary follows from part (*b*) of the theorem, since multiplication by z is bounded below on A^2.

PROOF OF THEOREM. (a). Let M be a cyclic subspace of A^2, and let f be a generator, so that $[f] = M$. Split f into components by writing

$$f = f_1 + f_2\,, \qquad \text{where} \;\; f_1 \in M \ominus (z-\lambda)M\,, \;\; f_2 \in (z-\lambda)M\,.$$

Now choose an arbitrary function $g \in M \ominus (z-\lambda)M$. Then $g \in [f]$, so that $\|Q_n f - g\|_2 \to 0$ for some sequence of polynomials Q_n. Write $Q_n f - g = h_{n1} + h_{n2}$, where

$$h_{n1} = Q_n(\lambda) f_1 - g \in M \ominus (z-\lambda)M\,,$$
$$h_{n2} = (Q_n - Q_n(\lambda)) f_1 + Q_n f_2 \in (z-\lambda)M\,.$$

Then $\|h_{n1}\|_2 \to 0$ and $\|h_{n2}\|_2 \to 0$ as $n \to \infty$, since

$$\|h_{n1}\|_2^2 + \|h_{n2}\|_2^2 = \|Q_n f - g\|_2^2 \to 0\,.$$

But the property $\|Q_n(\lambda) f_1 - g\|_2 \to 0$ implies that $g = \alpha f_1$ for some $\alpha \in \mathbb{C}$. Thus $M \ominus (z-\lambda)M$ is one-dimensional.

(b). First choose a function $f \in M \ominus (z-\lambda)M$, not the zero function. If $f(\lambda) = 0$, then $f(z) = (z-\lambda)F(z)$ for some analytic function F, which belongs to A^2 since multiplication by $(z-\lambda)$ is bounded below. But $f \in M$, which implies $F \in M$ by hypothesis, so that $f \in (z-\lambda)M$, a contradiction. Thus $f(\lambda) \neq 0$. Now let g be another function in $M \ominus (z-\lambda)M$. With the notation $\beta = g(\lambda)/f(\lambda)$, we then consider the function $h = g - \beta f$. Since $h \in M \ominus (z-\lambda)M$ and $h(\lambda) = 0$, it follows that $h(z) \equiv 0$. In other words, g is a scalar multiple of f, and we have shown that $M \ominus (z-\lambda)M$ is one-dimensional.

(c). By hypothesis, all elements of $M \ominus (z-\lambda)M$ are scalar multiples of a single function $f \in M \ominus (z-\lambda)M$. Note that $f(\lambda) \neq 0$, since not all functions in M vanish at λ. If $(z-\lambda)g(z)$ is in M for some analytic function g, then

$$(z-\lambda)g(z) = \alpha f(z) + (z-\lambda)g_1(z)$$

for some constant $\alpha \in \mathbb{C}$ and some function $g_1 \in M$. But this implies $\alpha f(\lambda) = 0$, which shows that $\alpha = 0$, since $f(\lambda) \neq 0$. Hence $g = g_1 \in M$. □

It follows from parts (a) and (c) that if M is a cyclic subspace of A^2 and $\lambda \in \mathbb{D}$ is a point at which some function in M does not vanish, then $f(z)/(z-\lambda) \in M$ for every $f \in M$ with $f(\lambda) = 0$.

The same arguments apply with little change to the Hardy space H^2. In particular, since by Beurling's theorem every invariant subspace of H^2 is cyclic, it follows that every invariant subspace has index 1. This is far from true in the Bergman space A^2, as we shall see in the next section. In fact, after extending the notion of index to A^p spaces for $0 < p < \infty$, we will see that an invariant subspace of A^p may have arbitrary index.

It is easy to see that one simple property of the invariant subspaces of H^p is not shared by those of A^p. According to Beurling's theorem, the intersection of two nonzero invariant subspaces M and N of H^p can never be the zero subspace. Indeed, if $M = [\varphi]$ and $N = [\psi]$ for inner functions φ and ψ, then $M \cap N = [\varphi \vee \psi] \neq \{0\}$. However, the invariant subspaces of A^p do not have this property. We know from work of Horowitz (Theorem 3 in Section 4.4) that there exist two disjoint A^p zero-sets whose union is not an A^p zero-set. Thus if M is the invariant subspace of A^p functions vanishing on one of those zero-sets, and N is the subspace of functions vanishing on the other, then $M \cap N$ contains only the zero-function.

§8.5. Invariant subspaces of higher index.

In sharp contrast with the situation for the Hardy space H^2, the invariant subspaces of the Bergman space A^2 need not be cyclic and may have index greater than one. In fact, through abstract considerations of operator algebras, Apostol, Bercovici, Foiaş, and Pearcy [1] were able to demonstrate the existence of invariant subspaces of A^2 with *arbitrary* index $n = 1, 2, 3, \ldots$. They even showed that an invariant subspace of A^2 can have infinite index; more precisely, it can have index $\aleph_0$. However, no actual examples of invariant subspaces with higher index had been found before Hedenmalm [3] recognized that Seip's results on sampling sequences (presented in Chapter 6) could be applied for that purpose. We shall now describe Hedenmalm's construction of an invariant subspace of A^2 with index 2. The point of departure is the observation (see Section 6.7, Theorem 9) that there exist disjoint A^2 zero-sets whose union is an A^2 sampling set.

THEOREM 5. *Let A and B be disjoint A^2 zero-sets whose union $A \cup B$ is an A^2 sampling set. Let M and N be the subspaces of functions in A^2 that vanish on A and B, respectively. Then $M \vee N = [M \cup N]$ is an invariant subspace of A^2 with index 2.*

Before passing to the proof, let us recall that the Cartesian product $\mathcal{H}_1 \times \mathcal{H}_2$ of two Hilbert spaces $\mathcal{H}_1$ and $\mathcal{H}_2$ is a Hilbert space with inner product

$$\langle (f_1, f_2), (g_1, g_2) \rangle = \langle f_1, g_1 \rangle + \langle f_2, g_2 \rangle ,$$

which defines the norm

$$\|(f, g)\| = \{\|f\|^2 + \|g\|^2\}^{\frac{1}{2}} .$$

If M and N are any subspaces of a Hilbert space $\mathcal{H}$ with $M \cap N = \{0\}$, then their sum

$$M + N = \{f + g \ : \ f \in M , \ g \in N\}$$

is isomorphic to the Cartesian product $M \times N$, since each element $h \in M + N$ has a *unique* representation $h = f + g$ with $f \in M$ and $g \in N$. However, the Cartesian product norm thus induced on a sum $f + g$ need not be equivalent to its ordinary norm $\|f + g\|$ as an element of $\mathcal{H}$. The key point of Hedenmalm's construction is that in the situation of Theorem 5, the two norms are equivalent.

PROOF OF THEOREM 5. Observe first that if M and N are the two subspaces defined in the theorem, then $M \cap N = \{0\}$. Indeed, if $f \in M \cap N$, then f vanishes on $A \cup B$, which is not a zero-set of A^2, so f must be the zero-function. Thus $M + N$ is isomorphic to $M \times N$. Furthermore, for $f \in M$ and $g \in N$, the norm $\|f + g\|$ is equivalent to the Cartesian product norm $\{\|f\|^2 + \|g\|^2\}^{\frac{1}{2}}$. To see this, note first the trivial inequality

$$\|f + g\|^2 \leq \|f + g\|^2 + \|f - g\|^2 = 2\{\|f\|^2 + \|g\|^2\}.$$

The reverse inequality comes from the fact that $A \cup B$ is a sampling set. Then for some positive constants C_1 and C_2,

$$\begin{aligned} C_1\{\|f\|^2 + \|g\|^2\} &\leq \sum_{z \in A \cup B} (1 - |z|^2)^2 \{|f(z)|^2 + |g(z)|^2\} \\ &= \sum_{z \in A \cup B} (1 - |z|^2)^2 |f(z) + g(z)|^2 \leq C_2 \|f + g\|^2, \end{aligned}$$

since $f(z) = 0$ for $z \in A$ and $g(z) = 0$ for $z \in B$. This shows that the Cartesian product norm induced on $M + N$ is equivalent to the ordinary norm. In particular, $M + N$ is closed under its ordinary norm, so that $M + N = M \vee N$.

Recall next that the dual space of a Cartesian product is the Cartesian product of the dual spaces. Specifically, the dual space of $M \times N$ is represented by ordered pairs of bounded linear functionals on M and N, respectively. Thus each functional $\Phi \in (M \times N)^*$ has the form

$$\Phi((f, g)) = \langle f, h \rangle + \langle g, k \rangle$$

for some functions $h \in M$ and $k \in N$, and

$$\|\Phi\|^2 = \|(h, k)\|^2 = \|h\|^2 + \|k\|^2.$$

Now

$$(M \times N) \ominus z(M \times N) = (M \times N) \ominus \{(zM) \times (zN)\} = (M \ominus zM) \times (N \ominus zN),$$

and $\dim(M \ominus zM) = \dim(N \ominus zN) = 1$, so any functional $\Phi \in (M \times N)^*$ that annihilates $z(M \times N)$ has the form

$$\Phi((f, g)) = \langle f, \alpha h \rangle + \langle g, \beta k \rangle$$

for some $h \in M$ and $k \in N$, and for arbitrary $\alpha, \beta \in \mathbb{C}$. In other words, the annihilator of $z(M \times N)$ in $(M \times N)^*$ is identified with a subspace spanned by $(h, 0)$ and $(0, k)$; hence it is two-dimensional. But by the equivalence of $M \times N$ with $M \vee N$, this proves that the subspace $(M \vee N) \ominus z(M \vee N)$ is two-dimensional. Thus $\operatorname{ind}(M \vee N) = 2$. □

The same ideas can be used to construct invariant subspaces of A^2 with arbitrary finite index, as Hedenmalm [3] indicated. Hedenmalm, Richter, and Seip [1] extended the construction to A^p. These generalizations are discussed in the next section.

§8.6. Generalizations to A^p.

The concept of index will now be extended to invariant subspaces of the Bergman spaces A^p for $0 < p < \infty$. This will allow us to generalize results in the previous two sections, expressed so far only in the context of the Hilbert space A^2. Moreover, a generalization of the construction just given will produce invariant subspaces of A^p with arbitrary finite index. The results of this section are due primarily to Richter [1].

Let M be an arbitrary (closed) invariant subspace of A^p. The *index* of M is defined to be $\operatorname{ind}(M) = \dim(M/zM)$, the dimension of the quotient space M/zM or the codimension of zM in M. It should be observed that this definition agrees with our previous definition of index in the Hilbert space case $p = 2$, namely $\operatorname{ind}(M) = \dim(M \ominus zM)$. Of course, the notion of orthogonality makes sense only in Hilbert space.

In the space A^2, Theorem 4 says that certain properties of the index extend to subspaces $(z-\lambda)M$ for $\lambda \in \mathbb{D}$. Underlying those relations is a general fact about stability of the index.

LEMMA 7. *Let M be any invariant subspace of A^p, where $0 < p < \infty$. Then for points $\lambda \in \mathbb{D}$, the codimension of $(z-\lambda)M$ in M is independent of λ.*

This is a general property of Banach spaces of analytic functions in which the operator of multiplication by $(z-\lambda)$ is bounded below, or equivalently has closed range. A proof can be given via Fredholm theory (*cf.* Richter [1]). We shall base the proof on the following theorem that can be found for instance in Kato [1], Theorem 5.31, where it appears in greater generality.

THEOREM A. *Let X be a Fréchet space, and let $T : X \to X$ be a bounded injective linear operator with closed range. Then there is a radius $r > 0$ such that the codimension of the range of $T - \lambda I$ in X is constant for $|\lambda| < r$.*

PROOF OF LEMMA 7. For $\varphi \in H^\infty$, let M_φ denote the operator of multiplication by φ. For $\lambda \in \mathbb{D}$, the operator $T = M_z - \lambda$ is injective, is bounded below, and has closed range. Let $n = \dim(M/(z-\lambda)M) \le \infty$. Let $r > 0$ be the constant from Theorem A, and choose $\zeta \in \mathbb{D}$ with $|\zeta - \lambda| < r$. By Theorem A,

$$\begin{aligned}\dim((M/(z-\zeta)M) &= \dim((M/(T-(\zeta-\lambda)I)(M)) \\ &= \dim((M/(z-\lambda)M) = n\,.\end{aligned}$$

This shows that $\dim((M/(z-\lambda)M)$ is a locally constant function of λ, and since $\mathbb{D}$ is connected, this implies the desired result. □

The *common zero-set* $Z(M)$ of a subspace M of A^p consists of those points $\lambda \in \mathbb{D}$ such that $f(\lambda) = 0$ for all $f \in M$. The set $Z(M)$ is not the whole disk, since we are assuming that M is not the zero subspace.

THEOREM 6. *For an invariant subspace M of A^p, where $0 < p < \infty$, the following statements are equivalent.*

(*i*) *M has index 1.*

(*ii*) *For some point $\lambda \in \mathbb{D} \setminus Z(M)$, the function*

$$g_\lambda(z) = \begin{cases} \frac{g(z)}{z-\lambda} & \text{if } z \neq \lambda \\ g'(\lambda) & \text{if } z = \lambda \end{cases}$$

belongs to M whenever $g \in M$ and $g(\lambda) = 0$.

(*iii*) *For every point $\lambda \in \mathbb{D} \setminus Z(M)$, the function g_λ belongs to M whenever $g \in M$ and $g(\lambda) = 0$.*

PROOF. (*i*) $\Longrightarrow$ (*iii*). Suppose that $\operatorname{ind}(M) = 1$, and choose an arbitrary point $\lambda \in \mathbb{D} \setminus Z(M)$. Then

$$M_\lambda = \{f \in M \ : \ f(\lambda) = 0\}$$

is a proper subspace of M with $\dim(M/M_\lambda) = 1$. We are to show that $g \in M_\lambda$ implies $g_\lambda \in M$, or equivalently that $M_\lambda \subset (z-\lambda)M$. But the reverse inclusion $(z-\lambda)M \subset M_\lambda$ holds trivially for every invariant subspace M, so we are to show that $M_\lambda = (z-\lambda)M$. By Lemma 7, our assumption that $\operatorname{ind}(M) = 1$ implies $\dim(M/(z-\lambda)M) = 1$. Since $(z-\lambda)M \subset M_\lambda$ and $\dim(M/M_\lambda) = 1$, this shows that $(z-\lambda)M = M_\lambda$.

(*iii*) $\Longrightarrow$ (*ii*). This is trivially true.

(*ii*) $\Longrightarrow$ (*i*). Suppose there is a point $\lambda \in \mathbb{D} \setminus Z(M)$ such that $g \in M_\lambda$ implies $g_\lambda \in M$. An equivalent assumption, as noted above, is that $M_\lambda = (z-\lambda)M$. But $\dim(M/M_\lambda) = 1$, so we infer that the codimension of $(z-\lambda)M$ in M is equal to 1. But by Lemma 7, this implies that $\operatorname{ind}(M) = 1$. □

COROLLARY 1. *Every cyclic subspace of A^p has index 1.*

PROOF. Let M be a cyclic subspace of A^p, and let f be a generator. Choose a point $\lambda \in \mathbb{D} \setminus Z(M)$ and a function $g \in M$ with $g(\lambda) = 0$. Note that $f(\lambda) \neq 0$, since all elements of $M = [f]$ vanish wherever f does, while $\lambda \notin Z(M)$. Because $g \in [f]$, there is a sequence of polynomials P_n for which $P_n f \to g$ in A^p norm. Then $P_n f(\lambda) \to g(\lambda) = 0$, and so $P_n(\lambda) \to 0$. Consequently, the function

$$h_n(z) = \big(P_n(z) - P_n(\lambda)\big) f(z) = Q_n(z) f(z)$$

has the properties $h_n(\lambda) = 0$ and $h_n \to g$. But $f(\lambda) \neq 0$, so it follows that $Q_n(\lambda) = 0$, and $R_n(z) = Q_n(z)/(z-\lambda)$ is a polynomial. Since multiplication by $(z-\lambda)$ is bounded below, we conclude that

$$\|R_n f - g_\lambda\|_p = \left\|\frac{h_n - g}{z-\lambda}\right\|_p \leq C\,\|h_n - g\|_p \to 0$$

as $n \to \infty$, so $g_\lambda \in [f] = M$. The proof that $\operatorname{ind}(M) = 1$ is completed by appeal to Theorem 6. □

COROLLARY 2. *Let $\{z_k\}$ be an A^p zero-set with $z_k \neq 0$, and let N^p be the subspace of functions in A^p that vanish on $\{z_k\}$ to at least the prescribed multiplicity. Then N^p has index 1.*

PROOF. Choose a point $\lambda \in \mathbb{D}$ that is not equal to z_k for any k. Then $\lambda \in \mathbb{D} \setminus Z(N^p)$. If $g \in N^p$ and $g(\lambda) = 0$, then $g_\lambda \in A^p$ and g_λ vanishes on $\{z_k\}$, so $g_\lambda \in N^p$. By Theorem 6, this shows that $\operatorname{ind}(N^p) = 1$. □

In Section 8.5 we produced an invariant subspace of A^2 with index 2. The same ideas can be used to construct such a subspace in A^p, where $0 < p < \infty$. In fact, without much additional effort we can construct an invariant subspace of A^p with index n, where n is an arbitrary positive integer. The construction again requires a type of reverse triangle inequality.

THEOREM 7. *For a fixed positive integer n, let $M_1, \ldots, M_n$ be invariant subspaces of A^p with index 1. If there is a positive constant C such that*

$$\|f_1 + \cdots + f_n\|_p \geq C(\|f_1\|_p + \ldots \|f_n\|_p) \tag{5}$$

for all $f_1 \in M_1, \ldots, f_n \in M_n$, then $M = M_1 + \cdots + M_n$ is an invariant subspace of A^p with index n.

PROOF. The inequality (5) implies that a function $f \in A^p$ can have at most one representation of the form $f = f_1 + \cdots + f_n$ with $f_k \in M_k$, and that M is a *closed* invariant subspace of A^p. It is clear that $f \in M$ implies $zf \in M$, so M is an invariant subspace. For each $k = 1, \ldots, n$, let $l_k \geq 0$ be the smallest integer such that $g^{(l_k)}(0) \neq 0$ for some $g \in M_k$. Then $\phi_k(g) = g^{(l_k)}(0)$ is a bounded linear functional on M_k with kernel zM_k. Define $\phi : M \to \mathbb{C}^n$ by

$$\phi(f) = (\phi_1(f_1), \ldots, \phi_n(f_n)),$$

where $f = f_1 + \cdots + f_n$ is the unique decomposition of a function $f \in M$ with $f_k \in M_k$ for $k = 1, \ldots, n$. Note that the inequality (5) guarantees that ϕ is continuous. Since ϕ maps M onto $\mathbb{C}^n$ and has kernel zM, we see that M/zM is isomorphic to $\mathbb{C}^n$. Therefore, $\dim(M/zM) = \dim(\mathbb{C}^n) = n$ and M has index n. □

The preceding proof is due to Hedenmalm, Korenblum, and Zhu [2].

Theorem 7 will give explicit examples of invariant subspaces with index n if we can produce invariant subspaces $M_1, \ldots, M_n$ of index 1 with the property (5). For this purpose we refer to Theorem 9 in Section 6.7. There we constructed n mutually disjoint sets $B_1, \ldots, B_n$ in $\mathbb{D}$ with the property that $B = B_1 \cup \cdots \cup B_n$ is a sampling set for A^p, while $B \setminus B_k$ is an A^p interpolation set (hence an A^p zero-set) for each k. Now take M_k to be the invariant subspace of A^p consisting of the functions that vanish on $B \setminus B_k$. Then M_k has index 1, by Corollary 2 of Theorem 6. To show that (5) holds, observe first that for each point $\zeta \in B$,

$$f(\zeta) = f_1(\zeta) + \cdots + f_n(\zeta) = f_k(\zeta)$$

if $\zeta \in B_k$, so we can write

$$|f_1(\zeta) + \cdots + f_n(\zeta)|^p = |f(\zeta)|^p = |f_1(\zeta)|^p + \cdots + |f_n(\zeta)|^p, \qquad \zeta \in B.$$

Now apply the lower sampling inequality to $f_1, \ldots, f_n$, add the results, and then apply the upper sampling inequality to $f = f_1 + \cdots + f_n$ to see that

$$K_1\big(\|f_1\|_p^p + \cdots + \|f_n\|_p^p\big) \le \sum_{\zeta \in B} (1 - |\zeta|^2)^2 \big(|f_1(\zeta)|^p + \cdots + |f_n(\zeta)|^p\big)$$

$$= \sum_{\zeta \in B} (1 - |\zeta|^2)^2 |f_1(\zeta) + \cdots + f_n(\zeta)|^p \le K_2 \|f_1 + \ldots f_n\|_p^p .$$

It follows that

$$\|f_1 + \cdots + f_n\|_p \ge C\left(\|f_1\|_p^p + \cdots + \|f_n\|_p^p\right)^{1/p} \ge \frac{C}{n}\left(\|f_1\|_p + \cdots + \|f_n\|_p\right).$$

Thus Theorem 7 shows that $M = M_1 + \cdots + M_n$ is an invariant subspace of A^p with index n.

Other methods are required to construct an invariant subspace of infinite index. This is done by Hedenmalm, Richter and Seip [1]. See also the discussion in the book of Hedenmalm, Korenblum and Zhu [2].

Borichev [2] and later Abakumov and Borichev [1] have found different approaches to the construction of invariant subspaces of arbitrary (possibly infinite) index in certain weighted sequence spaces. Some of their results specialize to the Bergman space A^2.

One of the most famous problems in mathematics is the invariant subspace problem. If T is a bounded linear operator on a separable Hilbert space of dimension greater than 1, must it have a proper invariant subspace? In other words, must there exist a subspace M, not the zero subspace or the whole space, such that $T(M) \subset M$? It turns out that this question is *equivalent* to a special problem about the lattice of invariant subspaces of the Bergman space A^2 under the operator of multiplication by z. Specifically, if M and N are invariant subspaces of A^2 such that $N \subset M$ and

$\dim(M \ominus N)$ is infinite, must there be another invariant subspace L of A^2 such that $N \subsetneq L \subsetneq M$? An affirmative answer to this question would mean an affirmative answer to the invariant subspace problem and conversely. For further discussion of this topic, the reader may consult Hedenmalm, Richter and Seip [1].

CHAPTER 9

Structure of Invariant Subspaces

In the previous chapter we examined the subspaces of the Bergman space A^p invariant under multiplication by polynomials. We found them not so nicely structured as their counterparts for the Hardy space H^p. We now turn to more positive results. The focus of this chapter is a theorem of Alexandru Aleman, Stefan Richter, and Carl Sundberg that gives a basic structural property of the invariant subspaces of A^2 and can be viewed as a version of Beurling's theorem for the Bergman space. The results of Aleman–Richter–Sundberg lead to an appropriate A^p analogue of the factorization of H^p functions into products of inner and outer functions. A more immediate consequence is that canonical divisors generate their corresponding zero-based subspaces in A^2. We give Serguei Shimorin's relatively simple proof of the central theorem, which recasts it as a special case of a more general theorem about linear operators in Hilbert space. The chapter concludes with a proof that every cyclic subspace of A^p is generated by its canonical extremal function.

§9.1. Description of generators.

Beurling's theorem states that every invariant subspace $M \neq \{0\}$ of H^2 has the form $M = [\varphi]$ for some inner function φ. An alternate statement, more amenable to generalization, is that $M = [M \ominus zM]$. To prove this, let φ be an inner function that generates M. Then $\varphi \in (zM)^\perp$, since

$$\langle z\varphi f, \varphi\rangle = \frac{1}{2\pi}\int_0^{2\pi} e^{i\theta}|\varphi(e^{i\theta})|^2 f(e^{i\theta})\, d\theta = \frac{1}{2\pi}\int_0^{2\pi} e^{i\theta} f(e^{i\theta})\, d\theta = 0$$

for every $f \in H^2$, because $|\varphi(e^{i\theta})| = 1$ almost everywhere. This shows that $\varphi \in M \ominus zM$. But $\operatorname{ind} M = 1$, since M is cyclic (see remarks after proof of Theorem 4 in Chapter 8), so $[M \ominus zM] = [\varphi] = M$.

As we have already seen in Section 8.4, the invariant subspaces of the Bergman space A^2 need not be singly generated and may have arbitrary index. Nevertheless, Aleman, Richter, and Sundberg [1] discovered that a complete set of generators of each invariant subspace M of A^2 is always to be found in the orthocomplement of zM within M. Their theorem can be regarded as an analogue of Beurling's theorem for the Bergman space. It can be stated as follows.

THEOREM 1. *Every invariant subspace M of A^2 has the property $M = [M \ominus zM]$.*

COROLLARY. *Let G be the canonical divisor of an A^2 zero-set $\{z_k\}$ and let N^2 be the subspace of functions in A^2 that vanish on $\{z_k\}$. Then $N^2 = [G]$.*

DEDUCTION OF COROLLARY. As indicated in Section 5.1, a variational argument shows that

$$\int_{\mathbb{D}} \overline{G(z)}\, h(z)\, d\sigma = 0 \qquad \text{for all} \quad h \in N^2 \quad \text{with} \quad h(0) = 0 .$$

In other words, G is orthogonal to every function $zf(z)$ with $f \in N^2 = M$. Thus $G \in M \ominus zM$. But $\dim(M \ominus zM) = 1$, so $[G] = [M \ominus zM] = M$. □

The same argument applies to any invariant subspace M of A^2 with index 1. Then, as indicated in Section 5.7, the canonical extremal function φ of M has the similar property

$$\int_{\mathbb{D}} \overline{\varphi(z)}\, h(z)\, d\sigma = 0 \qquad \text{for all} \quad h \in M \quad \text{with} \quad h(0) = 0 .$$

This again says that φ is orthogonal to every function $zf(z)$ with $f \in M$, and it follows in the same way that $[\varphi] = [M \ominus zM] = M$. In particular, if M is a *cyclic* subspace of A^2, then M is generated by its canonical extremal function: $M = [\varphi]$.

This last property can be generalized to cyclic subspaces of any A^p space, where $0 < p < \infty$, although it is not clear how Theorem 1 might extend to that setting, since it is based on the Hilbert space concept of orthogonality. The following theorem is also due to Aleman, Richter, and Sundberg [1].

THEOREM 2. *For $0 < p < \infty$, let $M = [\psi]$ be an arbitrary cyclic subspace of A^p, generated by a function $\psi \in A^p$. Let φ be the canonical extremal function of M. Then $M = [\varphi]$.*

For $0 < p < 1$ it is an open question whether every invariant subspace of A^p has an extremal function, or whether a canonical extremal function when it exists is unique. Existence of an extremal function is an open question even for invariant subspaces of A^1, although uniqueness in A^1 is a consequence of strict convexity (see Section 2.2). For *cyclic* subspaces of A^p, however, the existence and uniqueness assertions are true even for $0 < p \le 1$, as we saw in Chapter 5 (Theorem 7). This fact is implicit in the statement of Theorem 2 above.

With the help of Theorem 2, the corollary to Theorem 1 can be generalized to A^p for $0 < p < \infty$. In other words, if G is the canonical divisor of an A^p zero-set $\{z_k\}$ and N^p is the subspace of functions in A^p that vanish on $\{z_k\}$, then $N^p = [G]$. However the more general result does not follow immediately from Theorem 2, because it must be shown *a priori* that N^p is cyclic. We will address this problem later.

Abkar [1] has given a different proof of the corollary (for $p = 2$) that makes no appeal to Theorem 1.

The proofs of Theorems 1 and 2 will be deferred to the end of this chapter. Meanwhile, we propose to show how the results can be applied to obtain a credible analogue for Bergman spaces of the classical "inner–outer" factorization of functions in Hardy spaces.

§9.2. Inner and outer functions for Bergman spaces.

Each function in the Hardy space H^p $(0 < p < \infty)$, has a unique factorization of the form $f = \varphi F$, where φ is an inner function and F is an H^p outer function. The canonical factorization plays a major role in the theory of H^p spaces, so it is desirable to find a counterpart for Bergman spaces. However, in searching for appropriate analogues of "inner" and "outer" functions in Bergman spaces, one is confronted by the fact that functions in Bergman spaces need not have boundary values. Thus the strategy is to reformulate the classical Hardy space definitions in more suggestive ways, without reference to boundary values.

Recall that a classical inner function is a function $\varphi \in H^\infty$ whose boundary function satisfies $|\varphi(e^{i\theta})| = 1$ almost everywhere. This concept can be recaptured by requiring that $\varphi \in H^p$ with norm $\|\varphi\|_{H^p} = 1$ and

$$\int_0^{2\pi} |\varphi(e^{i\theta})|^p e^{in\theta}\, d\theta = 0\,, \qquad n = 1, 2, \ldots .$$

To see this, observe that the property implies (by conjugation) the vanishing of all Fourier coefficients of $|\varphi(e^{i\theta})|^p$ with index $n \neq 0$, so $|\varphi(e^{i\theta})|$ is constant.

The last formulation suggests an appropriate analogue for Bergman spaces. Accordingly, an A^p inner function was defined in Chapter 5 as a function $\varphi \in A^p$ of norm $\|\varphi\|_p = 1$ such that

$$\int_{\mathbb{D}} |\varphi(z)|^p z^n\, d\sigma = 0, \qquad n = 1, 2, \ldots .$$

We have seen that the canonical extremal function of every invariant subspace is an A^p inner function, for $0 < p < \infty$. Conversely, we shall see that every A^p inner function is (up to rotation) the canonical extremal function of some invariant subspace.

The next problem is to find a suitable definition of "outer function" for the Bergman space. The outer functions of H^p are the cyclic elements, and it is desirable that this property extend to A^p. It is not difficult to show, using the canonical inner-outer factorization, that $F \in H^p$ is an outer function if and only if for every $g \in H^p$ the inequality $|g(e^{i\theta})| \leq |F(e^{i\theta})|$

a.e. implies $|g(0)| \le |F(0)|$. But the inequality between boundary functions is equivalent to requiring that $\|Qg\|_{H^p} \le \|QF\|_{H^p}$ for all polynomials Q. This prompted Korenblum [6] to say that a function $f \in A^p$ *dominates* a function $g \in A^p$, written $g \prec f$, if $\|Qg\|_p \le \|Qf\|_p$ for all polynomials Q. A function $F \in A^p$ is then defined to be an A^p *outer function* if every $g \in A^p$ with $g \prec F$ has the property $|g(0)| \le |F(0)|$.

The following theorems show that these are the "right" definitions of inner and outer functions for the Bergman space.

THEOREM 3. *For $0 < p < \infty$, a function in A^p is cyclic if and only if it is an A^p outer function.*

THEOREM 4. *For $0 < p < \infty$, every function $f \in A^p$ has a factorization $f = \varphi F$, where φ is an A^p inner function and F is an A^p outer function.*

Korenblum [6] showed that every cyclic function in A^p is an A^p outer function. The converse was proved by Aleman, Richter, and Sundberg [1], who also proved Theorem 4. Unfortunately, the inner–outer factorization of Theorem 4 is not unique. Borichev and Hedenmalm [1,2] have constructed an example of a function in A^2 with two distinct factorizations.

Proofs of Theorem 1 (due to Shimorin) and Theorem 2 will be presented at the end of this chapter. Meanwhile, we propose to show how the results can be used to prove Theorems 3 and 4. For this purpose we need more information about extremal functions. We have already recalled that an extremal function φ of any invariant subspace of A^p is an A^p inner function. Thus it is an expansive multiplier in the sense that $\|\varphi Q\|_p \ge \|Q\|_p$ for every polynomial Q. (See Chapter 5, Theorem 3.) We will also need the following lemma, which says essentially that A^p inner functions are contractive divisors.

LEMMA 1. *If φ is an A^p inner function and $f \in [\varphi]$, then $f/\varphi \in A^p$ and $\|f/\varphi\|_p \le \|f\|_p$.*

PROOF OF LEMMA. The expansive multiplier property of inner functions says that $\|\varphi Q\|_p \ge \|Q\|_p$ for all polynomials Q. Now if $f \in [\varphi]$, there is a sequence of polynomials Q_n for which $\varphi Q_n \to f$ in norm. Thus in particular,

$$\|\varphi Q_n - \varphi Q_m\|_p = \|\varphi(Q_n - Q_m)\|_p \to 0$$

as $n, m \to \infty$. But

$$\|Q_n - Q_m\|_p \le \|\varphi(Q_n - Q_m)\|_p ,$$

so this implies that $\{Q_n\}$ is a Cauchy sequence, hence that it converges in norm to some function $g \in A^p$. But then it follows that the sequence $\{\varphi Q_n\}$ converges pointwise in $\mathbb{D}$ to the function φg. On the other hand, it is also true that $\{\varphi Q_n\}$ converges pointwise to f, since $\|\varphi Q_n - f\|_p \to 0$. Thus $f = \varphi g$, and so $f/\varphi = g \in A^p$. Furthermore, $\|Q_n\|_p \le \|\varphi Q_n\|_p$ while $\|Q_n\|_p \to \|g\|_p$ and $\|\varphi Q_n\|_p \to \|f\|_p$. Thus $\|g\|_p \le \|f\|_p$, or $\|f/\varphi\|_p \le \|f\|_p$. □

PROOF OF THEOREM 3. Suppose first that a function $F \in A^p$ is cyclic. Then $F(z) \neq 0$ in $\mathbb{D}$, and there is a sequence $\{Q_n\}$ of polynomials such that $\|Q_n F - 1\|_p \to 0$ as $n \to \infty$. But norm convergence implies pointwise convergence, so it follows that $Q_n(0)F(0) \to 1$. Suppose now that $g \in A^p$ and $g \prec F$. Then $Q_n(0)g(0) \to g(0)/F(0)$ and

$$|Q_n(0)g(0)| \leq \|Q_n g\|_p \leq \|Q_n F\|_p \to 1$$

as $n \to \infty$, so that $|g(0)/F(0)| \leq 1$. This proves that F is an A^p outer function.

Conversely, let F be an A^p outer function, and let $M = [F]$ be the invariant subspace it generates. Let φ be the extremal function of M. Then φ is an A^p inner function and $[\varphi] = M$, by Theorem 2. Since $QF \in [\varphi]$ for every polynomial Q, Lemma 1 shows that $QF/\varphi \in A^p$ and $\|Q(F/\varphi)\|_p \leq \|QF\|_p$. This says that $F/\varphi \prec F$, which implies $|(F/\varphi)(0)| \leq |F(0)|$, by the definition of an A^p outer function. But $\varphi \in [F]$, so $(F/\varphi)(0) \neq 0$, and it follows that $F(0) \neq 0$. Thus the above inequality shows that $\varphi(0) \geq 1$. But $\|\varphi\|_p = 1$, so we conclude that $\varphi(z) \equiv 1$. Consequently, $[F] = [\varphi] = A^p$, and F is cyclic. □

PROOF OF THEOREM 4. Let φ be the (uniquely determined) extremal function of $[f]$, so that $[\varphi] = [f]$ by Theorem 2. Thus $f \in [\varphi]$, and Lemma 1 says that the function $F = f/\varphi$ belongs to A^p. Since $\varphi \in [f]$, it is seen that $Q_n f \to \varphi$ in A^p norm for some sequence of polynomials Q_n. But $(Q_n f - \varphi) \in [\varphi]$, so Lemma 1 gives the inequality

$$\|Q_n F - 1\|_p \leq \|Q_n f - \varphi\|_p \to 0,$$

proving that $1 \in [F]$. But this implies that $[F] = A^p$, so F is cyclic and hence is an A^p outer function. □

As a consequence of Theorem 3, we can say that for $0 < p < q < \infty$, every A^q outer function is also an A^p outer function. This is not immediately apparent from the definition, but it follows easily from the equivalent property of cyclicity. (See the remarks at the beginning of Section 8.3.) A classical outer function for H^p is always an A^p outer function. In fact, we have shown (Chapter 8, Corollary 1 to Theorem 2) that an H^p function is an A^{2p} outer function if and only if its (classical) inner factor is purely singular and the associated singular measure places no mass on any Carleson set.

One further application of Lemma 1 will now be noted. It shows that the family of extremal functions of invariant subspaces is the same as the family of inner functions.

PROPOSITION 1. *For $0 < p < \infty$, every normalized A^p inner function is the canonical extremal function of the invariant subspace it generates.*

PROOF. Recall that "normalized" means $\varphi(0) > 0$ if $\varphi(0) \neq 0$, and $\varphi^{(m)}(0) > 0$ if φ has a zero of order m at the origin. Let f be the canonical extremal function of the subspace $[\varphi]$ generated by an A^p inner function φ. Because $f \in [\varphi]$, we can infer from Lemma 1 that $g = f/\varphi \in A^p$ and $\|g\|_p \leq \|f\|_p = 1$. But $|h(0)| \leq \|h\|_p$ for every function $h \in A^p$, with equality only if $h(z) \equiv h(0)$. Thus it follows that $f^{(m)}(0) \leq \varphi^{(m)}(0)$, in violation of the extremal property of f unless $f^{(m)}(0) = \varphi^{(m)}(0)$, or $g(0) = 1$, which implies that $f(z) \equiv \varphi(z)$. □

Alternatively, we can appeal to Theorem 2 to see that $[f] = [\varphi]$, so that $\varphi \in [f]$ and $\|\varphi/f\|_p \leq 1$ by a similar application of Lemma 1, since every extremal function is an A^p inner function. This together with the inequality $\|f/\varphi\|_p \leq 1$ implies that $(f/\varphi)(0) = 1$, so that $(f/\varphi)(z) \equiv 1$ and $f = \varphi$.

§9.3. Generalization of the main theorem.

The original proof of Theorem 1, as given by Aleman, Richter, and Sundberg, was long and technical. However, Serguei Shimorin [7] was able to show that the result is actually a special case of a much more general theorem about bounded linear operators in Hilbert space, which applies for instance to certain weighted Bergman spaces. Furthermore, Shimorin's proof of the more general theorem is shorter and simpler, although it is by no means easy. It should be remarked that Stefan Richter [2] had previously obtained results of the same general type.

Before stating Shimorin's theorem, we need to introduce some notation. Given a linear operator T on a Banach space X and an arbitrary subset $E \subset X$, we denote by $[E]_T$ the smallest T-invariant subspace of X containing E. A subspace Y of X is said to be *T-invariant* if $TY \subset Y$. Here is Shimorin's theorem.

THEOREM 5. *Let T be a linear operator on a Hilbert space $\mathcal{H}$ with the properties*

$$(i) \quad \|Tx + y\|^2 \leq 2(\|x\|^2 + \|Ty\|^2) \qquad \text{for all } x, y \in \mathcal{H};$$

$$(ii) \quad \bigcap_{n=1}^{\infty} T^n\mathcal{H} = \{0\}.$$

Then $\mathcal{H} = [\mathcal{H} \ominus T\mathcal{H}]_T$.

It is not immediately clear that the Aleman–Richter–Sundberg theorem actually follows from Shimorin's theorem. To see that it does, let $\mathcal{H}$ be a subspace of A^2 invariant under the operator T of multiplication by z. Then the hypothesis (ii) is clearly satisfied, since an analytic function not identically zero can have at most a zero of finite order at the origin. A calculation is required to check the hypothesis (i). It must be shown that (with abuse of notation)

$$\|zf(z)+g(z)\|_2^2 \le 2(\|f(z)\|_2^2 + \|zg(z)\|_2^2)$$

for every pair of functions f and g in A^2. Writing

$$f(z) = \sum_{n=0}^{\infty} a_n z^n \qquad \text{and} \qquad g(z) = \sum_{n=0}^{\infty} b_n z^n ,$$

and recalling from Chapter 1 that

$$\|f\|_2^2 = \sum_{n=0}^{\infty} \frac{|a_n|^2}{n+1} ,$$

we see that the condition to be verified is

$$|b_0|^2 + \sum_{n=1}^{\infty} \frac{|a_{n-1}+b_n|^2}{n+1} \le 2\left\{ \sum_{n=1}^{\infty} \frac{|a_{n-1}|^2}{n} + \frac{|b_0|^2}{2} + \sum_{n=1}^{\infty} \frac{|b_n|^2}{n+2} \right\} .$$

But this follows at once from the inequality

$$\frac{|a+b|^2}{n+1} \le 2\left(\frac{|a|^2}{n} + \frac{|b|^2}{n+2} \right) ,$$

valid for arbitrary complex numbers a and b and any real number $n > 0$. Indeed, this last inequality reduces to the form

$$2n(n+2)\text{Re}\{\overline{a}b\} \le (n+2)^2|a|^2 + n^2|b|^2 ,$$

which is equivalent to the inequality $|(n+2)a - nb|^2 \ge 0$. Thus Theorem 1 is a consequence of Theorem 5.

Before passing to the proof of Theorem 5, we recall some basic facts about positive operators. Any bounded linear operator A on a Hilbert space $\mathcal{H}$ has an *adjoint operator* A^* defined by $\langle Ax, y\rangle = \langle x, A^*y\rangle$ for all $x, y \in \mathcal{H}$. A bounded linear operator A is said to be *positive* if $\langle Ax, x\rangle \ge 0$ for all $x \in \mathcal{H}$. A positive operator on a *complex* Hilbert space is self-adjoint. The spectrum of a positive operator is nonnegative. If B is another bounded linear operator on $\mathcal{H}$, the notation $A \le B$ means that $B - A$ is positive. Every positive operator A has a unique positive square root, written as $A^{\frac{1}{2}}$. This is a positive operator B such that $B^2 = A$. If a positive operator A is invertible, then the inverse operator A^{-1} is also positive, so it has a square root $A^{-\frac{1}{2}} = (A^{-1})^{\frac{1}{2}}$. In this case, $A^{\frac{1}{2}}$ is also invertible and its inverse coincides with the square root of A^{-1}. For any bounded linear operator A, the operator A^*A is clearly positive, so it has a positive square root $(A^*A)^{\frac{1}{2}}$.

For a discussion of positive operators, see the books of Rudin [3] or Reed and Simon [1].

It will be useful to record some preliminary observations about linear operators T with the property (i). First, we can take $y = 0$ to see that T is bounded. We can take $x = 0$ to see that $\|y\|^2 \leq 2\|Ty\|^2$, so that T is one-to-one and has closed range. Next we observe that the operator T^*T is invertible. To see that T^*T is one-to-one, suppose $T^*Tx = 0$ for some $x \in \mathcal{H}$. Then $0 = \langle T^*Tx, x\rangle = \langle Tx, Tx\rangle = \|Tx\|^2$, so $Tx = 0$. But T is one-to-one, so this implies $x = 0$. Hence T^*T is one-to-one. To see that T^*T maps $\mathcal{H}$ *onto* itself, suppose on the contrary that there is an element $y \neq 0$ in $\mathcal{H}$ such that $\langle T^*Tx, y\rangle = 0$ for all $x \in \mathcal{H}$. Then in particular

$$\|Ty\|^2 = \langle Ty, Ty\rangle = \langle T^*Ty, y\rangle = 0\,,$$

so that $Ty = 0$ and $y = 0$, again because T is one-to-one. This contradiction shows that the range of T^*T is the whole space $\mathcal{H}$, so the inverse operator $(T^*T)^{-1}$ is defined on $\mathcal{H}$.

We can now state a lemma that will be a key ingredient in the proof of Theorem 5.

LEMMA 2. *If T is a linear operator on a Hilbert space $\mathcal{H}$ that satisfies the condition (i) of Theorem 5, then the operator $S = T(T^*T)^{-1}$ is concave in the sense that*

$$\|S^2x\|^2 + \|x\|^2 \leq 2\|Sx\|^2 \qquad \textit{for all } \; x \in \mathcal{H}\,.$$

PROOF. It is easily seen that the operator $T(T^*T)^{-\frac{1}{2}}$ is an isometry. Thus with the substitution $y = (T^*T)^{-\frac{1}{2}}z$, the condition (i) takes the form

$$\|Tx + (T^*T)^{-\frac{1}{2}}z\|^2 \leq 2(\|x\|^2 + \|z\|^2)\,.$$

In terms of the operator $L : \mathcal{H} \times \mathcal{H} \mapsto \mathcal{H}$ defined by

$$L(x, z) = Tx + (T^*T)^{-\frac{1}{2}}z\,,$$

this says that $\|L\| \leq \sqrt{2}$, so that $LL^* \leq 2I$, where $L^* : \mathcal{H} \mapsto \mathcal{H} \times \mathcal{H}$ is the adjoint of L. Here $\mathcal{H} \times \mathcal{H}$ has the product norm $\|(x, y)\| = \{\|x\|^2 + \|y\|^2\}^{\frac{1}{2}}$ derived from the inner product

$$\langle (x_1, y_1), (x_2, y_2)\rangle = \langle x_1, x_2\rangle + \langle y_1, y_2\rangle\,,$$

and it can be verified that

$$L^* = \left(T^*, (T^*T)^{-\frac{1}{2}}\right).$$

Thus the inequality $LL^* \leq 2I$ takes the form

$$TT^* + (T^*T)^{-1} \leq 2I\,.$$

Substituting the operator $S = T(T^*T)^{-1}$ by the relation $T = S(S^*S)^{-1}$, we find

$$S(S^*S)^{-1}(S^*S)^{-1}S^* + S^*S \leq 2I\,.$$

But if an operator A is positive, then so is S^*AS. Thus the last inequality can be recast in the form

$$I + (S^*)^2S^2 \leq 2S^*S\,,$$

which is easily seen to be the desired inequality

$$\|x\|^2 + \|S^2x\|^2 \leq 2\|Sx\|^2, \qquad x \in \mathcal{H}\,. \qquad \square$$

LEMMA 3. *If U is a bounded linear concave operator on a Hilbert space $\mathcal{H}$, then the restriction of U to the subspace*

$$\widehat{\mathcal{H}} = \bigcap_{n=1}^{\infty} U^n\mathcal{H}$$

is an isometry, $U\widehat{\mathcal{H}} = \widehat{\mathcal{H}}$, and $U^*\widehat{\mathcal{H}} = \widehat{\mathcal{H}}$.

PROOF. To say that U is *concave* means that

$$\|U^2x\|^2 + \|x\|^2 \leq 2\|Ux\|^2 \qquad \text{for all } x \in \mathcal{H}\,.$$

This implies that $Ux \neq 0$ unless $x = 0$, so U is one-to-one on $\mathcal{H}$. Moreover, $\|Ux\| \geq \|x\|$ for all $x \in \mathcal{H}$. Indeed, the hypothesis says that the sequence $\{\|U^nx\|^2\}$ is concave:

$$\|U^{n+2}x\|^2 - \|U^{n+1}x\|^2 \leq \|U^{n+1}x\|^2 - \|U^nx\|^2, \qquad n = 0, 1, 2, \ldots.$$

From this it follows that the sequence is nondecreasing. To see this, suppose on the contrary that $\|U^{k+1}x\| < \|U^kx\|$ for some $x \in \mathcal{H}$ and some integer $k \geq 0$. Then by the concavity property, $\|U^{n+1}x\| < \|U^nx\|$ for all $n \geq k$. But $\|U^nx\| \geq 0$, so the sequence is eventually monotone and bounded, hence convergent. Consequently,

$$0 = \lim_{n\to\infty} \left(\|U^{n+1}x\|^2 - \|U^nx\|^2\right) \leq \|U^{k+1}x\|^2 - \|U^kx\|^2 < 0\,,$$

which is a contradiction. Thus the sequence $\{\|U^nx\|\}$ is nondecreasing. In particular, $\|Ux\| \geq \|x\|$.

We now show that U is an isometry on $\widehat{\mathcal{H}}$. If $x \in \widehat{\mathcal{H}}$, then $x = U^ny_n$ for some $y_n \in \mathcal{H}$, for each $n = 0, 1, 2, \ldots$. Thus $Uy_n = y_{n-1}$, since $U^{n-1}(Uy_n) =$

$U^{n-1}y_{n-1}$ and U is one-to-one. This shows that the sequence $\{\|y_n\|^2\}$ is concave; *i.e.*,

$$\|y_n\|^2 + \|y_{n+2}\|^2 \le 2\|y_{n+1}\|^2, \qquad n = 0, 1, 2, \dots .$$

But $\|y_n\|^2 = \|Uy_{n+1}\|^2 \ge \|y_{n+1}\|^2$, by what we have already shown, so the nonnegative sequence $\{\|y_n\|^2\}$ is both concave and nonincreasing. It is therefore constant, and so in particular $\|y_1\| = \|y_0\|$, or $\|Ux\| = \|x\|$ for all $x \in \widehat{\mathcal{H}}$.

It is clear from the definition of $\widehat{\mathcal{H}}$ that $U\widehat{\mathcal{H}} = \widehat{\mathcal{H}}$. We shall now show that $U^*\widehat{\mathcal{H}} = \widehat{\mathcal{H}}$. Since U is an isometry on $\widehat{\mathcal{H}}$, it is clear that

$$\langle (U^*U - I)x, x\rangle = 0 \qquad \text{for all } x \in \widehat{\mathcal{H}}.$$

On the other hand, we know that $\|Ux\| \ge \|x\|$ for every $x \in \mathcal{H}$, which says that $U^*U - I \ge 0$ on $\mathcal{H}$. It now follows from the spectral theorem for self-adjoint operators that $U^*U - I = 0$ on $\widehat{\mathcal{H}}$. Thus $U^*Ux = x$ for every $x \in \widehat{\mathcal{H}}$. But $U\widehat{\mathcal{H}} = \widehat{\mathcal{H}}$, so this proves that $U^*\widehat{\mathcal{H}} = \widehat{\mathcal{H}}$. Briefly, the conclusion is that the restriction of U to $\widehat{\mathcal{H}}$ is a unitary operator. □

Proof of Theorem 5. For convenience, let $E = \mathcal{H} \ominus T\mathcal{H}$. Observe first that E is the kernel of T^*, since $\langle Tx, y\rangle = 0$ for all $x \in \mathcal{H}$ if and only if $T^*y = 0$. Next note that the adjoint of $S = T(T^*T)^{-1}$ is $S^* = (T^*T)^{-1}T^*$, so $S^*T = I$. This shows that $TS^*x = x$ for every $x \in T\mathcal{H}$. On the other hand, TS^* annihilates $E = (T\mathcal{H})^\perp$, since $T^*E = \{0\}$ and so $S^*E = \{0\}$. Since $\mathcal{H} = E \oplus T\mathcal{H}$, this shows that TS^* is the orthogonal projection onto $T\mathcal{H}$. Therefore, $I - TS^* = P_E$, the orthogonal projection onto E.

We claim now that the kernel of $(S^*)^n$ has the form

$$\ker(S^*)^n = E + TE + \cdots + T^{n-1}E = \bigvee_{k=0}^{n-1} T^kE\,, \qquad n = 1, 2, \dots .$$

To prove this, suppose first that $x \in T^kE$ for some k with $0 \le k \le n-1$. Then $x = T^ky$ for some $y \in E$, and so

$$(S^*)^nx = (S^*)^nT^ky = (S^*)^{n-k}y = 0\,,$$

because $S^*T = I$, $S^*E = \{0\}$, and $n - k > 0$. This shows that

$$\bigvee_{k=0}^{n-1} T^kE \subset \ker(S^*)^n.$$

To verify the reverse inclusion, observe that

$$T^kP_E(S^*)^k = T^k(S^*)^k - T^{k+1}(S^*)^{k+1},$$

since $P_E = I - TS^*$, and therefore

$$I - T^n(S^*)^n = \sum_{k=0}^{n-1} T^k P_E (S^*)^k, \qquad n = 1, 2, \dots .$$

Thus if $x \in \ker(S^*)^n$, we can write

$$x = \big[I - T^n(S^*)^n\big]x = \sum_{k=0}^{n-1} T^k P_E (S^*)^k x \,,$$

which shows that $x \in \bigvee_{k=0}^{n-1} T^k E$. This proves the reverse inclusion and establishes the claimed expression for $\ker(S^*)^n$.

Next combining Lemmas 2 and 3, we see that both S and S^* map the subspace

$$\widehat{\mathcal{H}} = \bigcap_{n=1}^{\infty} S^n \mathcal{H}$$

onto itself. It follows that $T = S(S^*S)^{-1}$ and $T^* = (S^*S)^{-1}S^*$ also map $\widehat{\mathcal{H}}$ onto $\widehat{\mathcal{H}}$. Since T is one-to-one, we infer that the restriction of T to $\widehat{\mathcal{H}}$ is invertible. Thus any element $x \in \widehat{\mathcal{H}}$ can be written in the form $x = T^n(T^{-n}x)$ for every $n = 1, 2, \dots$, which implies that

$$\widehat{\mathcal{H}} \subset \bigcap_{n=1}^{\infty} T^n \mathcal{H} = \{0\} \,,$$

by the hypothesis (ii) of the theorem. This shows that $\widehat{\mathcal{H}} = \{0\}$.

Now choose an arbitrary element $x \in \mathcal{H}$ and write $x = x_n + y_n$, where $y_n \in S^n\mathcal{H}$ and

$$x_n \in \big(S^n\mathcal{H}\big)^{\perp} = \ker(S^*)^n = \bigvee_{k=0}^{n-1} T^k E \,.$$

Since $\|x\|^2 = \|x_n\|^2 + \|y_n\|^2$, the sequence $\{y_n\}$ is bounded and therefore has a weakly convergent subsequence $\{y_{n_k}\}$. In other words, there is some element $y \in \mathcal{H}$ such that $\langle y_{n_k}, x\rangle \to \langle y, x\rangle$ as $k \to \infty$, for every $x \in \mathcal{H}$. We claim that $y = 0$. If not, then $y \notin \widehat{\mathcal{H}} = \{0\}$ and so $y \notin S^m\mathcal{H}$ for some m. Let z be the orthogonal projection of y into $\big(S^m\mathcal{H}\big)^{\perp}$. Then $\langle y, z\rangle = \|z\|^2 \neq 0$ while $\langle y_{n_k}, z\rangle = 0$ for all $n_k \geq m$. This contradicts the weak convergence, so we conclude that the weak limit $y = 0$. But the Banach–Saks theorem says that whenever a sequence converges weakly, then the arithmetic means of some subsequence converge strongly to the same limit. (See Riesz–Nagy [1], p. 80.) Thus we can extract a further subsequence, which we again denote by $\{y_{n_k}\}$, such that

$$\eta_k = \frac{1}{k}\big(y_{n_1} + y_{n_2} + \cdots + y_{n_k}\big) \to 0 \qquad \text{as } k \to \infty \,.$$

Finally, we can write

$$x = \frac{1}{k}\sum_{j=1}^{k}(x_{n_j} + y_{n_j}) = \xi_k + \eta_k\,, \qquad k = 1, 2, \dots\,,$$

where

$$\xi_k = \frac{1}{k}(x_{n_1} + x_{n_2} + \cdots + x_{n_k}) \in \bigvee_{j=0}^{n_k - 1} T^j E \subset [E]_T\,.$$

Since $\eta_k \to 0$, we conclude that $\xi_k \to x$ and so $x \in [E]_T$. This shows that $\mathcal{H} = [E]_T = [\mathcal{H} \ominus T\mathcal{H}]_T$, as the theorem asserts. □

§9.4. Cyclic subspaces of A^p.

The main goal of these final two sections is to prove Theorem 2, that every cyclic subspace of A^p is generated by its canonical extremal function. The proof will make heavy use of the integral formula

$$\|\varphi f\|_p^p = \|f\|_p^p + \pi \int_{\mathbb{D}}\int_{\mathbb{D}} \Gamma(z,\zeta)\,\Delta\big(|\varphi(z)|^p\big)\,\Delta\big(|f(\zeta)|^p\big)\,d\sigma(z)d\sigma(\zeta)\,,$$

proved in Chapter 5 for an arbitrary A^p inner function φ and any polynomial f. Here

$$\Gamma(z,\zeta) = \frac{1}{16}\left\{|z-\zeta|^2 \log\left|\frac{z-\zeta}{1-\bar{\zeta}z}\right|^2 + (1-|\zeta|^2)(1-|z|^2)\right\}$$

is the biharmonic Green function for the unit disk. (See Section 1.6 for its construction.) Specifically, the proof of Theorem 2 will be based on a structural formula for the subspace $[\varphi]$ generated by an A^p inner function φ, expressed in terms of the biharmonic Green function of the disk. The result (Theorem 6 below) is of independent interest and will be derived in this section. The actual proof of Theorem 2 will be deferred to the next section.

We begin with some preliminary results involving the biharmonic Green function. The first lemma is a general measure-theoretic principle that we record here for convenient reference.

LEMMA 4. *Suppose $0 < p < \infty$ and let μ be a finite measure on $\mathbb{D}$. If $f_n \in L^p(\mu)$ and $f_n(z) \to f(z)$ almost everywhere (with respect to μ) in $\mathbb{D}$, and if*

$$\lim_{n\to\infty} \int_{\mathbb{D}} |f_n(z)|^p\,d\mu(z) = \int_{\mathbb{D}} |f(z)|^p\,d\mu(z) < \infty\,,$$

then

$$\lim_{n\to\infty} \int_{\mathbb{D}} |f_n(z) - f(z)|^p\,d\mu(z) = 0\,.$$

For a proof, see Duren [5], p. 21 or Novinger [2].

The next lemma sharpens the property $\Gamma(z,\zeta) > 0$, already noted in Chapter 1, and allows us to approximate $\Gamma(z,\zeta)$ by a simpler function.

LEMMA 5. *Let $\Gamma(z,\zeta)$ be the biharmonic Green function of the unit disk, and let*

$$\widetilde{\Gamma}(z,\zeta) = \frac{1}{16}\frac{(1-|\zeta|^2)^2(1-|z|^2)^2}{|1-\overline{\zeta}z|^2}.$$

Then $\frac{1}{2}\widetilde{\Gamma}(z,\zeta) \le \Gamma(z,\zeta) \le \widetilde{\Gamma}(z,\zeta)$ for all points $z,\zeta \in \mathbb{D}$.

PROOF. In view of the standard identity

$$1-|\varphi_\zeta(z)|^2 = \frac{(1-|\zeta|^2)(1-|z|^2)}{|1-\overline{\zeta}z|^2}, \qquad \varphi_\zeta(z) = \frac{\zeta - z}{1-\overline{\zeta}z},$$

we can write

$$\Gamma(z,\zeta) = \widetilde{\Gamma}(z,\zeta)\, F\big(1-|\varphi_\zeta(z)|^2\big),$$

where $F(x) = [(1-x)\log(1-x)+x]/x^2$. But it can be shown that $F(x)$ increases from $\frac{1}{2}$ to 1, so that $\frac{1}{2} < F(x) < 1$ for $0 < x < 1$ and the result follows. □

LEMMA 6. *Let $f(z)$ be analytic in $\mathbb{D}$ and let $f_r(z) = f(rz)$ denote its dilations, where $0 < r < 1$. Then for each fixed $\zeta \in \mathbb{D}$ and $0 < p < \infty$,*

$$\lim_{r\to 1}\int_{\mathbb{D}} \Gamma(z,\zeta)\,\Delta\big(|f_r(z)|^p\big)\,d\sigma(z) = \int_{\mathbb{D}} \Gamma(z,\zeta)\,\Delta\big(|f(z)|^p\big)\,d\sigma(z).$$

PROOF. Note first that $\Delta\big(|f(z)|^p\big) = p^2|f(z)|^{p-2}|f'(z)|^2$ is locally integrable in $\mathbb{D}$, even near a zero of f, which creates a singularity when $p < 2$. Thus the left-hand integral is finite for each $r < 1$. But Lemma 5 implies

$$\frac{1}{32}(1-|\zeta|)^2(1-|z|^2)^2 \le \Gamma(z,\zeta) \le \frac{1}{16}(1-|\zeta|)^2(1+|z|^2)^2,$$

since

$$\frac{1}{(1+|\zeta|)^2} \le \frac{1}{|1-\overline{\zeta}z|^2} \le \frac{1}{(1-|\zeta|)^2};$$

so the right-hand integral is finite if and only if

$$\int_{\mathbb{D}} (1-|z|^2)^2\Delta\big(|f(z)|^p\big)\,d\sigma(z)$$

is finite. Changing variables and applying the monotone convergence theorem, we find that

$$\begin{aligned}\lim_{r\to 1}\int_{\mathbb{D}}(1-|z|^2)^2\Delta\big(|f_r(z)|^p\big)\,d\sigma(z) &= \lim_{r\to 1}\frac{1}{r^4}\int_{r\mathbb{D}}(r^2-|z|^2)^2\Delta\big(|f(z)|^p\big)\,d\sigma(z)\\ &= \int_{\mathbb{D}}(1-|z|^2)^2\Delta\big(|f(z)|^p\big)\,d\sigma(z).\end{aligned}$$

Because $\Delta(|f_r(z)|^p)$ converges pointwise to $\Delta(|f(z)|^p)$ as $r \to 1$, it now follows from Lemma 4 that

$$\lim_{r\to 1} \int_{\mathbb{D}} (1-|z|^2)^2 \left|\Delta(|f(z)|^p) - \Delta(|f_r(z)|^p)\right| d\sigma(z) = 0,$$

which implies Lemma 6. □

LEMMA 7. *If $h \in L^1$ and $h(\zeta) \geq 0$ almost everywhere in $\mathbb{D}$, then for each $z \in \mathbb{D}$,*

$$r^3 \int_{\mathbb{D}} \Gamma(z,\zeta) h(r\zeta)\, d\sigma(\zeta) \leq 2 \int_{\mathbb{D}} \Gamma(z,\zeta) h(\zeta)\, d\sigma(\zeta), \qquad 0 < r < 1.$$

PROOF. A calculation of the derivative shows that for fixed $z, \zeta \in \mathbb{D}$, the expression $r\widetilde{\Gamma}(z,\zeta/r)$ is an increasing function of r in the interval $|\zeta| < r < 1$. Therefore, by the positivity of the integrand,

$$r^3 \int_{\mathbb{D}} \widetilde{\Gamma}(z,\zeta) h(r\zeta)\, d\sigma(\zeta) = \int_{r\mathbb{D}} r\widetilde{\Gamma}(z,\zeta/r) h(\zeta)\, d\sigma(\zeta)$$

$$\leq \int_{r\mathbb{D}} \widetilde{\Gamma}(z,\zeta) h(\zeta)\, d\sigma(\zeta) \leq \int_{\mathbb{D}} \widetilde{\Gamma}(z,\zeta) h(\zeta)\, d\sigma(\zeta).$$

The result now follows from Lemma 5. □

LEMMA 8. *Let $\{z_k\}$ be an A^p zero-set with $z_k \neq 0$, and let G be its canonical divisor. Let G_n be the canonical divisor of the set $\{z_1, z_2, \dots, z_n\}$, and let H_n be the canonical divisor of the remaining set $\{z_{n+1}, z_{n+2}, \dots\}$. Then $G_n(z) \to G(z)$ and $H_n(z) \to 1$ as $n \to \infty$, uniformly on compact subsets of $\mathbb{D}$.*

PROOF. Note first the distinction between $G_n H_n$ and G. Both have the same zero-set $\{z_k\}$, and both are contractive divisors of analytic functions that vanish there, but $G_n H_n$ need not have unit norm. It was proved in Chapter 5 (Corollary to Theorem 6) that $\|G_n - G\|_p \to 0$ as $n \to \infty$. Hence $G_n(z) \to G(z)$ uniformly on compact subsets of $\mathbb{D}$. On the other hand, $\{H_n(z)\}$ is locally bounded in $\mathbb{D}$, since $\|H_n\| = 1$ for all n, so by Montel's theorem these functions form a normal family. Thus some subsequence $\{H_{n_j}(z)\}$ converges locally uniformly to a function $H(z)$ analytic in $\mathbb{D}$, and $H(z) \neq 0$ in $\mathbb{D}$, by Hurwitz' theorem. By Fatou's lemma, $H \in A^p$ and $\|H\|_p \leq 1$. Thus $H(0) \leq 1$. If we can show that $H(0) = 1$, it will follow that $H(z) \equiv 1$. But $\|G/G_n\|_p \leq \|G\|_p = 1$ and G/G_n vanishes on $\{z_{n+1}, z_{n+2}, \dots\}$, so $H_n(0) \geq G(0)/G_n(0)$ by the extremal property of H_n. Since $G_n(0) \to G(0)$, it follows that $H(0) \geq 1$. Combining this with the reverse inequality $H(0) \leq 1$, we see that $H(0) = 1$ and so $H(z) \equiv 1$ in $\mathbb{D}$. The argument shows that every subsequence of $\{H_n(z)\}$ has a further subsequence that converges locally uniformly to 1. This implies that $H_n(z) \to 1$ locally uniformly in $\mathbb{D}$. □

Recall now from Chapter 5 that for any nonnegative weight function $w \in L^1$, the space A^p_w consists of all functions f analytic in $\mathbb{D}$ for which

$$\|f\|^p_{p,w} = \int_{\mathbb{D}} |f(z)|^p w(z)\, d\sigma < \infty .$$

For weight functions of the form $w = |\psi|^q$ for $0 < q < \infty$ and some function $\psi \in A^q$, the space A^p_w is complete. The closure of the polynomials in the A^p_w norm is denoted by $\mathcal{P}^p_w$. We have observed in Section 5.7 that for any function $\psi \in A^p$, the subspace $[\psi]$ generated in A^p by the polynomial multiples of ψ consists of all functions $f = \psi h$ where $h \in \mathcal{P}^p_w$ and $w = |\psi|^p$. Thus we may write $[\psi] = \psi \mathcal{P}^p_w$.

For a nonconstant A^p inner function φ, the space $\mathcal{A}^p(\varphi)$ is defined to consist of all analytic functions f with the property

$$\|f\|^p_{\mathcal{A}^p(\varphi)} = \|f\|^p_p + \pi \int_{\mathbb{D}} \int_{\mathbb{D}} \Gamma(z,\zeta) \Delta\big(|f(z)|^p\big) \Delta\big(|\varphi(\zeta)|^p\big)\, d\sigma(z) d\sigma(\zeta) < \infty .$$

If f is a polynomial, or more generally a function analytic in $\overline{\mathbb{D}}$, the integral formula (Chapter 5, Theorem 2) asserts that $\|\varphi f\|_p = \|f\|_{\mathcal{A}^p(\varphi)}$. Recall that $\Gamma(z,\zeta) > 0$ for all $z, \zeta \in \mathbb{D}$, and $\Delta\big(|f(z)|^p\big) \geq 0$ for any analytic function f, so that $\|f\|_{\mathcal{A}^p(\varphi)} \geq \|f\|_p$. Note also that $\mathcal{A}^p(\varphi) \subset A^p$. It is not clear from the definition that $\mathcal{A}^p(\varphi)$ is a vector space, or that $\|\cdot\|_{\mathcal{A}^p(\varphi)}$ is a true norm, but this is a consequence of the following theorem.

THEOREM 6. *If $0 < p < \infty$ and φ is an A^p inner function, then $\mathcal{A}^p(\varphi) = \mathcal{P}^p_w$, where $w = |\varphi|^p$. Consequently, $[\varphi] = \varphi \mathcal{P}^p_w = \varphi \mathcal{A}^p(\varphi)$. Moreover, $\|f\|_{\mathcal{A}^p(\varphi)} = \|f\|_{p,w} = \|\varphi f\|_p$ for every function $f \in \mathcal{A}^p(\varphi)$, and $\varphi f_r \to \varphi f$ in A^p norm as $r \to 1$, where $f_r(z) = f(rz)$ denotes a dilation of f.*

Before embarking on the proof, we recall that the space $[\varphi]$ is invariant under multiplication by H^∞ functions. In other words, if $f \in [\psi]$ and $h \in H^\infty$, then $hf \in [\psi]$. For a proof, see the remarks preceding the proof of Theorem 7 in Chapter 5.

PROOF OF THEOREM 6. The first step is to show that $\mathcal{P}^p_w \subset \mathcal{A}^p(\varphi)$. If $f \in \mathcal{P}^p_w$, then $\|Q_n - f\|_{p,w} \to 0$, or equivalently $\|\varphi(Q_n - f)\|_p \to 0$ for some sequence of polynomials Q_n. Hence for arbitrary $\varepsilon > 0$ the expansive multiplier property of φ (Theorem 3 in Chapter 5) shows that

$$\|Q_n - Q_m\|_p \leq \|\varphi(Q_n - Q_m)\|_p < \varepsilon$$

for n, m sufficiently large, so $\{Q_n\}$ is a Cauchy sequence and $\|Q_n - g\|_p \to 0$ for some function $g \in A^p$. Consequently, $Q_n(z) \to g(z)$ locally uniformly in $\mathbb{D}$, which implies that $g = f$. Furthermore, the integral formula gives

$$\begin{aligned} \|\varphi Q_n\|^p_p &= \|Q_n\|^p_p + \pi \int_{\mathbb{D}} \int_{\mathbb{D}} \Gamma(z,\zeta) \Delta\big(|Q_n(z)|^p\big) \Delta\big(|\varphi(\zeta)|^p\big)\, d\sigma(z) d\sigma(\zeta) \\ &= \|Q_n\|^p_{\mathcal{A}^p(\varphi)}. \end{aligned}$$

Because the integrand is nonnegative, Fatou's lemma allows us to conclude that $\|\varphi f\|_p \geq \|f\|_{\mathcal{A}^p(\varphi)}$. In particular, $f \in \mathcal{A}^p(\varphi)$. This proves that $\mathcal{P}^p_w \subset \mathcal{A}^p(\varphi)$.

To prove the reverse inclusion $\mathcal{A}^p(\varphi) \subset \mathcal{P}^p_w$, suppose $f \in \mathcal{A}^p(\varphi)$. Since each dilation f_r is analytic in $\overline{\mathbb{D}}$, the integral formula gives

$$\begin{aligned}\|\varphi f_r\|_p^p &= \|f_r\|_p^p + \pi \int_{\mathbb{D}} \int_{\mathbb{D}} \Gamma(z,\zeta) \Delta\big(|f_r(\zeta)|^p\big) \Delta\big(|\varphi(z)|^p\big)\, d\sigma(\zeta) d\sigma(z) \\ &= \|f_r\|^p_{\mathcal{A}^p(\varphi)}\,.\end{aligned}$$

(The formula is stated in Chapter 5 (Theorem 2) only for polynomials f, but an approximation argument shows it holds for any function analytic in $\overline{\mathbb{D}}$, as remarked after the proof.) We begin by showing that $\|f_r\|_{\mathcal{A}^p(\varphi)} \to \|f\|_{\mathcal{A}^p(\varphi)}$ as $r \to 1$. Since $\|f_r\|_p \to \|f\|_p$ for any $f \in A^p$ (*cf.* Chapter 2, proof of Theorem 3), we need only consider the behavior of the double integral. We apply Lemma 7 with $h = \Delta(|f|^p) = p^2|f|^{p-2}|f'|^2$, so that $h_r = r^{-2}\Delta(|f_r|^p)$. Thus Lemma 7 gives

$$\int_{\mathbb{D}} \Gamma(z,\zeta)\Delta\big(|f_r(\zeta)|^p\big)\, d\sigma(\zeta) \leq 4 \int_{\mathbb{D}} \Gamma(z,\zeta)\Delta\big(|f(\zeta)|^p\big)\, d\sigma(\zeta)$$

for $\frac{1}{2} \leq r \leq 1$ and all $z \in \mathbb{D}$. Since $f \in \mathcal{A}^p(\varphi)$, Lemma 6 and the Lebesgue dominated convergence theorem permit us to conclude that

$$\begin{aligned}&\lim_{r\to 1} \int_{\mathbb{D}} \int_{\mathbb{D}} \Gamma(z,\zeta)\Delta\big(|f_r(\zeta)|^p\big)\, d\sigma(\zeta)\, \Delta\big(|\varphi(z)|^p\big)\, d\sigma(z) \\ &= \int_{\mathbb{D}} \int_{\mathbb{D}} \Gamma(z,\zeta)\Delta\big(|f(\zeta)|^p\big)\, d\sigma(\zeta)\, \Delta\big(|\varphi(z)|^p\big)\, d\sigma(z)\,.\end{aligned}$$

This shows that $\|f_r\|_{\mathcal{A}^p(\varphi)} \to \|f\|_{\mathcal{A}^p(\varphi)}$, and so $\|\varphi f_r\|_p \to \|f\|_{\mathcal{A}^p(\varphi)}$ as $r \to 1$. By Fatou's lemma, this implies $\varphi f \in A^p$ and $\|\varphi f\|_p \leq \|f\|_{\mathcal{A}^p(\varphi)}$.

So far we have shown that $f \in A^p_w$ for $w = |\varphi|^p$, and that $\|f\|_{p,w} \leq \|f\|_{\mathcal{A}^p(\varphi)}$. We wish to show that $f \in \mathcal{P}^p_w$. Let $\{z_k\}$ be the zero-set of f and let G be the corresponding canonical divisor in A^p. Let G_n and H_n be the canonical divisors of the zero-sets $\{z_1, z_2, \dots\}$ and $\{z_{n+1}, z_{n+2}, \dots\}$, respectively. For $0 < r < 1$, let G_{nr} and H_{nr} be the canonical divisors of the finite zero-sets $\frac{1}{r}\{z_1, \dots, z_n\} \cap \mathbb{D}$ and $\frac{1}{r}\{z_{n+1}, z_{n+2}, \dots\} \cap \mathbb{D}$. Note that G_{nr} and H_{nr} are not the same as the dilations $(G_n)_r$ and $(H_n)_r$, although their respective zero-sets coincide. We claim now that $G_{nr}(z) \to G_n(z)$ and $H_{nr}(z) \to H_n(z)$ as $r \to 1$, locally uniformly in $\mathbb{D}$. Indeed, since the functions H_{nr} have unit norm, they are locally bounded in $\mathbb{D}$, so they form a normal family. Hence some subsequence $\{H_{nr_j}\}$ converges locally uniformly to an analytic function h_n, where $r_j \to 1$ as $j \to \infty$. This function h_n vanishes on the set $\{z_{n+1}, z_{n+2}, \dots\}$, and $\|h_n\|_p \leq 1$ by Fatou's lemma. On the other hand, the dilation $(H_n)_r$ vanishes on the set $\frac{1}{r}\{z_{n+1}, z_{n+2}, \dots\} \cap \mathbb{D}$ and has norm $\|(H_n)_r\|_p \leq \|H_n\|_p = 1$, so

$$h_n(0) \leq H_n(0) = (H_n)_r(0) \leq H_{nr}(0)$$

by the extremal properties of H_n and H_{nr}. But $H_{nr_j}(0) \to h_n(0)$ as $j \to \infty$, so $h_n(0) = H_n(0)$. By the uniqueness of the canonical divisor, this implies that $h_n = H_n$. It now follows that $H_{nr}(z) \to H_n(z)$ locally uniformly in $\mathbb{D}$ as $r \to 1$. The same proof shows that $G_{nr}(z) \to G_n(z)$ locally uniformly.

We have already shown that $\|\varphi f_r\|_p \to \|f\|_{\mathcal{A}^p(\varphi)}$ as $r \to 1$. In particular, $\|\varphi f_r\|_p \le C$ for $\frac{1}{2} \le r < 1$. (In fact, the above application of Lemma 7 shows that this is true with $C = 4^{1/p}\|f\|_{\mathcal{A}^p(\varphi)}$.) Since f_r vanishes on $\frac{1}{r}\{z_1, z_2, \dots\} \cap \mathbb{D}$, the contractive properties of the canonical divisors G_{nr} and H_{nr} allow us to infer that

$$\left\| \frac{\varphi f_r}{G_{nr}H_{nr}} \right\|_p \le \|\varphi f_r\|_p \le C\,, \qquad \tfrac{1}{2} \le r < 1\,.$$

But $f_r/(G_{nr}H_{nr})$ has no zeros in $\mathbb{D}$, so this inequality can be recast in the form $\|F_r\|_{2,w} \le C^{p/2}$ for $\frac{1}{2} \le r < 1$, where

$$F_r = \left(\frac{f_r}{G_{nr}H_{nr}} \right)^{\frac{p}{2}} \qquad \text{and} \qquad w = |\varphi|^p\,.$$

Since the functions F_r have bounded norms, some subsequence $\{F_{r_j}\}$ is weakly convergent to a function $F \in A^2_w$, where $r_j \to 1$. In particular, $F_{r_j} \to F$ pointwise in $\mathbb{D}$. But $G_{nr} \to G_n$, $H_{nr} \to H_n$, and $f_r \to f$ pointwise in $\mathbb{D}$, so $F = \big(f/(G_nH_n)\big)^{p/2}$. A standard application of the Hahn–Banach theorem now shows that F belongs to the subspace spanned (in A^2_w norm) by the functions F_r. Indeed, if F does not belong to this subspace, then $\langle F_r, g\rangle = 0$ but $\langle F, g\rangle \ne 0$ for some function $g \in A^2_w$, which violates the weak convergence of F_r to F. Observe that for each $r < 1$, the function F_r is analytic in $\mathbb{D}$ and continuous in $\overline{\mathbb{D}}$, since G_{nr} and H_{nr} are canonical divisors of *finite* zero-sets and therefore have analytic continuations across the unit circle, with $|G_{nr}(z)| \ge 1$ and $|H_{nr}(z)| \ge 1$ for $z \in \mathbb{T}$. (See Chapter 5, Theorem 5 and Lemma 5.) Thus F_r can be approximated by polynomials, uniformly in $\mathbb{D}$. In particular, $F_r \in \mathcal{P}^2_w$ for each $r < 1$, which implies that $F \in \mathcal{P}^2_w$. This says that some sequence of polynomials Q_k converges to F in A^2_w norm, where $w = |\varphi|^p$. Now by a slight variant of the integral formula (*cf.* Chapter 5, remarks after the proof of Theorem 2), we have

$$\|Q_k\|^2_{2,w} = \|Q_k\|^2_2 + \pi \int_{\mathbb{D}} \int_{\mathbb{D}} \Gamma(z,\zeta)\Delta\big(|Q_k(z)|^2\big)\Delta\big(|\varphi(\zeta)|^p\big)\, d\sigma(z) d\sigma(\zeta)\,.$$

Letting $k \to \infty$ and applying Fatou's lemma, we conclude that

$$\|F\|^2_{2,w} \ge \|F^{2/p}\|^p_{\mathcal{A}^p(\varphi)}\,,$$

or

$$\|\varphi f_n\|_p \ge \|f_n\|_{\mathcal{A}^p(\varphi)}\,, \qquad \text{where} \qquad f_n = \frac{f}{G_nH_n}\,.$$

On the other hand, the integral formula can be applied to the dilation $(f_n)_r$ to give $\|\varphi(f_n)_r\|_p = \|(f_n)_r\|_{\mathcal{A}^p(\varphi)}$. We can then apply Lemma 7 and the Lebesgue dominated convergence theorem as above to conclude that $\|\varphi(f_n)_r\|_p \to \|f_n\|_{\mathcal{A}^p(\varphi)}$. Fatou's lemma then gives $\|\varphi f_n\|_p \le \|f_n\|_{\mathcal{A}^p(\varphi)}$. Combining this with the reverse inequality proved earlier, we conclude that $\|\varphi f_n\|_p = \|f_n\|_{\mathcal{A}^p(\varphi)}$ and so $\|\varphi(f_n)_r\|_p \to \|\varphi f_n\|_p$. Hence $\varphi(f_n)_r \to \varphi f_n$ in A^p norm as $r \to 1$, by Lemma 4.

This shows that $\varphi f_n \in [\varphi]$. It follows that $G_n \varphi f_n = \varphi f / H_n \in [\varphi]$, since G_n is the canonical divisor of a finite zero-set and therefore $G_n \in H^\infty$. Now observe that $\|\varphi f / H_n\|_p \le \|\varphi f\|_p$ by the contractive property of H_n. But Lemma 8 says that $H_n(z) \to 1$ as $n \to \infty$, for each point $z \in \mathbb{D}$. Consequently, $\|\varphi f / H_n\|_p \to \|\varphi f\|_p$ as $n \to \infty$, by Fatou's lemma. It now follows from Lemma 4 that $\varphi f / H_n \to \varphi f$ in A^p norm, so that $\varphi f \in [\varphi]$, or $f \in \mathcal{P}^p_w$. This completes the proof that $\mathcal{A}^p(\varphi) \subset \mathcal{P}^p_w$. Since we saw earlier that $\mathcal{P}^p_w \subset \mathcal{A}^p(\varphi)$, we conclude that $\mathcal{A}^p(\varphi) = \mathcal{P}^p_w$.

The remaining assertions of the theorem now follow by combining the results already obtained. We have shown that $\|f\|_{\mathcal{A}^p(\varphi)} \le \|f\|_{p,w}$ if $f \in \mathcal{P}^p_w$. We have also shown that if $f \in \mathcal{A}^p(\varphi)$, then $\|\varphi f_r\|_p \to \|f\|_{\mathcal{A}^p(\varphi)}$ as $r \to 1$, and so $\|f\|_{p,w} = \|\varphi f\|_p \le \|f\|_{\mathcal{A}^p(\varphi)}$. Thus $\|f\|_{p,w} = \|f\|_{\mathcal{A}^p(\varphi)}$ and therefore $\|\varphi f_r\|_p \to \|\varphi f\|_p$. By another appeal to Lemma 4, we infer that $\varphi f_r \to \varphi f$ in A^p norm, and the proof is complete. □

§9.5. Extremal functions as generators.

We now turn to the proof of Theorem 2. For $0 < p < \infty$, let $M = [\psi]$ be the invariant subspace of A^p generated by a function $\psi \in A^p$, and let φ be the canonical extremal function of M. We are to show that $[\varphi] = M$. Since $\varphi \in M$, it is clear that $[\varphi] \subset M$. If we can show that $\psi \in [\varphi]$, then it will follow that $M = [\psi] \subset [\varphi]$, and so $M = [\varphi]$.

Therefore, the proof reduces to showing that $\psi \in [\varphi]$. According to Theorem 6, the subspace $[\varphi]$ consists of all functions of the form φh, where $h \in \mathcal{A}^p(\varphi)$. Thus the problem is to show that $\psi/\varphi \in \mathcal{A}^p(\varphi)$. The connection between φ and ψ is displayed in the formula

$$\varphi(z) = J(0,0)^{-1/p} \psi(z) J(z,0)^{2/p},$$

proved in Theorem 7 of Chapter 5. Here $J(z,\zeta)$ is the kernel function of the Hilbert space $\mathcal{P}^2_w$ with weight $w = |\psi|^p$. Implicit in this representation of φ is the fact that $J(z,0) \ne 0$ in $\mathbb{D}$, as proved in Chapter 5 (Lemma 7). Thus ψ/φ is an analytic function with no zeros in $\mathbb{D}$.

The proof that $\psi/\varphi \in \mathcal{A}^p(\varphi)$ is long and technical. The first step is to establish an orthogonality property for difference quotients of the function $F = (\psi/\varphi)^{p/2}$. This is then applied to obtain an integral formula for $|F|^2 = |\psi/\varphi|^p$, which leads to an expression for $\|\psi\|_p^p$ as a sum of two integrals.

These integrals are approximated by passing to dilations, then manipulated with Green's formula to introduce the biharmonic Green function. The final step is to let the dilation parameter tend to 1, with the conclusion that $\|\psi/\varphi\|_{A^p(\varphi)} \le \|\psi\|_p$.

Here are the details. Define $F = (\psi/\varphi)^{p/2}$. Note that $F \in A^2_v$, where $v = |\varphi|^p$, since $\psi \in A^p$. Because the function $g(z) = J(0,0)^{-1/2}J(z,0)$ belongs to $\mathcal{P}^2_w$ and has unit norm, some sequence of polynomials Q_k converges in norm to g, and they can be chosen with norms $\|Q_k\|_{2,w} = 1$. Therefore, $Q_k(z) \to g(z)$ locally uniformly in $\mathbb{D}$. Since $g = (\varphi/\psi)^{p/2} = 1/F$ by the representation formula for φ, it follows that $Q_k(z)F(z) \to 1$ for each $z \in \mathbb{D}$. Moreover,

$$\begin{aligned} \|Q_kF\|^2_{2,v} &= \int_{\mathbb{D}} |Q_kF|^2|\varphi|^p\,d\sigma = \int_{\mathbb{D}} |Q_k|^2|\psi|^p\,d\sigma \\ &= \|Q_k\|^2_{2,w} = 1 = \|\varphi\|^p_p = \|1\|^2_{2,v}\,, \end{aligned}$$

so $\|Q_kF - 1\|_{2,v} \to 1$ as $k \to \infty$, by Lemma 4. This says that $1 \in [F]_v$, the closure in A^2_v of the polynomial multiples of F. On the other hand, by the representation of φ and the reproducing property of the kernel function $J(z,0)$ in the space $\mathcal{P}^2_w$, we have

$$\int_{\mathbb{D}} z^n F(z)|\varphi(z)|^p\,d\sigma = J(0,0)^{-1/2}\int_{\mathbb{D}} z^n\,\overline{J(z,0)}\,|\psi(z)|^p\,d\sigma = 0$$

for $n = 1, 2, \dots$, which says that 1 is orthogonal in the space A^2_v to all polynomial multiples QF with $Q(0) = 0$. Thus 1 is orthogonal to the subspace $z[F]_v$, and we can write $1 \in [F]_v \ominus z[F]_v$.

One can check easily that multiplication by F, viewed as an operator from A^2_w to A^2_v, is an isometry. We have observed that $[\psi] = \psi\mathcal{P}^p_w$, where $w = |\psi|^p$. Similarly, it is easy to see that $[F]_v = F\mathcal{P}^2_w$.

For fixed $\zeta \in \mathbb{D}$, define

$$H_\zeta(z) = \frac{F(z) - F(\zeta)}{z - \zeta}\,, \qquad z \ne \zeta\,,$$

and $H_\zeta(\zeta) = F'(\zeta)$. The function $g_\zeta(z) = (z-\zeta)H_\zeta(z) = F(z) - F(\zeta)$ belongs to $[F]_v$, since $1 \in [F]_v$. Thus it has the form $g_\zeta = Fh$ for some $h \in \mathcal{P}^2_w$. Also $h(\zeta) = 0$, since $g_\zeta(\zeta) = 0$ and $F(z) \ne 0$ for all $z \in \mathbb{D}$. It follows that h can be approximated in A^2_w norm by polynomials that vanish at ζ. (Compare the proof of Theorem 7 in Chapter 5.) Thus for each $\varepsilon > 0$ there is a polynomial $\widetilde{Q}$ such that $Q(z) = (z-\zeta)\widetilde{Q}(z)$ has the property $\|h - Q\|_{2,w} < \varepsilon$. Writing $h(z) = (z-\zeta)\widetilde{h}(z)$, we infer that $\widetilde{Q}$ approximates $\widetilde{h}$ in A^2_w norm, since multiplication by $(z-\zeta)$ is bounded below (*cf.* Section 2.9). Hence $\widetilde{h} \in \mathcal{P}^2_w$, and $H_\zeta = F\widetilde{h} \in F\mathcal{P}^2_w = [F]_v$. Because 1 is orthogonal to $z[F]_v$, this implies that

$$\int_{\mathbb{D}} zH_\zeta(z)|\varphi(z)|^p\,d\sigma = 0\,. \tag{1}$$

We are now prepared to prove that $\psi/\varphi \in \mathcal{A}^p(\varphi)$. As previously noted, this will prove Theorem 2. For fixed $\zeta \in \mathbb{D}$, we define

$$I_\zeta(z) = \frac{zF(z) - \zeta F(\zeta)}{z - \zeta}, \qquad z \neq \zeta,$$

and $I_\zeta(\zeta) = F(\zeta) + \zeta F'(\zeta)$. Thus $I_\zeta(z) = F(z) + \zeta H_\zeta(z)$, and

$$|I_\zeta(z)|^2 = |F(z)|^2 + 2\,\mathrm{Re}\{\zeta H_\zeta(z)\overline{F(z)}\} + |\zeta|^2 |H_\zeta(z)|^2 .$$

We will now integrate with respect to the parameter ζ. For $0 < r < 1$, let C_r be the circle $|\zeta| = r$. By the mean-value property of harmonic functions,

$$\begin{aligned}
&\frac{1}{2\pi r}\int_{C_r} (|I_\zeta(z)|^2 - |\zeta|^2 |H_\zeta(z)|^2)\,|d\zeta| \\
&= \frac{1}{2\pi r}\int_{C_r} (|F(z)|^2 + 2\,\mathrm{Re}\{\zeta H_\zeta(z)\overline{F(z)}\})\,|d\zeta| \\
&= |F(z)|^2 = |\psi(z)/\varphi(z)|^p, \qquad z \in \mathbb{D}.
\end{aligned}$$

Therefore, by Fubini's theorem,

$$\int_{\mathbb{D}} |\psi|^p\,d\sigma(z) = \frac{1}{2\pi r}\int_{C_r}\int_{\mathbb{D}} (|I_\zeta(z)|^2 - |\zeta|^2|H_\zeta(z)|^2)|\varphi(z)|^p\,d\sigma(z)|d\zeta|. \tag{2}$$

Now observe that $I_\zeta(z) = F(\zeta) + zH_\zeta(z)$ and so the inner integral takes the form

$$\begin{aligned}
&\int_{\mathbb{D}} (|I_\zeta(z)|^2 - |\zeta|^2|H_\zeta(z)|^2)|\varphi(z)|^p\,d\sigma(z) \\
&= \int_{\mathbb{D}} (|F(\zeta)|^2 + 2\,\mathrm{Re}\{zH_\zeta(z)\overline{F(\zeta}\} + (|z|^2 - |\zeta|^2)|H_\zeta(z)|^2)|\varphi(z)|^p\,d\sigma(z) \\
&= \left|\frac{\psi(\zeta)}{\varphi(\zeta)}\right|^p + \int_{\mathbb{D}} (|z|^2 - |\zeta|^2)|H_\zeta(z)|^2|\varphi(z)|^p\,d\sigma(z),
\end{aligned}$$

by the orthogonality property (1) and the relations $|F|^2 = |\psi/\varphi|^p$ and $\|\varphi\|_p = 1$. Inserting this expression into (2) and applying Fubini's theorem again, we find

$$\begin{aligned}
&\int_{\mathbb{D}} |\psi(z)|^p\,d\sigma(z) \\
&= \frac{1}{2\pi r}\int_{C_r}\left|\frac{\psi(\zeta)}{\varphi(\zeta)}\right|^p |d\zeta| + \frac{1}{2\pi r}\int_{C_r}\int_{\mathbb{D}} (|z|^2 - |\zeta|^2)|H_\zeta(z)|^2|\varphi(z)|^p\,d\sigma(z)|d\zeta| \\
&\geq \frac{1}{2\pi r}\int_{C_r}\left|\frac{\psi(\zeta)}{\varphi(\zeta)}\right|^p |d\zeta| + \frac{1}{2\pi r}\int_{r\mathbb{D}} |\varphi(z)|^p \int_{C_r} (|z|^2 - r^2)|H_\zeta(z)|^2\,|d\zeta|d\sigma(z).
\end{aligned} \tag{3}$$

The remainder of the proof consists of transforming and estimating the two integrals on the last line of (3) to obtain the inequality $\|\psi/\varphi\|_{A^p(\varphi)} \le \|\psi\|_p$ as $r \to 1$. The first integral can be transformed by Green's formula (see Section 1.5) with $u = |\psi/\varphi|^p$ and $v(\zeta) = \frac{1}{2r}(|\zeta|^2 - r^2)$. After a change of variables, we find

$$\begin{aligned}
&\frac{1}{2\pi r}\int_{C_r}\left|\frac{\psi(\zeta)}{\varphi(\zeta)}\right|^p |d\zeta| \\
&= \frac{1}{r^2}\int_{r\mathbb{D}}\left|\frac{\psi(\zeta)}{\varphi(\zeta)}\right|^p d\sigma(\zeta) + \frac{1}{4r^2}\int_{r\mathbb{D}}(r^2-|\zeta|^2)\,\Delta\left(\left|\frac{\psi(\zeta)}{\varphi(\zeta)}\right|^p\right) d\sigma(\zeta) \qquad (4) \\
&= \int_{\mathbb{D}}\left|\frac{\psi_r(\zeta)}{\varphi_r(\zeta)}\right|^p d\sigma(\zeta) + \frac{1}{4}\int_{\mathbb{D}}(1-|\zeta|^2)\,\Delta\left(\left|\frac{\psi_r(\zeta)}{\varphi_r(\zeta)}\right|^p\right) d\sigma(\zeta),
\end{aligned}$$

where $f_r(\zeta) = f(r\zeta)$ denotes a dilation.

To handle the last term in (3), we will need two more lemmas. Recall that Green's function of the unit disk is given by the formula

$$G(z,\zeta) = -\log\left|\frac{z-\zeta}{1-\bar{\zeta}z}\right|.$$

LEMMA 9. *Suppose $0 < r < 1$, $z \in r\mathbb{D}$, and $f \in C^2(\overline{r\mathbb{D}})$. Then*

$$\int_{r\mathbb{D}} G\left(\frac{z}{r},\frac{\zeta}{r}\right)\Delta f(\zeta)\,d\sigma(\zeta) = \frac{1}{\pi r}\int_{C_r}\frac{r^2-|z|^2}{|z-\zeta|^2}f(\zeta)\,|d\zeta| - 2f(z).$$

PROOF. Apply Green's formula, placing a small "safety circle" about the point z.

LEMMA 10. *Suppose $0 < r < 1$ and $\zeta \in \mathbb{D}$, and let φ be an A^p inner function. Then*

$$4\int_{\mathbb{D}}\Gamma(z,\zeta)\Delta(|\varphi_r(z)|^p)\,d\sigma(z) \le -2\int_{\mathbb{D}}G(z,\zeta)|\varphi_r(z)|^p\,d\sigma(z) + 1 - |\zeta|^2,$$

where $\Gamma(z,\zeta)$ is the biharmonic Green function of $\mathbb{D}$.

Deferring the proof of Lemma 10, we apply it to the second integral in the last line of (3). Using Lemma 9 with $f(\zeta) = |F(z) - F(\zeta)|^2$, and noting that $\Delta f = \Delta(|F|^2)$, we have

$$\frac{1}{2\pi r}\int_{C_r}\frac{|z|^2-r^2}{|z-\zeta|^2}|F(z)-F(\zeta)|^2\,|d\zeta| = -\frac{1}{2}\int_{r\mathbb{D}}G\left(\frac{z}{r},\frac{\zeta}{r}\right)\Delta(|F(\zeta)|^2)\,d\sigma(\zeta).$$

Thus by Fubini's theorem and Lemma 10,

$$
\begin{aligned}
\frac{1}{2\pi r}&\int_{r\mathbb{D}}|\varphi(z)|^p\int_{C_r}(|z|^2-r^2)|H_\zeta(z)|^2\,|d\zeta|d\sigma(z)\\
&=-\frac{1}{2}\int_{r\mathbb{D}}|\varphi(z)|^p\int_{r\mathbb{D}}G\Big(\frac{z}{r},\frac{\zeta}{r}\Big)\,\Delta\big(|F(\zeta)|^2\big)\,d\sigma(\zeta)d\sigma(z)\\
&=-\frac{1}{2}\int_{r\mathbb{D}}\int_{r\mathbb{D}}G\Big(\frac{z}{r},\frac{\zeta}{r}\Big)\,|\varphi(z)|^p\,d\sigma(z)\,\Delta\bigg(\bigg|\frac{\psi(\zeta)}{\varphi(\zeta)}\bigg|^p\bigg)\,d\sigma(\zeta)\\
&=-\frac{r^2}{2}\int_{\mathbb{D}}\int_{\mathbb{D}}G(z,\zeta)|\varphi_r(z)|^p\,d\sigma(z)\,\Delta\bigg(\bigg|\frac{\psi_r(\zeta)}{\varphi_r(\zeta)}\bigg|^p\bigg)\,d\sigma(\zeta)\\
&\geq r^2\int_{\mathbb{D}}\int_{\mathbb{D}}\Gamma(z,\zeta)\,\Delta\big(|\varphi_r(z)|^p\big)d\sigma(z)\,\Delta\bigg(\bigg|\frac{\psi_r(\zeta)}{\varphi_r(\zeta)}\bigg|^p\bigg)\,d\sigma(\zeta)\\
&\qquad-\frac{r^2}{4}\int_{\mathbb{D}}(1-|\zeta|^2)\,\Delta\bigg(\bigg|\frac{\psi_r(\zeta)}{\varphi_r(\zeta)}\bigg|^p\bigg)\,d\sigma(\zeta)\,.
\end{aligned}
\tag{5}
$$

Substitute (4) and (5) into (3) to obtain

$$
\begin{aligned}
\int_{\mathbb{D}}|\psi(z)|^p\,d\sigma(z)\geq&\int_{\mathbb{D}}\bigg|\frac{\psi_r(\zeta)}{\varphi_r(\zeta)}\bigg|^p\,d\sigma(\zeta)\\
&+r^2\int_{\mathbb{D}}\int_{\mathbb{D}}\Gamma(z,\zeta)\Delta\big(|\varphi_r(z)|^p\big)\,\Delta\bigg(\bigg|\frac{\psi_r(\zeta)}{\varphi_r(\zeta)}\bigg|^p\bigg)\,d\sigma(z)d\sigma(\zeta)\\
&+\frac{1}{4}(1-r^2)\int_{\mathbb{D}}(1-|\zeta|^2)\,\Delta\bigg(\bigg|\frac{\psi_r(\zeta)}{\varphi_r(\zeta)}\bigg|^p\bigg)\,d\sigma(\zeta)\,.
\end{aligned}
$$

Now let $r\to 1$ and apply Fatou's lemma to conclude that

$$
\|\psi/\varphi\|_p^p+\int_{\mathbb{D}}\Gamma(z,\zeta)\Delta\big(|\varphi(z)|^p\big)\Delta\bigg(\bigg|\frac{\psi(\zeta)}{\varphi(\zeta)}\bigg|^p\bigg)\,d\sigma(z)d\sigma(\zeta)\leq\|\psi\|_p^p\,,
$$

or $\|\psi/\varphi\|_{\mathcal{A}^p(\varphi)}\leq\|\psi\|_p$. Thus $\psi/\varphi\in\mathcal{A}^p(\varphi)$, and the proof of Theorem 2 is complete, except for Lemma 10.

For the proof of Lemma 10, one further lemma is required.

LEMMA 11. *Let φ be an A^p inner function, and let $\varphi_r(z)=\varphi(rz)$ be a dilation, where $0<r<1$. Then*

$$
\int_{\mathbb{D}}|\varphi_r(z)|^p h(z)\,d\sigma(z)\leq h(0)
$$

for every positive harmonic function h.

PROOF OF LEMMA 11. By Lemma 1 of Chapter 5, stated for canonical divisors but valid for any A^p inner function,

$$
\int_{\mathbb{D}}|\varphi(z)|^p h_r(z)\,d\sigma(z)=h_r(0)=h(0)\,,
$$

since h_r is a bounded harmonic function. Hence an application of Fatou's lemma shows that

$$\int_{\mathbb{D}} |\varphi(z)|^p h(z)\, d\sigma(z) \le h(0)\,.$$

But for fixed $z \in \mathbb{D}$ the function $\lambda \mapsto |\varphi(\lambda z)|^p$ is subharmonic, so

$$|\varphi_r(z)|^p = |\varphi(rz)|^p \le \frac{1}{2\pi}\int_{\mathbb{T}} P(\zeta, r)|\varphi(\zeta z)|^p\, |d\zeta|\,,$$

where $\mathbb{T}$ is the unit circle and

$$P(\zeta, r) = \frac{1-r^2}{|\zeta - r|^2}$$

is the Poisson kernel. Thus by Fubini's theorem,

$$\begin{aligned}\int_{\mathbb{D}} |\varphi_r(z)|^p h(z)\, d\sigma(z) &\le \frac{1}{2\pi}\int_{\mathbb{T}} P(\zeta, r)\int_{\mathbb{D}} |\varphi(\zeta z)|^p h(z)\, d\sigma(z) |d\zeta| \\ &\le \frac{1}{2\pi} h(0)\int_{\mathbb{T}} P(\zeta, r)\, |d\zeta| = h(0)\,. \qquad \square\end{aligned}$$

PROOF OF LEMMA 10. The explicit formula for $\Gamma(z,\zeta)$ is

$$\Gamma(z,\zeta) = \frac{1}{16}\left\{|z-\zeta|^2 \log\left|\frac{z-\zeta}{1-\bar{\zeta}z}\right|^2 + (1-|\zeta|^2)(1-|z|^2)\right\},$$

as shown in Chapter 1 (Section 1.6). A direct calculation gives

$$4\,\Delta_z \Gamma(z,\zeta) = -2\,G(z,\zeta) + \frac{(1-|\zeta|^2)(1-|\zeta z|^2)}{|1-\bar{\zeta}z|^2}\,. \tag{6}$$

Next observe that by Green's formula,

$$\int_{\mathbb{D}} \Gamma(z,\zeta)\Delta\big(|f(z)|^p\big)\, d\sigma(z) = \int_{\mathbb{D}} \Delta_z \Gamma(z,\zeta)\, |f(z)|^p\, d\sigma(z) \tag{7}$$

if f is analytic in $\overline{\mathbb{D}}$, since $\Gamma(z,\zeta) = \frac{\partial \Gamma}{\partial n}(z,\zeta) = 0$ on the boundary. (Use safety circles to deal with the singularities at ζ and the zeros of f.) Now apply (7) with $f = \varphi_r$ and use the formula (6) to obtain

$$\begin{aligned} 4\int_{\mathbb{D}} \Gamma(z,\zeta)\Delta\big(|\varphi_r(z)|^p\big)\, d\sigma(z) &= 4\int_{\mathbb{D}} \Delta_z\Gamma(z,\zeta)|\varphi_r(z)|^p\, d\sigma(z) \\ = -2\int_{\mathbb{D}} G(z,\zeta)|\varphi_r(z)|^p\, d\sigma(z) &+ (1-|\zeta|^2)\int_{\mathbb{D}} \frac{1-|\zeta z|^2}{|1-\bar{\zeta}z|^2}\, |\varphi_r(z)|^p\, d\sigma(z)\,.\end{aligned}$$

Therefore, since

$$\frac{1-|\zeta z|^2}{|1-\overline{\zeta}z|^2} = \operatorname{Re}\left\{\frac{1+\overline{\zeta}z}{1-\overline{\zeta}z}\right\}$$

is a positive harmonic function of z whose value at the origin is 1, the desired result follows from Lemma 11. $\square$

The proofs of Theorem 2 and the preliminary Theorem 6 are based on the original paper of Aleman, Richter, and Sundberg [1]. The organization of lemmas as presented here is derived in part from the book of Hedenmalm, Korenblum, and Zhu [2].

As an application of Theorem 2, it can now be shown that the canonical divisor of an arbitrary A^p zero-set generates the corresponding invariant subspace.

THEOREM 7. *For $0 < p < \infty$, let $\{z_k\}$ be any A^p zero-set and let N^p be the subspace of functions in A^p that vanish on $\{z_k\}$. If G is the canonical divisor of $\{z_k\}$, then $[G] = N^p$.*

This theorem is easily proved for *finite* zero-sets; an argument was outlined in the last section of Chapter 5. The generalization to arbitrary A^p zero-sets was established by Aleman, Richter, and Sundberg [1] as a consequence of Theorem 2. Duren, Khavinson, and Shapiro [1] had obtained the result, with a more elementary proof, for the special case of Blaschke sequences. Then the corresponding Blaschke product is also a generator of N^p. (See Theorem 6 in Chapter 4.)

Theorem 7 does not follow trivially from Theorem 2 because it is not evident that N^p is cyclic. The cyclicity of N^p is a corollary of a more general theorem that has independent interest.

THEOREM 8. *Suppose $0 < p < \infty$. Let $M_1 \supset M_2 \supset \dots$ be cyclic subspaces of A^p, and suppose the invariant subspace $M = \bigcap_{n=1}^{\infty} M_n \neq \{0\}$. Then M is cyclic. Moreover, if φ_n is the canonical extremal function of M_n, then $\{\varphi_n\}$ converges in A^p norm to the canonical extremal function of M.*

PROOF OF THEOREM 8. Suppose first that some function in M does not vanish at the origin. Since $\|\varphi_n\|_p = 1$ for all n, the sequence $\{\varphi_n\}$ is locally bounded in $\mathbb{D}$ and so by Montel's theorem it has a subsequence $\{\varphi_{n_k}\}$ that converges locally uniformly to an analytic function φ. By Theorem 2, the extremal function φ_n generates M_n, and so by Theorem 6 each function $f \in M_n$ has the form $f = \varphi_n h_n$ with $h_n \in \mathcal{A}^p(\varphi_n) \subset A^p$ and

$$\|h_n\|_p \leq \|h_n\|_{\mathcal{A}^p(\varphi_n)} = \|\varphi_n h_n\|_p = \|f\|_p .$$

Hence every function $f \in M$ can be written $f = \varphi_{n_k} h_{n_k}$ with $\|h_{n_k}\|_p \leq \|f\|_p$ for $k = 1, 2, \dots$. If $\varphi(z) \equiv 0$, then $f(z) \equiv 0$ and so $M = \{0\}$, contrary to hypothesis. Thus φ is not the zero function. Because the subspaces M_n are nested, we see for each index m that $\varphi_{n_k} \in M_m = [\varphi_m]$ for all $n_k \geq m$. In view of Theorem 6, this implies that $\varphi_{n_k}/\varphi_m \in \mathcal{A}^p(\varphi_m)$ for $n_k \geq m$, with

$$\|\varphi_{n_k}/\varphi_m\|_{\mathcal{A}^p(\varphi_m)} = \|\varphi_m(\varphi_{n_k}/\varphi_m)\|_p = \|\varphi_{n_k}\|_p = 1\,.$$

An application of Fatou's lemma now shows that $\varphi/\varphi_m \in \mathcal{A}^p(\varphi_m)$, and by Theorem 6 this says that $\varphi \in [\varphi_m] = M_m$. Since m was chosen arbitrarily, it follows that $\varphi \in M$. Fatou's lemma also shows that $\|\varphi\|_p \leq 1$. Moreover, φ is the canonical extremal function of M. If not, then $\varphi(0) < g(0)$ for some $g \in M$ with $\|g\|_p = 1$. Since $g \in M_n$ for all n, the extremal property of φ_n shows that $g(0) \leq \varphi_n(0)$. Thus

$$\varphi(0) < g(0) \leq \varphi_{n_k}(0) \to \varphi(0) \qquad \text{as } k \to \infty\,,$$

a contradiction that shows φ is the extremal function for M. By the uniqueness of the canonical extremal function, it follows that $\varphi_n \to \varphi$ locally uniformly in $\mathbb{D}$. Since $\|\varphi_n\|_p = 1$ and $\|\varphi\|_p = 1$, we can invoke Lemma 4 to infer that $\|\varphi_n - \varphi\|_p \to 0$ as $n \to \infty$.

If all functions of M vanish at the origin, one makes the usual modification, working with the smallest integer m for which $f^{(m)}(0) \neq 0$ for some function $f \in M$. The argument is similar and the details are omitted. □

PROOF OF THEOREM 7. In order to deduce the result from Theorem 2, we have only to show that N^p is cyclic. We can do this with the help of Theorem 8. Let N_n^p denote the subspace of functions in A^p that vanish on the finite zero-set $\{z_1, z_2, \ldots, z_n\}$. Then as previously noted, each subspace N_n^p is cyclic and is generated by its canonical divisor (and also by the corresponding finite Blaschke product). Since $N_1^p \supset N_2^p \supset \ldots$ and $N^p = \bigcap_{n=1}^\infty N_n^p$, Theorem 8 implies that N^p is cyclic, and Theorem 2 then asserts that $[G] = N^p$. □

References

Abakumov, E. and Borichev, A.

[1] Shift invariant subspaces with arbitrary indices in ℓ^p spaces, *J. Funct. Anal.* **188** (2002), 1–26.

Abkar, A.

[1] Norm approximation by polynomials in some weighted Bergman spaces, *J. Funct. Anal.* **191** (2002), 224–240.

Aharonov, D., Shapiro, H. S., and Shields, A. L.

[1] Weakly invertible elements in the space of square-summable holomorphic functions, *J. London Math. Soc.* **9** (1974), 183–192.

Aleman, A.

[1] Invariant subspaces with finite codimension in Bergman spaces, *Trans. Amer. Math. Soc.* **330** (1992), 531–544.

Aleman, A., Hedenmalm, H., Richter, S., and Sundberg, C.

[1] Curious properties of canonical divisors in weighted Bergman spaces, *Entire Functions in Modern Analysis (Tel Aviv, 1997), Israel Math. Conf. Proc.* **15** (Bar-Ilan University, Ramat Gan, 2001), pp. 1–10.

Aleman, A. and Richter, S.

[1] Some sufficient conditions for the division property of invariant subspaces in weighted Bergman spaces, *J. Funct. Anal.* **144** (1997), 542–556.

[2] Single point extremal functions in Bergman-type spaces, *Indiana Univ. Math. J.* **51** (2002), 581–605.

Aleman, A., Richter, S., and Sundberg, C.

[1] Beurling's theorem for the Bergman space, *Acta Math.* **177** (1996), 275–310.

[2] The majorization function and the index of invariant subspaces in the Bergman spaces, *J. Analyse Math.* **86** (2002), 139–182.

Amar, E.

[1] Suites d'interpolation pour les classes de Bergman de la boule et du polydisque de $\mathbb{C}^n$, *Canad. J. Math.* **30** (1978), 711–737.

[2] Extension de fonctions holomorphes et courants, *Bull. Sci. Math.* **107** (1983), 25–48.

Anderson, J. M.

[1] Bloch functions: The basic theory, *Operators and Function Theory* (S. C. Power, editor), D. Reidel, Boston, 1985, pp. 1–17.

Anderson, J. M., Clunie, J., and Pommerenke, Ch.

[1] On Bloch functions and normal functions, *J. Reine Angew. Math.* **270** (1974), 12–37.

Andersson, Mats Erik

[1] Integral means on Bergman spaces, *Complex Variables Theory Appl.* **32** (1997), 147–160.

Apostol, C., Bercovici, H., Foiaş, C., and Pearcy, C.

[1] Invariant subspaces, dilation theory, and the structure of the predual of a dual algebra, I, *J. Funct. Anal.* **63** (1985), 369–404.

Axler, S.

[1] Zero-multipliers of Bergman spaces, *Canad. Math. Bull.* **28** (1985), 237–242.

[2] The Bergman space, the Bloch space, and commutators of multiplication operators, *Duke Math. J.* **53** (1986), 315–332.

[3] Bergman spaces and their operators, *Surveys of Some Recent Results in Operator Theory*, Vol. I (J. B. Conway and B. B. Morrel, editors), Pitman Research Notes in Math. No. 171, John Wiley & Sons, New York, 1988, pp. 1–50.

Beller, E.

[1] Zeros of A^p functions and related classes of analytic functions, *Israel J. Math.* **22** (1975), 68–80.

Beller, E. and Horowitz, C.

[1] Zero sets and random zero sets in certain function spaces, *J. Analyse Math.* **64** (1994), 203–217.

Bergman, S.

[1] *The Kernel Function and Conformal Mapping* (Second Edition, American Mathematical Society, Providence, R. I., 1970).

Bergman, S. and Schiffer, M.

[1] Kernel functions and conformal mapping, *Compositio Math.* **8** (1951), 205–249.

[2] *Kernel Functions and Elliptic Differential Equations in Mathematical Physics* (Academic Press, New York, 1953).

Berman, R., Brown, L., and Cohn, W.

[1] Cyclic vectors of bounded characteristic in Bergman spaces, *Michigan Math. J.* **31** (1984), 295–306.

Berndtsson, B. and Ortega-Cerdà, J.

[1] On interpolation and sampling in Hilbert spaces of analytic functions, *J. Reine Angew. Math.* **464** (1995), 109–128.

Beurling, A.
[1] Ensembles exceptionels, *Acta Math.* **72** (1940), 1–13.
[2] On two problems concerning linear transformations in Hilbert space, *Acta Math.* **81** (1949), 239–255.
[3] Mittag-Leffler Lectures on Harmonic Analysis (1977–1978): *IV.* Balayage of Fourier–Stieltjes transforms; *V.* Interpolation for an interval on $\mathbb{R}^1$, *Collected Works of Arne Beurling* (Birkhäuser, Boston, 1989), Volume 2, pp. 341–365.

Boas, R. P.
[1] Some uniformly convex spaces, *Bull. Amer. Math. Soc.* **46** (1940), 304–311.

Bomash, G.
[1] A Blaschke-type product and random zero sets for Bergman spaces, *Arkiv for Matematik* **30** (1992), 45–60.

Borichev, A.
[1] Estimates from below and cyclicity in Bergman-type spaces, *Internat. Math. Res. Notices* **12** (1996), 603–611.
[2] Invariant subspaces of given index in Banach spaces of analytic functions, *J. Reine Angew. Math.* **505** (1998), 23–44.

Borichev, A. A. and Hedenmalm, H.
[1] Cyclicity in Bergman-type spaces, *Internat. Math. Res. Notices* **11** (1995), 253–262.
[2] Harmonic functions of maximal growth: invertibility and cyclicity in Bergman spaces, *J. Amer. Math. Soc.* **10** (1997), 761–796.

Bourdon, P.
[1] Cyclic Nevanlinna class functions in Bergman spaces, *Proc. Amer. Math. Soc.* **93** (1985), 503–506.
[2] Cellular-indecomposable operators and Beurling's theorem, *Michigan Math. J.* **33** (1986), 187–193.

Brown, L. and Korenblum, B.
[1] Cyclic vectors in $A^{-\infty}$, *Proc. Amer. Math. Soc.* **102** (1988), 137–138.

Brown, L. and Shields, A. L.
[1] Cyclic vectors in the Dirichlet space, *Trans. Amer. Math. Soc.* **285** (1984), 269–304.

Bruna, J. and Pascuas, D.
[1] Interpolation in $A^{-\infty}$, *J. London Math. Soc.* **40** (1989), 452–466.

Buckley, S. M., Koskela, P., and Vukotić, D.
[1] Fractional integration, differentiation, and weighted Bergman spaces, *Math. Proc. Cambridge Philos. Soc.* **126** (1999), 369–385.

Carleson, L.

[1] Sets of uniqueness for functions regular in the unit circle, *Acta Math.* **87** (1952), 325–345.

[2] An interpolation problem for bounded analytic functions, *Amer. J. Math.* **80** (1958), 921–930.

[3] Interpolations by bounded analytic functions and the corona problem, *Ann. of Math.* **76** (1962), 547–559.

Cima, J. A.

[1] The basic properties of Bloch functions, *Internat. J. Math. Math. Sci.* **2** (1979), 369–413.

Cima, J. A. and Derrick, W. R.

[1] Extremals for subspaces of the Bergman space, *Houston J. Math.* **20** (1994), 623–628.

Clarkson, J. A.

[1] Uniformly convex spaces, *Trans. Amer. Math. Soc.* **40** (1936), 396–414.

Coifman, R. R. and Rochberg, R.

[1] Representation theorems for holomorphic and harmonic functions in L^p, *Astérisque* **77** (1980), 1–66.

Cowen, C. and MacCluer, B.

[1] *Composition Operators on Spaces of Analytic Functions* (CRC Press, Boca Raton, 1995).

Djrbashian, A. E. and Shamoian, F. A.

[1] *Topics in the Theory of A^p_α Spaces* (B. G. Teubner, Leipzig, 1988).

Duffin, R. J.

[1] On a question of Hadamard concerning super-biharmonic functions, *J. Math. and Physics* **27** (1949), 253–258.

Duren, P. L.

[1] Smoothness of functions generated by Riesz products, *Proc. Amer. Math. Soc.* **16** (1965), 1263–1268.

[2] Extension of a theorem of Carleson, *Bull. Amer. Math. Soc.* **75** (1969), 143–146.

[3] On the multipliers of H^p spaces, *Proc. Amer. Math. Soc.* **22** (1969), 24–27.

[4] On the Bloch–Nevanlinna conjecture, *Colloq. Math.* **20** (1969), 295–297.

[5] *Theory of H^p Spaces* (Academic Press, New York, 1970; reprinted with supplement by Dover Publications, Mineola, N.Y., 2000).

[6] *Univalent Functions* (Springer–Verlag, New York, 1983).

Duren, P., Khavinson, D., and Shapiro, H. S.

[1] Extremal functions in invariant subspaces of Bergman spaces, *Illinois J. Math.* **40** (1996), 202–210.

Duren, P., Khavinson, D., Shapiro, H. S., and Sundberg, C.
[1] Contractive zero-divisors in Bergman spaces, *Pacific J. Math.* **157** (1993), 37–56.
[2] Invariant subspaces in Bergman spaces and the biharmonic equation, *Michigan Math. J.* **41** (1994), 247–259.

Duren, P. L., Romberg, B. W., and Shields, A. L.
[1] Linear functionals on H^p spaces with $0 < p < 1$, *J. Reine Angew. Math.* **238** (1969), 32–60.

Duren, P. and Schuster, A. P.
[1] Finite unions of interpolation sequences, *Proc. Amer. Math. Soc.* **130** (2002), 2609–2615.

Duren, P., Schuster, A. P., and Seip, K.
[1] Uniform densities of regular sequences in the unit disk, *Trans. Amer. Math. Soc.* **352** (2000), 3971–3980.

Duren, P., Schuster, A., and Vukotić, D.
[1] On uniformly discrete sets in the disk, *to appear.*

Duren, P. L. and Shields, A. L.
[1] Properties of H^p $(0 < p < 1)$ and its containing Banach space, *Trans. Amer. Math. Soc.* **141** (1969), 255–262.
[2] Coefficient multipliers of H^p and B^p spaces, *Pacific J. Math.* **32** (1970), 69–78.
[3] Restrictions of H^p functions to the diagonal of the polydisc, *Duke Math. J.* **42** (1975), 751–753.

Duren, P. L. and Taylor, G. D.
[1] Mean growth and coefficients of H^p functions, *Illinois J. Math.* **14** (1970), 419–423.

Engliš, M.
[1] A Loewner-type lemma for weighted biharmonic operators, *Pacific J. Math.* **179** (1997), 343–353.
[2] Weighted biharmonic Green functions for rational weights, *Glasgow Math. J.* **41** (1999), 239–269.

Epstein, B.
[1] *Orthogonal Families of Analytic Functions* (Macmillan, New York, 1965).

Fisher, S. D.
[1] *Function Theory on Planar Domains* (John Wiley & Sons, New York, 1983).

Folland, G.B.
[1] *Introduction to Partial Differential Equations* (Second Edition, Princeton University Press, Princeton, N. J., 1995).

Forelli, F.
[1] The isometries of H^p, *Canad. J. Math.* **16** (1964), 721–728.

Forelli, F. and Rudin, W.
[1] Projections on spaces of holomorphic functions in balls, *Indiana Univ. Math. J.* **24** (1974), 593–602.

Friedman, A.
[1] *Partial Differential Equations* (Holt, Rinehart and Winston, New York, 1969).

Garabedian, P. R.
[1] A partial differential equation arising in conformal mapping, *Pacific J. Math.* **1** (1951), 485–524.
[2] *Partial Differential Equations* (Second Edition, Chelsea Publ. Co., New York, 1986).

Garnett, J.
[1] *Bounded Analytic Functions* (Academic Press, New York, 1981).

Hansbo, J.
[1] Reproducing kernels and contractive divisors in Bergman spaces, *Zap. Nauchn. Sem. S.-Peterburg. Otdel. Mat. Inst. Steklov (POMI)* **232** (1996), *Issled. po Lineĭn. Oper. i Teor. Funktsiĭ* **24**, 174–198, 217 = *J. Math. Sci. (New York)* **92** (1998), no. 1, 3657–3674.

Hardy, G. H. and Littlewood, J. E.
[1] Some properties of conjugate functions, *J. Reine Angew. Math.* **167** (1931), 405–423.
[2] Some properties of fractional integrals, *II*, *Math. Z.* **34** (1932), 403–439.
[3] Notes of the theory of series (*XX*): Generalizations of a theorem of Paley, *Quart. J. Math., Oxford Ser.* **8** (1937), 161–171.

Hastings, W. W.
[1] A Carleson measure theorem for Bergman spaces, *Proc. Amer. Math. Soc.* **52** (1975), 237–241.

Hayman, W. K.
[1] On a conjecture of Korenblum, *Analysis (Munich)* **19** (1999), 195–205.

Hayman, W. K. and Korenblum, B.
[1] An extension of the Riesz-Herglotz formula, *Ann. Acad. Sci. Fenn. Ser. A.I. Math.* **2** (1976), 175–201.
[2] A critical growth rate for functions regular in a disk, *Michigan Math. J.* **27** (1980), 21–30.

Hedenmalm, H.

[1] A factorization theorem for square area-integrable analytic functions, *J. Reine Angew. Math.* **422** (1991), 45–68.

[2] A factoring theorem for a weighted Bergman space, *Algebra i Analiz* **4** (1992), no. 1, 167–176 = *St. Petersburg Math. J.* **4** (1993), 163–174.

[3] An invariant subspace of the Bergman space having the codimension two property, *J. Reine Angew. Math.* **443** (1993), 1–9.

[4] Spectral properties of invariant subspaces in the Bergman space, *J. Funct. Anal.* **116** (1993), 441–448.

[5] A factoring theorem for the Bergman space, *Bull. London Math. Soc.* **26** (1994), 113–126.

[6] A computation of Green functions for the weighted biharmonic operators $\Delta|z|^{-2\alpha}\Delta$, with $\alpha > -1$, *Duke Math. J.* **75** (1994), 51–78.

[7] Open problems in the function theory of the Bergman space, *Festschrift in Honour of Lennart Carleson and Yngve Domar* (Uppsala, 1995), pp. 153–169.

[8] Maximal invariant subspaces in the Bergman space, *Ark. Mat.* **36** (1998), 97–101.

[9] Boundary value problems for weighted biharmonic operators, *Algebra i Analiz* **8** (1996), no. 4, 173–192 = *St. Petersburg Math. J.* **8** (1997), 661–674.

[10] Recent progress in the function theory of the Bergman space, *Holomorphic Spaces (Berkeley, 1995)* (Cambridge Univ. Press, Cambridge, 1998), pp. 35–50.

[11] Recent developments in the function theory of the Bergman space, *European Congress of Mathematics (Budapest, 1996)*, Vol. *I*, *Progress in Math.* **168** (Birkhäuser, Basel, 1998), pp. 202–217.

Hedenmalm, H., Jakobsson, S., and Shimorin, S.

[1] A maximum principle à la Hadamard for biharmonic operators with applications to the Bergman spaces, *C. R. Acad. Sci. Paris* **328** (1999), 973–978.

[2] A biharmonic maximal principle for hyperbolic surfaces, *J. Reine Angew. Math.* **550** (2002), 25–75.

Hedenmalm, H., Korenblum, B., and Zhu, K.

[1] Beurling type invariant subspaces of the Bergman spaces, *J. London Math. Soc.* **53** (1996), 601–614.

[2] *Theory of Bergman Spaces* (Springer–Verlag, New York, 2000).

Hedenmalm, H., Richter, S., and Seip, K.

[1] Interpolating sequences and invariant subspaces of given index in the Bergman spaces, *J. Reine Angew. Math.* **477** (1996), 13–30.

Hedenmalm, H. and Zhu, K.

[1] On the failure of optimal factorization for certain weighted Bergman spaces, *Complex Variables Theory Appl.* **19** (1992), 165–176.

Hinkkanen, A.

[1] On a maximum principle in Bergman space, *J. Analyse Math* **79** (1999), 335–344.

Hörmander, L.

[1] *The Analysis of Linear Partial Differential Operators I* (Second Edition, Springer–Verlag, Berlin, 1990).

Horowitz, C.

[1] Zeros of functions in the Bergman spaces, Ph.D. thesis, University of Michigan, 1974.

[2] Zeros of functions in the Bergman spaces, *Duke Math. J.* **41** (1974), 693–710.

[3] Factorization theorems for functions in the Bergman spaces, *Duke Math. J.* **44** (1977), 201–213.

[4] Some conditions on Bergman space zero sets, *J. Analyse Math.* **62** (1994), 323–348.

Horowitz, C., Korenblum, B., and Pinchuk, B.

[1] Extremal functions and contractive divisors in A^{-n}, *Ann. Scuola Norm. Sup. Pisa* **23** (1996), 179–191.

[2] Sampling sequences for $A^{-\infty}$, *Michigan Math. J.* **44** (1997), 389–398.

Horowitz, C. and Oberlin, D. M.

[1] Restriction of H^p functions to the diagonal of $\mathbb{U}^n$, *Indiana Univ. Math. J.* **24** (1975), 767–772.

Janas, J.

[1] A note on invariant subspaces under multiplication by z in Bergman spaces, *Proc. Roy. Irish Acad.* **83A** (1983), 157–164.

Jevtić, M., Massaneda, X., and Thomas, P. J.

[1] Interpolating sequences for weighted Bergman spaces of the ball, *Michigan Math. J.* **43** (1996), 495–517.

Kabaĭla, V.

[1] Refinement of some theorems on interpolation in the class H_δ, *Vilniaus Univ. Darbai.* **33**, *Mat.-Fiz.* **9** (1960), 15–19. (in Lithuanian; Russian summary)

[2] Interpolation sequences for the H_p classes in the case $p < 1$, *Litovsk. Mat. Sb.* **3** (1963), no. 1, 141–147. (in Russian)

Kato, T.

[1] *Perturbation Theory for Linear Operators* (Second Edition, Springer–Verlag, New York, 1976).

Khavinson, D. and Shapiro, H. S.

[1] Invariant subspaces in Bergman spaces and Hedenmalm's boundary value problem, *Ark. Mat.* **32** (1994), 309–321.

Khavinson, D. and Stessin, M. I.

[1] Certain linear extremal problems in Bergman spaces of analytic functions, *Indiana Univ. Math. J.* **46** (1997), 933–974.

Kim, A. K.

[1] Horowitz products in Bergman spaces, *Abstracts Amer. Math. Soc.* **20** (1999), 71.

Kolaski, C. J.

[1] Isometries of weighted Bergman spaces, *Canad. J. Math.* **34** (1982), 910–915.

Koosis, P.

[1] *Introduction to H_p Spaces* (Cambridge Univ. Press, London, 1980; Second Edition, 1999).

Korenblum, B.

[1] An extension of the Nevanlinna theory, *Acta Math.* **135** (1975), 187–219.

[2] A Beurling-type theorem, *Acta Math.* **138** (1977), 265–293.

[3] Cyclic elements in some spaces of analytic functions, *Bull. Amer. Math. Soc.* **5** (1981), 317–318.

[4] Transformation of zero sets by contractive operators in the Bergman space, *Bull. Sci. Math.* **114** (1990), 385–394.

[5] A maximum principle for the Bergman space, *Publ. Mat.* **35** (1991), 479–486.

[6] Outer functions and cyclic elements in Bergman spaces, *J. Funct. Anal.* **115** (1993), 104–118.

Korenblum, B., O'Neil, R., Richards, K., and Zhu, K.

[1] Totally monotone functions with applications to the Bergman space, *Trans. Amer. Math. Soc.* **337** (1993), 795–806.

Korenblum, B. and Richards, K.

[1] Majorization and domination in the Bergman space, *Proc. Amer. Math. Soc.* **117** (1993), 153–158.

Köthe, G.

[1] *Topological Vector Spaces I* (Springer–Verlag, Berlin, 1983).

Krosky, M. and Schuster, A. P.

[1] Multiple interpolation and extremal functions in the Bergman spaces, *J. Analyse Math.* **85** (2001), 141–156.

LeBlanc, E.

[1] A probabilistic zero set condition for the Bergman space, *Michigan Math. J.* **37** (1990), 427–438.

Lieb, E. H. and Loss, M.

[1] *Analysis* (Second Edition, American Mathematical Society, Providence, R. I., 2001).

Littlewood, J. E.

[1] *Lectures on the Theory of Functions* (Oxford University Press, London, 1944).

Loewner, Ch.

[1] On generation of solutions of the biharmonic equation in the plane by conformal mappings, *Pacific J. Math.* **3** (1953), 417–436.

Luecking, D. H.

[1] Inequalities on Bergman spaces, *Illinois J. Math.* **25** (1981), 1–11.

[2] A technique for characterizing Carleson measures on Bergman spaces, *Proc. Amer. Math. Soc.* **87** (1983), 656–660.

[3] Forward and reverse Carleson inequalities for functions in Bergman spaces and their derivatives, *Amer. J. Math.* **107** (1985), 85–111.

[4] Zero sequences for Bergman spaces, *Complex Variables Theory Appl.* **30** (1996), 345–362.

[5] Sampling measures for the Bergman space on the unit disk, *Math. Ann.* **316** (2000), 659–679.

Lyubarskii, Y. I. and Sodin, M.

[1] Analogues of sine type functions for convex domains, Preprint No. 17, Inst. Low Temperature Phys. Engrg., Ukrainian Acad. Sci., Kharkov (1986) (in Russian).

MacGregor, T. H. and Stessin, M. I.

[1] Weighted reproducing kernels in Bergman spaces, *Michigan Math. J.* **41** (1994), 523–533.

MacGregor, T. H. and Zhu, K.

[1] Coefficient multipliers between Bergman and Hardy spaces, *Mathematika* **42** (1995), 413–426.

Matero, J.

[1] On Korenblum's maximum principle for Bergman space, *Arch. Math. (Basel)* **64** (1995), 337–340.

McCarthy, J. E.

[1] Coefficient estimates on weighted Bergman spaces, *Duke Math. J.* **76** (1994), 751–760.

McCullough, S. and Richter, S.

[1] Bergman-type reproducing kernels, contractive divisors, and dilations, *J. Funct. Anal.* **190** (2002), 447–480.

McDonald, G. and Sundberg, C.

[1] Toeplitz operators on the disc, *Indiana Univ. Math. J.* **28** (1979), 595–611.

McKenna, P. J.

[1] Discrete Carleson measures and some interpolation problems, *Michigan Math. J.* **24** (1977), 311–319.

Nakamura, A., Ohya, F., and Watanabe, H.

[1] On some properties of functions in weighted Bergman spaces, *Proc. Fac. Sci. Tokai Univ.* **15** (1979), 33–44.

Nehari, Z.

[1] *Conformal Mapping* (New York, McGraw-Hill, 1952; reprinted by Dover Publications, New York, 1975).

Neville, C. W.

[1] A short proof of an inequality of Carleson's, *Proc. Amer. Math. Soc.* **65** (1977), 131–132.

Nicolau, A.

[1] Finite products of interpolating Blaschke products, *J. London Math. Soc.* **50** (1994), 520–531.

Nikolskii, N.

[1] *Treatise on the Shift Operator* (English translation, Springer–Verlag, Berlin, 1986).

Novinger, W. P.

[1] Holomorphic functions with infinitely differentiable boundary values, *Illinois J. Math.* **15** (1970), 80–90.

[2] Mean convergence in L^p spaces, *Proc. Amer. Math. Soc.* **34** (1972), 627–628.

Nowak, M. and Waniurski, P.

[1] Random zero sets for Bergman spaces, *Math. Proc. Cambridge Philos. Soc.* **134** (2003), 337–345.

Oleinik, V. L.

[1] Embedding theorems for weighted classes of harmonic and analytic functions, *Zap. Nauchn. Sem. Leningrad. Otdel. Mat. Inst. Steklov. (LOMI)* **47** (1974), 120–137 (in Russian) = *J. Soviet Math.* **9** (1978), 228–243.

Ortega-Cerdà, J.

[1] Sampling measures, *Publ. Mat.* **42** (1998), 559–566.

Osipenko, K. Yu. and Stessin, M. I.

[1] On recovery problems in Hardy and Bergman spaces, *Mat. Zametki* **49** (1991), no. 4, 95–104 (in Russian) = *Math. Notes* **49** (1991), 395–401.

[2] On optimal recovery of holomorphic functions in the unit ball in $\mathbb{C}^n$, *Constr. Approx.* **8** (1992), 141–159.

Pommerenke, Ch.

[1] On Bloch functions, *J. London Math. Soc.* **2** (1970), 689–695.

[2] *Boundary Behaviour of Conformal Maps* (Springer–Verlag, Berlin–Heidelberg, 1992).

Ransford, T.

[1] *Potential Theory in the Complex Plane* (Cambridge University Press, Cambridge, 1995).

Reed, M. and Simon, B.

[1] *Functional Analysis* (Academic Press, San Diego, 1983).

Richter, S.

[1] Invariant subspaces in Banach spaces of analytic functions, *Trans. Amer. Math. Soc.* **304** (1987), 585–616.

[2] Invariant subspaces of the Dirichlet shift, *J. Reine Angew. Math.* **386** (1988), 205–220.

Riesz, F. and Sz.-Nagy, B.

[1] *Functional Analysis* (Frederick Ungar, New York, 1955).

Roberts, J. W.

[1] Cyclic inner functions in the Bergman spaces and weak outer functions in H^p, $0 < p < 1$, *Illinois J. Math.* **29** (1985), 25–38.

Rochberg, R.

[1] Interpolation by functions in Bergman spaces, *Michigan Math. J.* **29** (1982), 229–236.

[2] Decomposition theorems for Bergman spaces and their applications, in *Operators and Function Theory* (S. C. Power, editor; D. Reidel Publ. Co., Boston, 1985), pp. 225–277.

Rudin, W.

[1] *Function Theory in Polydiscs* (Benjamin, New York, 1969).

[2] *Principles of Mathematical Analysis* (Third Edition, McGraw–Hill, New York, 1976).

[3] *Functional Analysis* (Second Edition, McGraw–Hill, New York, 1991).

[4] L^p–isometries and equimeasurability, *Indiana Univ. Math. J.* **25** (1976), 215–228.

[5] *Function Theory in the Unit Ball of* $\mathbb{C}^n$ (Springer–Verlag, New York, 1980).

Saitoh, S.

[1] *Integral Transforms, Reproducing Kernels and Their Applications* (Pitman Research Notes in Math. No. 369, Addison Wesley Longman, Essex, England, 1997).

Schuster, A. P.

[1] Sets of sampling and interpolation in Bergman spaces, *Proc. Amer. Math. Soc.* **125** (1997), 1717–1725.

[2] Sampling and interpolation in Bergman spaces, Ph.D. thesis, University of Michigan, 1997.
[3] On Seip's description of sampling sequences for Bergman spaces, *Complex Variables Theory Appl.* **42** (2000), 347–367.

Schuster, A. and Seip, K.
[1] A Carleson-type condition for interpolation in Bergman spaces, *J. Reine Angew. Math.* **497** (1998), 223–233.
[2] Weak conditions for interpolation in holomorphic spaces, *Publ. Mat.* **44** (2000), 277–293.

Schuster, A. P. and Varolin, D.
[1] Sampling sequences for Bergman spaces for $p < 1$, *Complex Variables Theory Appl.* **47** (2002), 243–253.
[2] Sampling and interpolation on Riemann surfaces, *to appear.*

Schwick, W.
[1] On Korenblum's maximum principle, *Proc. Amer. Math. Soc.* **125** (1997), 2581–2587.

Seip, K.
[1] Reproducing formulas and double orthogonality in Bargmann and Bergman spaces, *SIAM J. Math. Anal.* **22** (1991), 856–876.
[2] Regular sets of sampling and interpolation for weighted Bergman spaces, *Proc. Amer. Math. Soc.* **117** (1993), 213–220.
[3] Beurling type density theorems in the unit disk, *Invent. Math.* **113** (1994), 21–39.
[4] On a theorem of Korenblum, *Ark. Mat.* **32** (1994), 237–243.
[5] On Korenblum's density condition for the zero sequences of $A^{-\alpha}$, *J. Analyse Math.* **67** (1995), 307–322.
[6] Developments from nonharmonic Fourier series, *Proceedings of International Congress of Mathematicians, Berlin 1998*, Vol. *II* (*Documenta Mathematica 1998*, Extra Vol. *II*), pp. 713–722.
[7] *Interpolation and Sampling in Spaces of Analytic Functions* (American Mathematical Society, Providence, R. I., *to appear*).

Shapiro, H. S.
[1] Weakly invertible elements in certain function spaces and generators in ℓ^1, *Michigan Math. J.* **11** (1964), 161–165.
[2] Weighted polynomial approximation and boundary behaviour of analytic functions, *Contemporary Problems in the Theory of Analytic Functions* (International Conference in Erevan, Armenia, 1965; Izdat. "Nauka", Moscow, 1966), pp. 326–335.
[3] Some remarks on weighted polynomial approximation of holomorphic functions, *Mat. Sbornik* **73** (115) (1967), 320–330 (in Russian) = *Math. USSR Sb.* **2** (1967), 285–294.

[4] Comparative approximation in two topologies, *Approximation Theory* (Banach Center Publications, Vol. 4, PWN—Polish Scientific Publishers, Warsaw 1979), pp. 225–232.

Shapiro, H. S. and Shields, A. L.

[1] On some interpolation problems for analytic functions, *Amer. J. Math.* **83** (1961), 513–532.

[2] On the zeros of functions with finite Dirichlet integral and some related function spaces, *Math. Z.* **80** (1962), 217–229.

Shapiro, H. S. and Tegmark, M.

[1] An elementary proof that the biharmonic Green function of an eccentric ellipse changes sign, *SIAM Review* **36** (1994), 99–101.

Shapiro, J. H.

[1] Linear functionals on non-locally convex spaces, Ph.D. thesis, University of Michigan, 1969.

[2] Mackey topologies, reproducing kernels, and diagonal maps on Hardy and Bergman spaces, *Duke Math. J.* **43** (1976), 187–202.

[3] Zeros of functions in weighted Bergman spaces, *Michigan Math. J.* **24** (1977), 243–256.

[4] Zeros of random functions in Bergman spaces, *Ann. Inst. Fourier (Grenoble)* **29** (1979), 159–171.

[5] Cyclic inner functions in Bergman spaces (unpublished notes, University of Wisconsin, 1980; *www.math.msu.edu/~shapiro*).

[6] *Composition Operators and Classical Function Theory* (Springer-Verlag, New York, 1993).

Shields, A.

[1] Weighted shift operators and analytic function theory, *Topics in Operator Theory* (C. M. Pearcy, editor), American Mathematical Society, Providence, R. I., 1974, pp. 49–128.

[2] Cyclic vectors in Banach spaces of analytic functions, *Operators and Function Theory* (S. C. Power, editor; D. Reidel Publ. Co., Boston, 1985), pp. 315–349.

Shimorin, S. M.

[1] Factorization of analytic functions in weighted Bergman spaces, *Algebra i Analiz* **5** (1993), no. 5, 155–177 (in Russian) = *St. Petersburg Math. J.* **5** (1994), 1005–1022.

[2] On a family of conformally invariant operators, *Algebra i Analiz* **7** (1995), no. 2, 133–158 (in Russian) = *St. Petersburg Math. J.* **7** (1996), 287–306.

[3] The Green function for the weighted biharmonic operator $\Delta(1-|z|^2)^{-\alpha}\Delta$, and factorization of analytic functions, *Zap. Nauchn. Sem. S.-Peterburg. Otdel. Math. Inst. Steklov. (POMI)* **222** (1995), *Issled. po Lineĭn. Oper. i Teor. Funktsiĭ* **23**, 203–221 (in Russian) = *J. Math. Sci. (New York)* **87** (1997), no. 5, 3912–3924.

[4] Single-point extremal functions in weighted Bergman spaces, *Probl. Mat. Anal.* No. 15 (1995), 241–252 (in Russian) = *J. Math. Sci. (New York)* **80** (1996), no. 6, 2349–2356.

[5] The Green functions for weighted biharmonic operators of the form $\Delta w^{-1}\Delta$ in the unit disk, *Probl. Mat. Anal.* (in Russian) = *J. Math. Sci. (New York)* **92** (1998), no. 6, 4404–4411.

[6] Approximative spectral synthesis in the Bergman space, *Duke Math. J.* **101** (2000), 1–39.

[7] Wold-type decompositions and wandering subspaces for operators close to isometries, *J. Reine Angew. Math.* **531** (2001), 147–189.

[8] An integral formula for weighted Bergman reproducing kernels, *Complex Variables Theory Appl.* **47** (2002), 1015–1028.

[9] On Beurling-type theorems in weighted ℓ^2 and Bergman spaces, *Proc. Amer. Math. Soc.* **131** (2003), 1777–1787.

Shvedenko, S. V.

[1] On the Taylor coefficients of functions from Bergman spaces in the polydisc, *Dokl. Akad. Nauk SSSR* **283** (1985), 325–328 (in Russian) = *Soviet Math. Dokl.* **32** (1985), 118–121.

[2] On a theorem of Shapiro and Shields on the zeros of analytic functions of restricted growth, *Izv. Vyssh. Uchebn. Zaved. Mat.* **2001**, no. 9, 75–79 (in Russian) = *Russian Math. (Izv. VUZ)* **45** (2001), no. 9, 71–75.

Stessin, M.

[1] An extension of a theorem of Hadamard and domination in the Bergman space, *J. Funct. Anal.* **115** (1993), 212–226.

Sundberg, C.

[1] Analytic continuability of Bergman inner functions, *Michigan Math. J.* **44** (1997), 399–407.

Szegő, G.

[1] Remark on the preceding paper of Charles Loewner, *Pacific J. Math.* **3** (1953), 437–446.

Taylor, B. A. and Williams, D. L.

[1] Ideals in rings of analytic functions with smooth boundary values, *Canad. J. Math.* **22** (1970), 1266–1283.

Thomson, J. E.

[1] Approximation in the mean by polynomials, *Ann. of Math.* **133** (1991), 477–508.

[2] Bounded point evaluations and polynomial approximation, *Proc. Amer. Math. Soc.* **123** (1995), 1757–1761.

Vukotić, D.

[1] A sharp estimate for A^p functions in $\mathbb{C}^n$, *Proc. Amer. Math. Soc.* **117** (1993), 753–756.

[2] On the coefficient multipliers of Bergman spaces, *J. London Math. Soc.* **50** (1994), 341–348.

[3] Linear extremal problems for Bergman spaces, *Expositiones Math.* **14** (1996), 313–352.

[4] The isoperimetric inequality and a theorem of Hardy and Littlewood, *Amer. Math. Monthly* **110** (2003), 532–536.

Watanabe, H.

[1] Some properties of functions in Bergman space A^p, *Proc. Fac. Sci. Tokai Univ.* **13** (1977), 39–54.

[2] Some inequalities of means and its applications to Bergman spaces, *Proc. Fac. Sci. Tokai Univ.* **28** (1993), 1–13.

Weir, R.

[1] Canonical divisors and invariant subspaces in weighted Bergman spaces, Ph.D. thesis, University of Michigan, 2001.

[2] Canonical divisors in weighted Bergman spaces and hypergeometric functions, *Proc. Amer. Math. Soc.* **130** (2002), 707–713.

[3] Zeros of extremal functions in weighted Bergman spaces, *Pacific J. Math.* **208** (2003), 187–199.

[4] Construction of Green functions for weighted biharmonic operators, *to appear.*

Wojtaszczyk, P.

[1] On multipliers into Bergman spaces and Nevanlinna class, *Canad. Math. Bull.* **33** (1990), 151–161.

Zaharjuta, V. P. and Judovič, V. I.

[1] The general form of a linear functional on H_p', *Uspekhi Mat. Nauk* **19** (1964), no. 2 (116), 139–142. (in Russian)

Zhu, K.

[1] *Operator Theory in Function Spaces* (Marcel Dekker, New York, 1990).

[2] Duality of Bloch spaces and norm convergence of Taylor series, *Michigan Math. J.* **38** (1991), 89–101.

[3] Interpolating sequences for the Bergman space, *Michigan Math. J.* **41** (1994), 73–86.

[4] Evaluation operators on the Bergman space, *Math. Proc. Cambridge Philos. Soc.* **117** (1995), 513–523.

[5] A sharp estimate for extremal functions, *Proc. Amer. Math. Soc.* **128** (2000), 2577–2583.

Zygmund, A.

[1] *Trigonometric Series* (Cambridge University Press, London, 1959).

Index